Patterns in
plant development

SECOND EDITION

Taylor A. Steeves

Ian M. Sussex

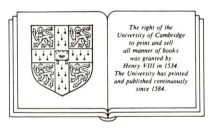

The right of the
University of Cambridge
to print and sell
all manner of books
was granted by
Henry VIII in 1534.
The University has printed
and published continuously
since 1584.

D0209462

CAMBRIDGE UNIVERSITY PRESS

Cambridge
New York   New Rochelle
Melbourne   Sydney

Published by the Press Syndicate of the University of Cambridge
The Pitt Building, Trumpington Street, Cambridge CB2 1RP
32 East 57th Street, New York, NY 10022, USA
10 Stamford Road, Oakleigh, Melbourne 3166, Australia

First published 1989

Printed in the United States of America

Library of Congress Cataloging-in-Publication Data
Steeves, Taylor A., 1926–
Patterns in plant development.
Includes bibliographies.
1. Plants – Development.   2. Plant morphogenesis.
I. Sussex, Ian M., 1927–   . II. Title.
QK665.S79   1989   581.3'32   88–1209

British Library Cataloguing in Publication Data
Steeves, Taylor A., 1926–
 Patterns in plant development. – 2nd ed.
1. Plants. Development
I. Title   II. Sussex, Ian M.
581.3

ISBN 0 521 24688 1 hard covers
ISBN 0 521 28895 9 paperback

Dedicated to
R. H. Wetmore
and to the memory of
C. W. Wardlaw

Contents

Preface

This volume is a revised edition of a book first published, under the same title, in 1972 and now several years out of print. In recognition of the impressive body of developmental research that has been reported since the publication of the original volume, this edition has been substantially modified and modestly enlarged. The point of view of the original, however, has been retained. It is, as the title implies, structural and organismal. We have attempted to document the developmental process as the plant undergoes it, beginning with the zygote and the formation of the embryo, continuing with the development of the primary body and completing the picture with a treatment of secondary growth. We have not, therefore, undertaken to analyze phenomena like cell growth, meristematic activity, or polarity as topics in themselves, although certain phenomena, notably differentiation and the potency of differentiated cells, have been given special treatment. It may be argued that this approach could fail to reveal fundamental generalizations about development. Nevertheless, our goal was to show how the plant develops as an organism and we have attempted to adhere to it.

In the more than fifteen years that have elapsed since the original edition was published, there have been phenomenal advances in the fields of cellular and molecular biology, and these discoveries are being applied with ever-increasing intensity to the interpretation of plant development. One may reasonably ask, therefore, whether the structural and organismal approach to development has become obsolete. The effort expended in revising this book signifies that our answer is emphatically "No." It is undeniably true that the enhanced understanding of processes at the cellular level is providing answers to developmental questions that seemed almost beyond our reach only a few years ago. It is equally true, however, that the fundamental developmental questions are posed by the formation of an integrated organism within which the multitudes of diverse cells occupy their places and perform their specific functions in the context of distinctive patterns. Indeed the events that occur at the cellular level reveal their developmental significance in the patterns within which they occur, that is, in the morphogenesis that forms a functioning organism. This essential point was well made more than one hundred years ago in the aphorism attributed to Anton deBary: "The plant forms cells, not cells the plant." Thus we believe that the explosion in molecular biology will achieve its most beneficial impact upon the understanding of development if it is based upon a profound appreciation of development at the organismal level.

On the other hand, we have not undertaken to give just a descriptive account of structural changes in development for such a treatment would grossly misrepresent the field of developmental morphology as it is to-day and has been for several decades. This field of investigation has become causal in its interpretations and experimental in its methods. Thus, while we have not attempted to analyze control mechanisms as physiological or biochemical processes, we have been very much concerned with them as they regulate developmental patterns. We have attempted as far as possible to limit the quantity of purely descriptive material by selecting particular examples that have been well documented and illustrated in published accounts and using them to present particular developmental patterns. The variability in pattern is then introduced briefly following the detailed account of the "type." We hope that this technique has enabled us to present sufficient detail for a meaningful picture without too great a burden of description. As in the earlier edition, we have found it necessary to limit the treatment to the vascular plants and thus, regrettably, to exclude significant work dealing with algae, fungi, and bryophytes.

In this revision much new material has been introduced. Wherever possible the exciting findings of molecular biology have been drawn upon to aid in the interpretation of morphogenetic events. This is particularly evident in the account of the embryonic phase of development. Substantial reference has been made to the application of the technique of clonal analysis, which, although not new, has only lately been recognized as a powerful tool in the difficult task of tracing cell lineages in plants. At the same time it is recognized that plant morphologists have made major strides in interpreting the organization of whole plants, particularly those of large stature, and the developmental significance of architectural analysis is stressed in this edition.

This book is intended to be an introduction to plant development, and we believe that it should be intelligible to any student or other potential user who has completed an introductory university course in biology or botany. Without preparing a text on plant anatomy, we have tried to provide enough fundamental information upon which to base an account of developmental phenomena. At the same time, in expanding each topic, we have tried to put the reader in contact with current problems and interpretations so that direct progression to the literature is possible.

Although we assume full responsibility for the contents of this book, the influence of many others has played an important part in its development. We shall always be indebted to the two distinguished botanists to whom this second edition, like the first, is dedicated. Their pioneering investigations in experimental morphogenesis, and the vision of this field that they expounded, inspired and influenced a generation of developmental botanists, particularly those of us who were fortunate enough to be their students. Many colleagues and students have also contributed to this work in numerous discussions and debates on the problems of development. We are particularly grateful to our students whose penetrating questions, unencumbered by preconceived notions, have so often provided a stimulus to clearer thinking. Finally, we hope that the result of our efforts will justify the confidence of the many colleagues who have urged us to undertake this revision.

Acknowledgments

We are most appreciative of those who contributed in particular ways to the preparation of this volume. Mrs. S. F. Rowley skillfully produced the manuscript from our nearly illegible script and patiently accommodated numerous editorial revisions. Ms. Sharon Pulvermacher and Ms. Colleen Shepstone are to be complimented for their fine artistic work, as are Mr. Dennis Dyck and Ms. P. J. Rennie for many of the included photographs. Finally we are indebted to Dr. Margaret Steeves for advice and assistance on many occasions, and especially for the preparation of the index.

1

*Development in
the vascular plants*

The vascular plant, like all sexually reproducing organisms, begins its existence as a single cell, the fertilized egg or zygote. Proliferation of this cell leads to the formation of an embryo within which, at an early stage, organs and tissues begin to be formed. Early in embryogeny two distinctive regions are set apart, approximately at opposite poles, that subsequently retain the capacity for continued growth. One of these, designated the *shoot apical meristem*, functions to produce an expanding shoot system by the continued formation of tissues and the initiation of a succession of leaf and bud primordia. The other, the *root apical meristem*, similarly forms an expanding root system. Furthermore, the development of these open-ended systems is repetitive; the same kinds of tissues and organs are produced in continuing succession.

The activity of the apical meristems results in the production of a continuously elongating body, which has been called the *primary body* of the plant. In many cases this primary body constitutes the whole plant. In other cases, particularly in those plants with an extended life-span, there is an additional component of development that leads to an increase in girth of the axis. This results from the activity of two additional meristems that are initiated in the postembryonic stage: the *vascular cambium*, which contributes additional cells to the conducting system, and the *cork cambium*, which produces a protective tissue replacing the original epidermis. These meristems and the tissues they produce constitute what is designated the *secondary body* of the plant. The secondary body does not constitute an entire plant in that it is composed of a few tissue types only and includes no organs. In some cases, however, it comes to constitute the bulk of the plant body.

There is a distinct difference between this pattern of development and that which characterizes the higher animals. In the animal embryo the fundamental plan of the adult body is laid down so that all the organs and tissues are present at least in rudimentary form. Postembryonic development of the animal consists of the enlargement of the body and then its maintenance in a functionally efficient state. Thus, whereas higher animals have characteristic numbers of organs, the indeterminate shoot has an indefinite number of leaves. On the other hand, this contrast in development between plants and animals should not be overemphasized. There are in the animal body many groups of cells that continue to proliferate throughout the life of the organism, giving rise to such structures as skin, blood, hair, claws, and various epithelial layers. Proliferating cell groups of this sort are very similar to the secondary plant meristems, but there is less similarity to the apical meristems. In these there is clearly a greater developmental potential in that they give rise to a wider range of structures and to a higher level of organization. In fact, they produce the organized plant body. Thus, the differences between plant and animal development are significant, but fundamental similiarities should be recognized.

The possible significance of the plant mode of development may be appreciated by considering the cellular structure of plants and their overall immobility. Plant cells are surrounded by a relatively rigid wall and are tightly cemented together within the framework of the tissues. In the animal body, cell replacement, which seems to be essential for the maintenance of functional efficiency of at least certain tissues, can be accomplished within the histological framework. In the plant this obviously is impossible, and a comparable replacement is effected by the continual addition of new cells at the growing tips of shoots and roots by the primary meristems and laterally by the secondary meristems. In a very real sense, plants assimilate at their most recently formed tips. The absorption of water and mineral salts is accomplished at or near growing root tips, and shoot apical meristems continually replenish the complement of photosynthetic leaves, which are shed often after one growing season and at most after several years. These expanding tips are connected by a vascular conducting system that may be renewed by cells derived from the vascular cambium. Moreover, continued growth endows the immobile plant with a measure of responsiveness to its environment, which is seen in the tropistic movements of the plant organs and in the continued advance of roots into unoccupied regions of the soil. It could be argued that, to a limited extent, development plays a role in plant survival analogous to that of behavior in animals.

In the preceding paragraphs we have been using the word *development* rather freely without stating precisely what we mean by it. Certainly, before plunging into the subject matter of this book about development, we have an obligation to state what we intend to encompass under this heading.

Development is the sum total of events that contribute to the progressive elaboration of the body of an organism. A more restrictive definition might unfortunately eliminate from consideration some essential

aspects of the development process. On the other hand, a broad interpretation, such as we have given, might seem to include all or most of
the physiological processes of living organisms. This is especially true
in plants, in which, in a very real sense, all physiological phenomena
throughout the life of the organism seem to be channeled into the progressive elaboration of its body. One might justifiably ask, then, whether
such phenomena as photosynthesis or the absorption of water are to be
regarded as part of development. Certainly, their absence would preclude development; but they are far removed from the actual mechanism by which the plant body is elaborated.

In order to avoid the impossible requirement of discussing the entire
biology of plants, it will be necessary to concentrate upon those phenomena that directly participate in the formation of the plant body.
Even in this limited sense, development encompasses numerous processes, such as cell division, cell enlargement, protein synthesis, the
elaboration of cell wall materials, quantitative and qualitative alterations in cell organelles, among many others. It is, however, convenient
to recognize two major aspects of development and to analyze developmental processes in terms of these two categories. These are growth
and differentiation.

In an organism *growth* is an irreversible increase in size, and it is
accomplished by a combination of cell division and cell enlargement.
Cell division does not of itself constitute growth and in fact may occur
without any increase in the overall size of the structure involved. Cell
enlargement alone does constitute growth; this is particularly evident
in plants in which there is a considerable net increase in cell size in
maturing regions. Nonetheless, with few exceptions the continued growth
of an organism requires the production of new cells and their enlargement, and these two processes are closely associated in space and time.
Growth by itself will not lead to the formation of an organized body,
but rather, at least in theory, to a homogeneous assemblage of cells.
Clearly, the formation of an organized body implies that cells and groups
of cells in different regions of the body have become structurally distinguishable and functionally distinctive. The changes that occur in these
cells and groups of cells and bring about their distinctiveness constitute
what is known as *differentiation*. Some biologists prefer to distinguish
between those changes that lead to distinctive histological patterns –
designated as cell differentiation, or histodifferentiation – and those that
set apart major segments of the body or organs – designated as organogenesis. Because there is no reason to suppose that mechanisms underlying these two types of change are fundamentally different, it seems
preferable to consider both as aspects of a general phenomenon of differentiation. There are cases in which growth can occur without differ-

entiation, and differentiation without growth, but it is almost always true that these two phenomena occur in intimate association. The development of an organized body depends upon the integrated activity of the two.

Although a book like this one ought to contain enough information and interpretation to be useful by itself, it is desirable that readers, particularly students, have access to the original studies in the field. For this reason, at the end of each chapter a selected bibliography is given, and reference is made to it in the context of the chapter. Although these lists are not comprehensive, they contain key reference works that offer the opportunity for a wider grasp of plant development. In addition, there are many extremely useful general reference works that are basic sources of information and points of view. A selected list of these references follows.

REFERENCES

Balls, M., and F. S. Billett. 1973. *The Cell Cycle in Development and Differentiation.* Cambridge: Cambridge University Press.

Barlow, P. W., and D. J. Carr. 1984. *Positional Controls in Plant Development.* Cambridge: Cambridge University Press.

Bernier, G., J. Kinet, and R. M. Sachs. 1981. *The Physiology of Flowering.* Vol. 1 and 2. Boca Raton, Fla.: CRC Press.

Bewley, J. D., and M. Black. 1985. *Seeds: Physiology of Development and Germination.* New York: Plenum.

Burgess, J. 1985. *An Introduction to Plant Cell Development.* Cambridge: Cambridge University Press.

Cutter, E. G. 1971. *Plant Anatomy: Experiment and Interpretation. Part 2: Organs.* London: Arnold.

———— 1978. *Plant Anatomy: Experiment and Interpretation. Part 1: Cells and Tissues.* 2d ed. London: Arnold.

Dale, J. E., and F. L. Milthorpe. 1983. *The Growth and Functioning of Leaves.* Cambridge: Cambridge University Press.

Esau, K. 1965. *Vascular Differentiation in Plants.* New York: Holt, Rinehart and Winston.

———— 1977. *Anatomy of Seed Plants.* 2d ed. New York: Wiley.

Francis, D., and J. A. Bryant. 1985. *The Cell Division Cycle in Plants.* Cambridge: Cambridge University Press.

Hallé, F., R. A. A. Oldeman, and P. B. Tomlinson. 1978. *Tropical Trees and Forests: An Architectural Analysis.* Berlin: Springer-Verlag.

Johri, B. M. 1982. *Experimental Embryology of Vascular Plants.* New York: Springer-Verlag.

———— 1984. *Embryology of Angiosperms.* New York: Springer-Verlag.

O'Brien, T. P., and M. E. McCully. 1981. *The Study of Plant Structure: Principles and Selected Methods.* Melbourne, Australia: Termarcarphi.

Philipson, W. R., J. M. Ward, and B. G. Butterfield. 1971. *The Vascular Cambium: Its Development and Activity.* London: Chapman and Hall.

Raghavan, V. 1986. *Embryogenesis in Angiosperms: A Developmental and Experimental Study.* Cambridge: Cambridge University Press.

Roberts, L. W. 1976. *Cytodifferentiation in Plants: Xylogenesis as a Model System.* Cambridge: Cambridge University Press.

Sinnott, E. W. 1960. *Plant Morphogenesis.* New York: McGraw-Hill.

Tilney-Bassett, R. A. E. 1986. *Plant Chimeras.* London: Arnold.

Torrey, J. G., and D. T. Clarkson. 1975. *The Development and Function of Roots.* New York: Academic Press.

Vasil, I. K. 1984. *Cell Culture and Somatic Cell Genetics of Plants.* Orlando, Fla.: Academic Press.

Wardlaw, C. W. 1968. *Morphogenesis in Plants: A Contemporary Study.* 2d ed. London: Methuen.

Wareing, P. F., and I. D. J. Phillips. 1981. *Growth and Differentiation in Plants.* 3d ed. New York: Pergamon Press.

Williams, R. F. 1975. *The Shoot Apex and Leaf Growth: A Study in Quantitative Biology.* Cambridge: Cambridge University Press.

Yeoman, M. M. 1976. *Cell Division in Higher Plants.* New York: Academic Press.

CHAPTER 2

Embryogenesis: beginnings of development

We saw in the previous chapter that it is characteristic for morphogenetic events to continue throughout the life-span of most plants. This is in marked contrast to animal development, in which there is a concentration of morphogenetic phenomena in the embryonic stages. Nonetheless, like the animal, the vascular plant begins life as a single cell, the fertilized egg, and passes through an embryonic phase during which the fundamental body plan is laid down. Although it may be argued that all plants that develop from a single cell into a multicellular state pass through an embryonic phase, historically the term *embryo* has been restricted to those groups in which the early stages are enclosed within parental tissue and are presumed to be nutritionally dependent upon the parent organism. On this basis the bryophytes and the vascular plants often are designated the *Embryophyta*. In the bryophytes and the lower vascular plants there is no interruption of growth to mark the end of the embryonic phase, which is therefore rather ill defined. On the other hand, in the seed plants, embryonic development is considered to be terminated at the maturation of the seed, and this leads to a sharp distinction between the embryo and all postgermination stages.

Throughout the Embryophyta, as well as in some lower groups, plant development from a zygote alternates in the life cycle with development of a second plant body from a single-celled spore. Alternation of generations poses interesting morphogenetic problems because of the contrasting morphology of the two phases, each developed from a single cell but under different conditions. In discussing embryogenesis we shall be concerned primarily with the early or embryonic stages of development of the diploid zygote into a vascular sporophyte. The contrasting development of the haploid spore into the gametophyte will not be considered except as it sheds some light upon the factors that control the different development of the two generations.

PATTERNS OF EMBRYO DEVELOPMENT

In view of the fundamental similarity of somatic organization in the sporophytes of the principal groups of vascular plants, it is startling to

6

discover the diversity of embryonic patterns that lead to this organization (Wardlaw, 1955). The conclusion is unavoidable that these patterns
in themselves have limited morphogenetic significance. Nevertheless,
the various patterns must be recognized, because any functional generalizations must be compatible with them. The classification of embryo
types based on the sequence of early cell divisions is a complex field
with a voluminous literature and is one that has had an important bearing upon taxonomic and phylogenetic interpretations.

In this chapter we shall not examine examples of all these types; rather
we shall describe several that reveal the range of embryonic diversity,
with the hope of arriving at some general understanding of the principles of embryogenesis.

Embryo development in angiosperms

An excellent account of embryo development in a flowering plant has
been given by Miller and Wetmore (1945) for *Phlox drummondii* (Figs.
2.1, 2.2). The first cell division is at right angles to the axis of the embryo
sac and divides the zygote into a smaller *terminal cell* and a larger *basal
cell*. The two cells each divide again in the same plane, giving rise to a
four-celled filament. Divisions continue in each of these cells. The cell
that lies nearest to the micropyle – the aperture through which the pollen tube grows prior to fertilization – divides in the same plane, producing a short, filamentous organ called the *suspensor*. The remaining
three cells initiate the *embryo proper*. These cells undergo both longitudinal and transverse divisions and give rise to a globular mass in which
the cells are arranged in regular tiers. At this stage the embryo, consisting of fewer than forty cells, is only four days old and is less than a
quarter of a millimeter in length. By the fifth day the first evidence of
histodifferentiation within the previously homogeneous globular embryo proper is detected. Divisions in the surface cells become progressively restricted to the anticlinal plane – that is, perpendicular to the
surface – resulting in the appearance of a superficial layer called the
protoderm. Shortly thereafter, internal differentiation becomes evident
in a central column of densely staining, narrow, elongated *procambium*
cells surrounded by a cylinder of vacuolated cells. Thus, at this early
stage, the three principal tissue systems of the plant (*dermal, vascular,*
and *fundamental*) have been initiated.

By the sixth day the first suggestion of the shoot apical meristem can
be detected in the spherical embryo as an area of small, densely staining
cells continuous with the central procambial core and at the pole of the
embryo opposite the suspensor. A day or two later the two cotyledons
appear as a result of localized concentrations of growth on either side
of the shoot meristem but not in it, and the embryo passes into what

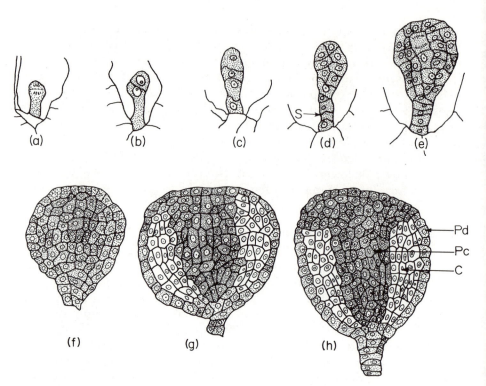

Figure 2.1. *Embryo development in* Phlox drummondii. *(a) First division of the zygote. (b–e) Stages of the embryogeny in the first four days after fertilization. (f–h) Later stages showing differentiation of protoderm, procambium, and cortical parenchyma. The shoot apex is first distinguishable in (g). Key: C, cortical parenchyma; Pc, procambium; Pd, protoderm; S, suspensor. ×225. (Miller and Wetmore, 1945.)*

has been called the heart stage. Procambium continuous with that of the central core of the embryo axis extends into the cotyledons. As the cotyledons enlarge, the axis of the embryo also elongates, and at the end of the axis opposite the shoot apical meristem, periclinal divisions – that is, in a plane parallel to the surface – initiate a root cap beneath which the apical meristem of the primary root may be detected. The mature embryo is thus bipolar, with shoot and root apical meristems located at opposite extremities of its axis. The mature embryo, like the full-term animal embryo, possesses the fundamental organization of the adult body, but unlike the animal embryo, in which the major organs are present at least in rudimentary form, the shoot and root systems are represented only by their apical meristems.

On the basis of the planes of early cell divisions and the contributions

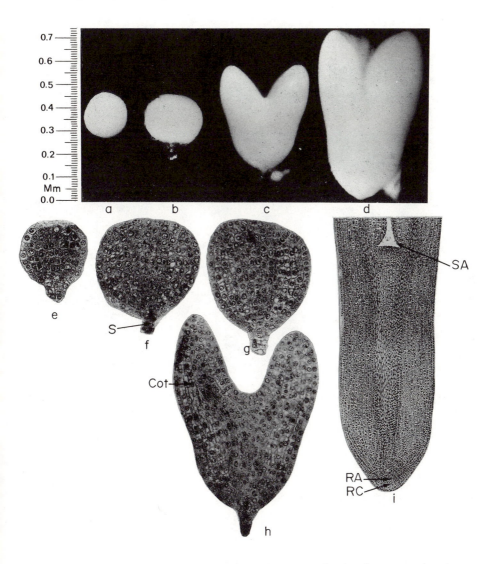

Figure 2.2. *Embryo development in* Phlox drummondii. *(a–d) Dissected embryos showing globular, heart, and early torpedo stages. (e–h) Sections of embryos showing cellular detail of comparable stages. (e), (f), and (g) correspond to stages (f), (g), and (h) in Fig. 2.1. (i) Fully developed embryo showing shoot and root apices. Key:* Cot, *cotyledon;* RA, *root apex;* RC, *root cap;* S, *suspensor;* SA, *shoot apex; (a–d) ×60, (e–g) ×160, (h) ×95, (i) ×25. (Miller and Wetmore, 1945.)*

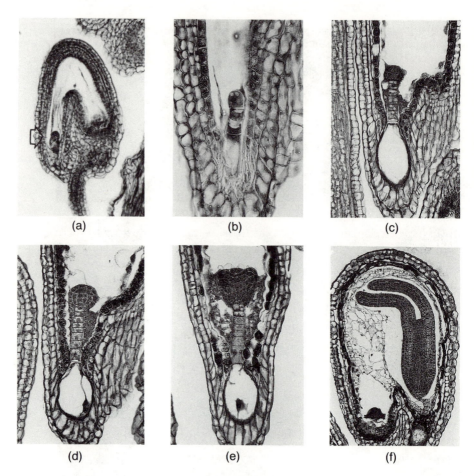

Figure 2.3. *Embryo development in* Capsella bursa-pastoris. *(a) Zygote (arrow) enclosed in ovule. (b) Four-celled, filamentous embryo. (c) Early globular stage. (d) Late globular stage showing protoderm and fully developed suspensor. (e) Heart stage. (f) Maturing, bipolar embryo. (a) ×140, (b) ×250, (c–e) ×140, (f) ×42.*

that their derivatives make to the development of the embryo, plant embryologists have recognized several patterns of embryonic development in the angiosperms (Maheshwari, 1950). The embryogeny of *Capsella bursa-pastoris* (shepherd's purse), which has been the subject of a number of detailed studies, shows a relatively precise sequence of early divisions and also contrasts with *Phlox* in features of later development (Fig. 2.3). Following the first transverse division of the zygote, the basal cell and its derivatives divide transversely to form a filamentous suspensor of six to ten cells. Meanwhile the terminal cell undergoes a longitudinal division and each of the derivative cells divides longitudinally in a plane at right angles to the prior division forming a terminal quartet of cells. Each of these cells then divides transversely resulting in a cluster of eight cells. The octants each undergo a perclinal division, which establishes the protoderm at this very early stage of development. The subsequent events that convert this globular structure into a bipolar, dicotyledonous embryo are much the same as those already described for *Phlox*. However, the limited space within the developing ovule results in a curvature that folds the cotyledons back along the axis of the embryo. In contrast to its rather limited development in *Phlox*, the suspensor of *Capsella* is considerably elaborated. In particular its basal-most cell becomes greatly enlarged and is embedded in the surrounding tissues of the ovule. On the other hand, the cell at the opposite end of the suspensor filament is incorporated as part of the embryo proper where it contributes to the initiation of the primary root. At about the late heart stage the suspensor begins to senesce, suggesting that its role as an embryonic organ has been completed.

Over the years it has become increasingly apparent that, though many species have fixed patterns of early segmentation, other species have either variable or obscure patterns. In *Daucus* (carrot), for example, the pattern of cleavage has been shown to be variable, and in *Gossypium* (cotton) it is without regularity. Further variations in the pattern of development occur in later stages of embryogeny. There appear to be differences in the extent of cellular proliferation in the globular stage before the onset of histogenesis and organogenesis. For example, in the primitive genus *Degeneria*, a tropical tree, there is a massive globular stage that precedes any differentiation. This contrasts with *Capsella*, in which the protoderm is initiated when the globular part of the embryo consists of only eight cells.

Embryos of the monocotyledons, which up to the globular stage closely resemble embryos of dicotyledons and conform to the same types, develop a single, prominent cotyledon in what appears to be a terminal position, the shoot apex apparently occupying a lateral position on the embryo. There is uncertainty as to whether the single cotyledon is truly terminal in position or whether it is initiated laterally as in the dicotyledons and by growth displacement assumes a terminal position. Supporting the latter view is the fact that in some dicotyledons only one of the two cotyledons develops, and this shows a terminal displacement. Embryo development in *Zea mays* (maize) of the grass family (Poaceae), which has been the subject of a substantial amount of analytical and experimental investigation, will serve as an example for the monocotyledons, although not necessarily a typical one (Fig. 2.4).

Ten to twelve hours after fertilization, a transverse division in the

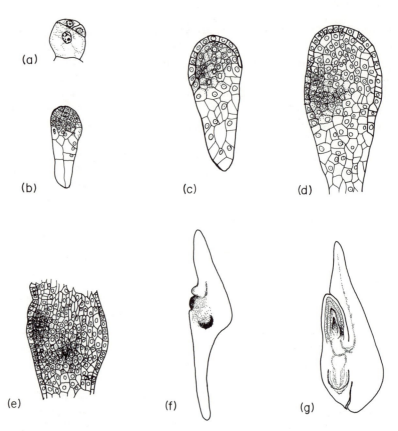

Figure 2.4. *Embryo development in* Zea mays. *(a) Three-celled embryo showing first division of terminal cell. (b) Six-day embryo showing embryo proper and suspensor. (c) Seven-day embryo showing delimitation of protoderm in embryo proper. (d) Nine-day embryo showing increased cytoplasmic density (on left) in region where meristems will arise. (e) Eleven-day embryo showing superficial position of shoot apical meristem and internal origin of root apical meristem. (f) Thirteen-day embryo showing shoot and root apical meristems, suspensor below and scutellum above. (g) Forty-four day, fully developed embryo. (a–d) ×140, (e) ×125, (f) ×42, (g) ×8.5. (Randolph, 1936.)*

zygote separates a small, lenticular terminal cell from a larger basal cell (Randolph, 1936). The first division of the terminal cell is in the vertical plane, but subsequent divisions throughout the embryo are irregular both in orientation and in sequence. Irregularity of early divisions is apparently unusual in grasses, although it is generally characteristic of later stages. The result of these divisions is the formation of a club-shaped embryo in which the cells of the embryo proper, where mitotic activity is higher, are small and densely protoplasmic but are not sharply set off from the larger, more vacuolated cells of the basal suspensor. The first evidence of differentiation within the embryo proper is the delimitation of the protoderm at about seven days after fertilization. A day or two later the appearance of a group of densely cytoplasmic cells in a lateral position below the tip foreshadows the location of the apical meristems of the embryo. Within the next few days the shoot apical meristem develops from the more distal, superficial part of this region, and the meristem of the primary root differentiates internally from its basal end near the suspensor. A strand of procambium differentiates through the embryo axis between the two meristems. The remainder of the embryo proper is involved in the formation of the scutellum. Growth in both axial and lateral directions results in the formation of this shieldlike organ, which is usually interpreted as the single cotyledon. The suspensor continues to enlarge up to about twenty days after fertilization, by which time the shoot apex has initiated several leaves. This prolongation of suspensor activity contrasts with the two cases previously described. Maize and other grasses are unusual in the extent of apical meristem activity during embryogeny. The shoot apex has initiated up to six leaves, a short length of primary root has been formed by the root apical meristem, and several adventitious roots have been initiated by the time that developmental arrest brings the embryonic phase to conclusion.

Plant embryologists have had a long-standing interest in ascertaining whether there is any constant relationship between organs of the fully formed embryo and the early segmentation pattern. Questions of this nature can be answered with confidence only if cells can be marked in such a way that their derivatives can be identified at later stages in development. In maize a large number of mutants have been described in which chlorophyll or other pigmentation is altered so that cells expressing the mutation can be identified. If a plant that carries such a mutation as a recessive and does not express it, is irradiated during development, individual cells may be altered in such a way that the mutation is expressed. Subsequently, the derivatives of those cells can be identified. It has been suggested, on the basis of histological analysis, that the two cells that result from the first longitudinal division of the terminal cell of the maize embryo each give rise to one-half of the embryo and subsequently of the plant. If maize ears carrying embryos with an unexpressed mutation for scutellum coloration are irradiated when the terminal cell is dividing or has recently divided, analysis of the resulting embryos reveals colored scutellum sectors that occupy variable proportions of that organ. This pattern does not support the concept of the early separation of the embryo into halves. Rather, it is more consistent with descriptions of maize embryogeny that have reported a lack of regularity in the early divisions (Poethig et al., 1986).

Embryo development in gymnosperms

Although there is considerable diversity in embryonic development among the gymnosperms, they all differ in characteristic ways from the angiosperms that have been considered. The embryogeny of *Ginkgo biloba*, an ancient and taxonomically isolated species, has been investigated extensively and will serve as a type for illustration (Fig. 2.5). As in all gymnosperms, the egg is large – 300 to 500 micrometers (μm) in diameter – and is contained within an archegonium, the female sex organ, at the micropylar end of a cellular female gametophyte. After fertilization the zygote nucleus divides, but no cell wall arises to separate the daughter nuclei. These nuclei and their progeny divide repeatedly until about 256 nuclei are distributed uniformly throughout the cytoplasmic mass of the embryo. At this time, or after one further general division, wall formation occurs in such a way as to partition the embryo into uninucleate cells of equal size. The cells farthest from the micropyle divide more rapidly in the ensuing period than do those at the micropylar end, and the embryo at this time exhibits a marked axial gradient in cell size. These small cells, occupying about one-third of the embryonic volume, give rise upon further development to the organized embryo with shoot and root apices and cotyledons. The larger cells, which make up the other two-thirds of the embryo, divide more slowly, are vacuolated, and, according to some, constitute a suspensor (Wardlaw, 1955).

The phenomenon of free nuclear division at the beginning of embryo development is characteristic of the gymnosperms and does not occur in other plant embryos. The free nuclear stage may be even more extensive than in *Ginkgo*, as in the cycad *Dioon edule*, where more than a thousand nuclei have been noted in the coenocytic embryo. Conversely, this stage may be relatively brief, as in the conifers, in which the number of free nuclei may range from 32 or 64 in members of the family Araucariaceae to 2 in the Cupressaceae. In *Sequoia* (redwood) the first division of the zygote is followed by cell wall formation, so there is no free nuclear stage in the embryo of this species.

Another peculiarity of gymnosperm embryogeny is the widespread occurrence of multiple embryos arising from a single zygote. Cleavage polyembryony is best developed among the conifers and may be illustrated in its most elaborate form in the genus *Pinus* (pine) (Fig. 2.6). The multiple embryos arise in characteristic positions, and in order to appreciate their development it is necessary to study the early stages of pine embryogeny. The four nuclei of the free nucleate stage move to the inner end of the embryo, opposite the micropyle, and there become arranged in a single tier. The nuclei then divide synchronously, producing

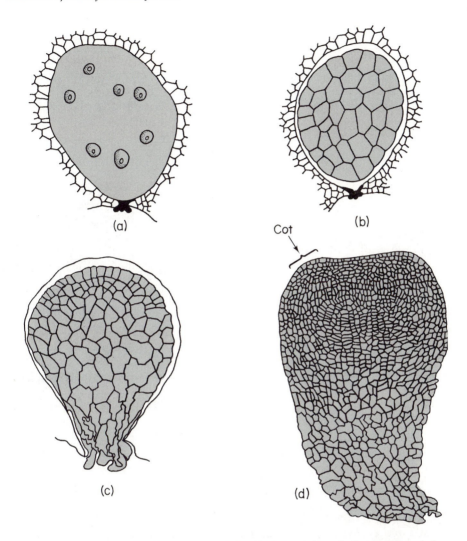

Figure 2.5. *Embryo development in* Ginkgo biloba. *(a) Early free nuclear embryo. (b) Wall formation complete throughout the embryo. (c) Cellular embryo at a later stage of development showing the axial gradient of cell size. (d) An early stage of cotyledon (Cot) development. (a,b) ×80, (c) ×90, (d) ×70. ([a,b] D. A. Johansen, Plant Embryol., Waltham: Chronica Botanica, 1950. [c,d] H. G. Lyon, Minn. Bot. Stud. 3:275, 1904.)*

a second tier of four nuclei, and this division is immediately followed by wall formation. The inner tier is completely enclosed by walls and the outer tier is open toward the micropyle. Each tier divides again so that there are now four tiers of four cells each, the outermost still open

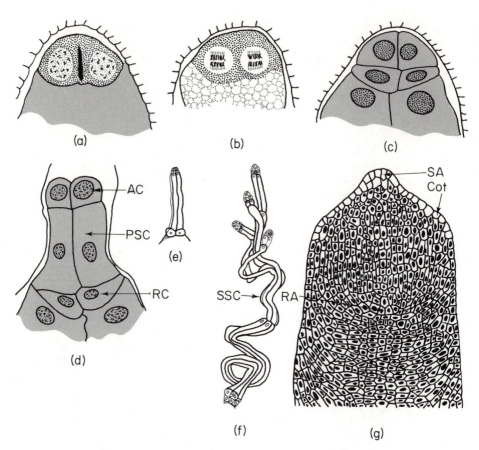

Figure 2.6. *Embryo development in* Pinus. *(a) Free nuclear stage showing two of the four nuclei at the end of the embryo opposite the micropyle. (b–d) Successive stages in the development of the four cell tiers. The suspensor cells are beginning to elongate in (d). (e,f) Stages in elongation of the suspensors and development of multiple embryos. (g) The embryo showing the differentiation of organs and tissues. Key: AC, apical cells; Cot, cotyledon; PSC, primary suspensor cells; RA, root apex; RC, rosette cells SA, shoot apex; SSC, secondary suspensor cells. (a)* ×130, *(b–d)* ×165, *(g)* ×110. *([a–d] D. A. Johansen,* Plant Embryol., *Waltham: Chronica Botanica, 1950. [e,f] J. T. Buchholz,* Trans. Ill. Acad. Sci. 23:117, 1931. *[g] Spurr, 1949.)*

toward the micropyle. The cells of the innermost tier are called the apical cells, those of the second tier the primary suspensor cells, and those of the next tier the rosette cells. The suspensor cells begin to elongate, and their growth pushes the apical cells into the tissue of the female gametophyte. The apical cells separate laterally from one another, meanwhile cutting off secondary suspensor cells, which also elongate.

Each apical cell, by dividing in various planes, builds up a multicellular mass that develops into a bipolar embryo. Even more striking is the fact that the rosette cells resume growth, and by a process similar to that occurring in the apical cell derivatives, each gives rise to a separate embryo. Thus, up to eight embryos may result from the development of a single zygote (Wardlaw, 1955).

The later stages of embryogeny that have been investigated in pine in many ways resemble those of angiosperms. The first embryonic organ to appear is the root apex, with a massive root cap adjacent to the suspensor. At an early stage histodifferentiation begins, as evidenced by the appearance of vacuolated pith parenchyma and elongated procambial cells. Finally the shoot apex, surrounded by the numerous cotyledons, arises at the top of the embryo opposite the root apex (Spurr, 1949).

Embryo development in lower vascular plants

In the lower vascular plants there is a bewildering array of embryonic types that have been described in more or less detail by numerous authors. In the majority of ferns the gametophytic phase of the life cycle is a distinct, free-living organism of diminutive size dependent upon its own photosynthesis for the energy required, not only for its own development, but for that of the embryo as well. Under these conditions one might expect to find the pattern of embryonic development in these organisms very different from those found in the embryos of the seed plants. In fact, the general pattern of embryogeny is not markedly dissimilar from embryogenic patterns of many higher plants. Much of the classical work on fern embryology has been concerned with early cleavage patterns and the possible organogenetic significance of this segmentation. In the leptosporangiate ferns (Fig. 2.7), with a few exceptions, the first division of the zygote is parallel to the long axis of the archegonium (see Fig. 3.1). The second division is perpendicular to the first and usually at right angles to the archegonial axis, but sometimes, for example in *Todea barbara*, parallel to it. In either case, the four quadrants divide again, usually synchronously, to form the octet stage embryo. The subsequent development of organs has been related to the original quadrants, each giving rise to a single organ, the shoot, the first leaf, the foot (an embryonic organ having a presumed absorptive function), and the root. The first division beyond the octet stage was presumed to set off the apical cells of the root, shoot, and first leaf.

The precise relationship between the quadrants and the embryonic organs has often been questioned, and several investigations have thrown doubt upon its general applicability. Ward (1954) has reported that in

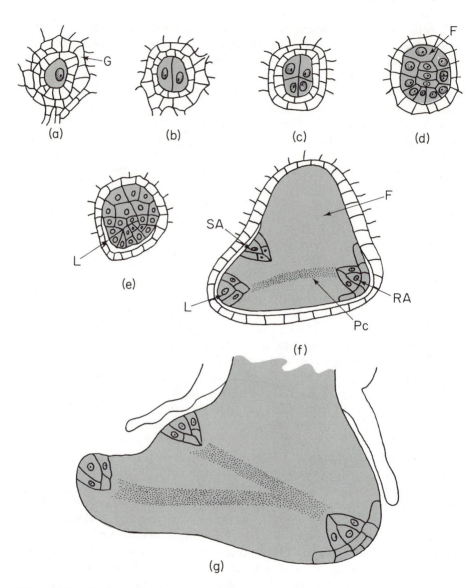

Figure 2.7. *Embryo development in* Phlebodium (Polypodium) aureum. *(a–c) Early developmental stages in which cell division in the embryo proceeds without evident differentiation. (d) Cells in the upper half of the embryo enlarge to form the foot. (e) Origin of the first leaf in the embryo. (f,g) Later stages of embryo development showing the origin and growth of the shoot and root apices, and the position of the procambium. Key: F, foot; G, gametophyte cells; L, leaf; Pc, procambium; RA, root apex; SA, shoot apex. (a–f) ×180, (g) ×140. (Drawn from Ward, 1954).*

the embryogeny of *Phlebodium* (*Polypodium*) *aureum* (Fig. 2.7) organs become recognizable only after many cell divisions and much growth. Cell divisions beyond the octet stage produce a globular mass within which differentiation begins, and the first evidence of this is in the part of the embryo that will give rise to the foot. The foot appears to origi-nate from much of the tissue derived from the two upper quadrants of the early embryo and not from one of these alone as the classical inter-pretation had suggested. Development of the foot involves the enlarge-ment and vacuolation of cells that remain thin-walled and intrude be-tween some of the adjacent cells of the gametophyte. Soon thereafter the first leaf appears as an outgrowth from the region referable to the lower anterior quadrant, in accordance with the classical description. The root arises next from derivatives of the lower posterior quadrant, and its origin, like that of the leaf, appears to conform to the traditional interpretation. The final embryonic organ to develop is the shoot apex, which appears between the leaf and the foot and is not referable to an entire quadrant. The four embryonic organs appear at about one-day intervals. Whereas classical fern embryology implied that organs are delimited at a very early stage in relation to segmentation of the zygote, this study and others like it show the prior formation of a multicellular mass within which the organs are delimited sequentially.

Some additional features of fern embryogeny have been revealed in a study by DeMaggio (1961) on *Todea*. In this fern the first evidence of differentiation in the embryo is the enlargement and vacuolation of cells in the foot region, as was the case in *Phlebodium*. In the remainder of the embryo, differentiation into an outer, surface layer and a central group of very small cells then takes place. It is only after this differen-tiation that the other embryonic organs have their origins. The leaf is the next organ to appear, but unlike the sequence in *Phlebodium*, the shoot apex arises before the root apex.

The practice of interpreting embryogenesis in relation to early cleav-age patterns has been characteristic of studies of other groups of lower vascular plants as well, and in the absence of modern reinvestigations, one questions this type of interpretation on the basis of comparison with the ferns. In both *Lycopodium* and *Selaginella* (Lycopsida) the first divi-sion of the zygote is transverse and cuts off a cell adjacent to the arche-gonial neck, which develops into a suspensor of limited extent. In *Ly-copodium* the suspensor cell only rarely divides, but in *Selaginella* it often becomes multicellular. A foot ordinarily develops in the region of the embryo adjacent to the suspensor, and it is usually massive in *Lycopo-dium* but variable in extent of development in *Selaginella*. Another in-teresting feature of both of these genera is the delayed development of an embryonic root that appears long after the shoot apex and first leaf are well advanced. The delay in root initiation in the primitive Lycopsida is of interest in relation to the complete absence of roots from both the embryo and the adult plant of both genera of the Psilopsida (*Psilotum* and *Tmesipteris*).

In all of the lower vascular plants, in contrast to seed plants, there is no interruption of growth to mark the end of a distinct embryonic phase. The embryo develops directly into a juvenile plant that bursts out of the enclosing parental gametopytic tissues and in a relatively short time becomes independent. In some species of *Lycopodium* the embryo ruptures the surrounding gametophytic tissues before differentiation of embryonic organs and grows out as a green parenchymatous structure, the protocorm, on which a shoot apex is subsequently differentiated.

CELLULAR CHANGES DURING EMBRYOGENESIS

The egg cell in angiosperms is a relatively small cell, ordinarily 50 to 100 μm in length. It has been described at the ultrastructural level as having characteristics suggestive of metabolic quiescence. The cytoplasm contains a relatively sparse endoplasmic reticulum, few ribosomes, and a small number of inactive Golgi bodies, mitochondria, and plastids that are more or less randomly distributed. Typically, a large vacuole occupies the micropylar end of the cell. As might be expected, among the relatively few species that have been examined some variability in the expression of these features has been reported.

Fertilization marks the onset of extensive change and reorganization of the egg cell cytoplasm, as has been described by Jensen (1968) in a study of *Gossypium* and by Schulz and Jensen (1968) in *Capsella* (Fig. 2.8). There are marked changes of appearance and distribution of cell organelles in the zygote. Ribosomes become aggregated into long, helical polysomes that are associated with plastids and mitochondria. Later, new ribosomes begin to be synthesized, but these aggregate into shorter polysomes that can be distinguished from those formed from maternal ribosomes on the basis of length. Plastids and mitochrondria accumulate around the nucleus, and Golgi bodies increase in number and activity. The effect of these changes is to convert the seemingly metabolically inactive egg cell into the metabolically active zygote.

During early stages of embryo development there are distinct changes in cell size that appear consistently in the various groups of vascular plants. The early cleavages of the embryo result in successively smaller cells because the embryo as a whole either increases slowly in size during the process, may not increase in size at all, or sometimes actually decreases because of reduction in vacuolar volume. In the embryo of *Gossypium* (Fig. 2.9) Pollock and Jensen (1964) have shown that cell size

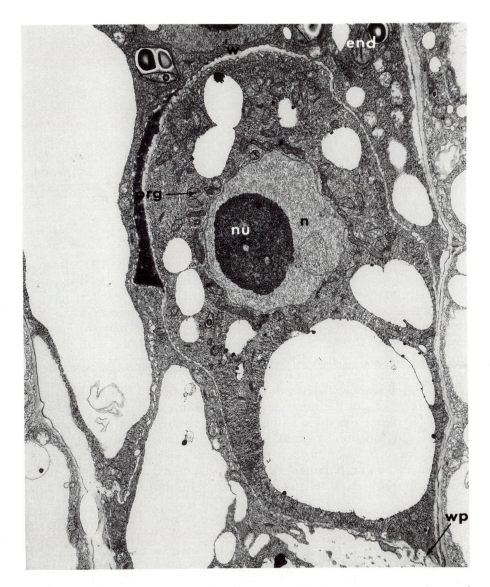

Figure 2.8. *Ultrastructure of the zygote of* **Capsella**. *The zygote is an elongated cell with the nucleus situated at the terminal end and a large vacuole at the basal end. Other smaller vacuoles occur throughout the cytoplasm. Plastids and mitochondria are concentrated around the nucleus. The cell wall separating the zygote from the surrounding endosperm is becoming thick, and at the basal end of the zygote there are wall projections that increase the absorbing surface. Key: end, endosperm; n, nucleus; nu, nucleolus; org, cytoplasmic organelles concentrated near the nucleus; w, wall of the zygote; wp, wall projections. ×6,000. (Schulz and Jensen, 1968.)*

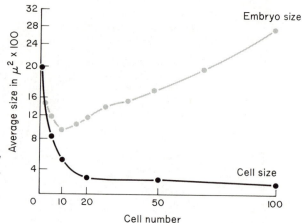

Figure 2.9. *Changes in cell size and embryo size during early development of the embryo of* Gossypium. *(Pollock and Jensen, 1964).*

in the embryo proper decreases progressively until, at the 100-cell stage, average cell size is about one-twentieth that of the zygote. Cell size then remains relatively constant until about the 1,000-cell stage, when it is further reduced by a half. Cell size continues to be reduced for a variable length of time in embroys of different species and may then remain constant during further growth until changes associated with histodifferentiation appear. Comparable reductions of cell size are a well-known feature of cleavage in animal embryos, in which it has been suggested that the significance of size reduction is to establish a desirable balance between the size of the nucleus and the volume of the associated cytoplasm. No doubt the same value may be assigned to this phenomenon in plants. In addition, it would seem that the onset of histogenesis and organogenesis in an embryonic mass of small size may require that the protoplasm be compartmentalized, so that it may be essential to increase the number of cells rapidly with a minimum increase of total mass.

The reduction in cell size in the embryo proper accentuates the distinctions between this region and the suspensor. The cells of the suspensor remain relatively large, may even increase in size, and are highly vacuolated (Fig. 2.10). It has also been reported in a number of cases that the content of nuclear DNA increases in these cells through endopolyploidy, polyteny, or other mechanisms (D'Amato, 1984). At the ultrastructural level the cytoplasm is seen to be less dense as a result of a lower concentration of ribosomes, but this does not mean that suspensor cells are metabolically inactive (Jensen, 1974). Rather, they are seen as highly active cells that function for a limited time in an exclusively embryonic organ in contrast to the organ- and tissue-forming cells of the embryo proper. An interesting characteristic found in the external walls of one or more cells of the suspensors of a number of species is the

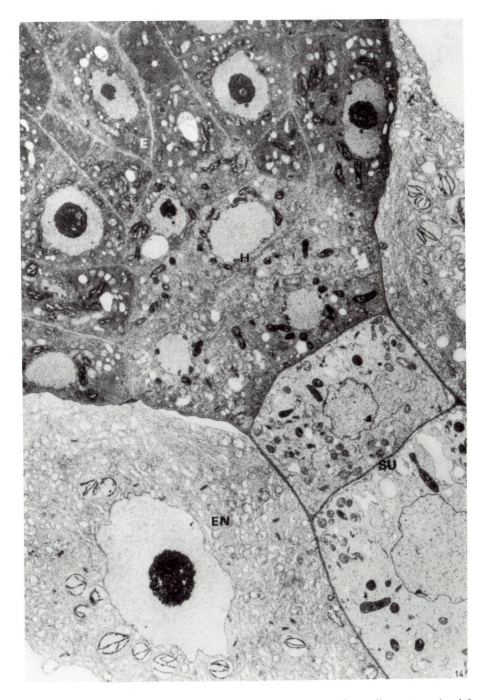

Figure 2.10. *An electron micrograph of a heart-stage embryo illustrating the difference in cytoplasmic density between the embryo proper (E) and the suspensor (SU). Key: EN, endosperm; H, region in which the root apical meristem will arise. ×3,600. (R. Schulz and W. A. Jensen, J. Ultrastr. Research 22:376, 1968).*

development of wall projections of the sort recognized in transfer cells elsewhere in the plant, which function in short-distance transport. This feature may be correlated with the fact that, although *plasmodesmata* (intercellular protoplasmic connections) are abundant in internal walls throughout both the embryo proper and the suspensor, they are conspicuously absent in the walls that separate the embryo from any surrounding tissue.

In the seed plants generally there is a feature of embryo development that does not occur in the lower vascular plants; the arrest of development and the entry into a metabolically quiescent state. Associated with this final phase of embryogeny are a number of cellular changes. As water is progressively withdrawn from the embryos, vacuoles become subdivided into smaller units, and into these, reserve materials are deposited. Some of these reserves are highly specific proteins the synthesis of which is restricted to this phase of the life cycle. Starch or oil are also common seed reserves. As noted earlier, the suspensor senesces during the latter part of embryonic development and has largely disappeared by the time of the changes just described in the embryo proper.

GENERAL COMMENT

Several general observations emerge from a comparative examination of embryonic development in the vascular plants. In spite of apparent great diversity, there is a strong suggestion of fundamental similarity of pattern in the establishment of the basic developmental plan of the embryo. The obvious differences that occur are those relating to early cleavage patterns before the onset of histogenesis or organogenesis, those concerned with accessory structures pertaining to embryonic survival, and those resulting from differential growth rates among the various embryonic tissues and organs. Thus, though the free nuclear stage of the gymnosperm embryo contrasts sharply with the cleavage patterns of other groups, though the precocious initiation and enlargement of the first leaf in fern embryos is a distinctive and consistent feature, and though the suspensor is variable in presence and in form in different groups, these features are somewhat peripheral to the basic development of the embryo, though each has its own morphogenetic interest. Thus, diverse patterns of embryogeny seem to converge upon a common point, the production of an embryo having the rudiments of the shoot and root systems.

From the morphogenetic point of view, one of the most interesting conclusions arising from the study of embryonic development is that the developmental relationships of organs and tissues to meristems that one has come to recognize as general from the study of adult plant development do not necessarily hold for embryonic stages. In the adult plant, leaf primordia arise only in relation to the activity of the shoot apical meristem but in many embryos the emergence of embryonic leaves may be unrelated to a localized meristem. Similarly, there are numerous instances in which histodifferentiation in embryos is initiated independently of any organized meristem of root, stem, or leaf, whereas in the adult plant the meristem commonly has been regarded as the initiator and organizer of these events. This leads to the not surprising conclusion that the regulation of morphogenetic processes at the time of their inception in the embryo is rather different from the regulation of their continuation in the expanding plant. In fact, the developmental progression of the embryo is essentially a sequential one, whereas post-embryonic development is dominated by repetitive events. It also calls attention to the possibility that the environment in which the embryo develops has an important bearing upon the control of morphogenetic processes. It is to these questions that attention will be turned in the next chapter.

REFERENCES

D'Amato, F. 1984. Role of polyploidy in reproductive organs. In *Embryology of Angiosperms*, ed. B. M. Johri, 519–66. Berlin: Springer-Verlag.

DeMaggio, A. E. 1961. Morphogenetic studies on the fern *Todea barbara*. (L.) Moore. II. Development of the embryo. *Phytomorphology* 11:64–79.

Jensen, W. A. 1968. Cotton embryogenesis: The zygote. *Planta* 79:346–66.

———. 1974. Reproduction in flowering plants. In *Dynamic Aspects of Plant Ultrastructure*, ed. A. W. Robards, 481–531. New York: McGraw-Hill.

Maheshwari, P. 1950. *An Introduction to the Embryology of Angiosperms*. New York: McGraw-Hill.

Miller, H. A., and R. H. Wetmore. 1945. Studies in the developmental anatomy of *Phlox drummondii* Hook. I. The embryo. *Am. J. Botany* 32:588–99.

Poethig, R. S., E. H. Coe, and M. M. Johri. 1986. Cell lineage patterns in maize embryogenesis: A clonal analysis. *Devel. Biol.* 117:392–404.

Pollock, E. G., and W. A. Jensen. 1964. Cell development during early embryogenesis in *Capsella* and *Gossypium*. *Am. J. Botany* 51:915–21.

Randolph, L. F. 1936. The developmental morphology of the caryopsis in maize. *J. Agric. Res.* 53:881–916.

Schulz, R., and W. A. Jensen. 1968. *Capsella* embryogenesis: The egg, zygote and young embryo. *Am. J. Botany* 55:807–19.

Spurr, A. R. 1949. Histogenesis and organization of the embryo in *Pinus strobus*. L. *Am. J. Botany* 36:629–41.

Ward, M. 1954. The development of the embryo of *Phlebodium aurem*. J. Sm. *Phytomorphology* 4:18–26.

Wardlaw, C. W. 1955. *Embryogenesis in Plants*. New York: Wiley.

3

Analytical and experimental studies of embryo development

The particular conditions under which development of a vascular plant begins in the embryo obviously hold great interest for the student of morphogenesis. The question that clearly requires investigation is whether the sequential pattern that emerges during embryogeny is to be regarded as an expression of the inherent capacity of the zygote, as the result of specific regulation from the environment, or as the manifestation of subtle interaction between the two. Although descriptive accounts of embryogenesis are helpful in exploring this problem, experimental and analytical techniques can provide a different and more penetrating analysis of these possibilities. Experimental embryology has been an extremely valuable discipline in elucidating problems of animal morphogenesis, but the plant counterpart of this field has played a limited role in the understanding of plant morphogenesis. A major factor contributing to this deficiency has been the relative inaccessibility of the plant embryo at the formative stages, with the result that the botanical work that most closely corresponds to experimental animal embryology has been done with the apical meristems of the adult plant. Nonetheless, there are several techniques by which the development of the embryo may be probed, and these have led to the acquisition of a body of information from which a meaningful interpretation of embryogeny is beginning to emerge.

Throughout the embryophytes, fertilization and embryogenesis occur in specialized structures that appear to provide a distinctive environment. It is important to consider these structures and the environment they create before turning to the sequential events of embryogeny.

THE ENVIRONMENT OF THE EMBRYO

The morphological environment

In the ferns, as in other lower vascular plants, the egg is produced on a gametophtyte that is entirely separated from the sporophyte. This is a small photosynthetic thallus in most ferns but in other lower groups it may be saprophytic and in some cases develops largely within a spore and is dependent upon food reserves contained within the spore. It would appear, therefore, that the supply of nutrients for the developing embryo in these plants must be somewhat limited. The egg itself is located within the basal portion of a closed flasklike structure known as an archegonium (Fig. 3.1) which opens at egg maturity in the presence of water, allowing a multiflagellated sperm to enter and effect fertilization. Following fertilization, as the zygote begins to divide, there are also divisions in the adjacent cells of the gametophyte, which give rise to a multilayered jacket. It is believed that this structure places physical constraints upon the growing embryo that could have consequences for its development.

In the gymnosperms the female gametophyte is a large, nutrient-filled structure enclosed by tissues of the parent sporophyte because the spore from which it arises is not shed (Fig. 3.2). The ovule is thus a combination of sporophytic and gametophytic tissues. The early development of the gametophyte proceeds by repeated nuclear divisions, and the large cytoplasmic mass is subsequently separated into cells by wall formation. Ultimately, archegonia are formed at one end of the gametophyte in the vicinity of the micropyle. Although the archegonia are somewhat reduced in structure compared to those of the ferns, the eggs are very large. The sperm are produced by a separate male gametophyte, and this is contained within a pollen grain that is transferred directly to the ovule and drawn into it through the micropyle. In some ancient gymnosperms (e.g., *Ginkgo*) motile sperm are released within the ovule, but in the modern conifers the sperm are nonmotile and are delivered to the archegonia by the growth of a pollen tube. After fertilization, embryogenesis begins in the archegonium, but the development and elongation of a suspensor soon force the embryo proper into the nutrient-filled tissue of the female gametophyte, where its development is completed.

As in the gymnosperms, the angiosperm ovule is composed of both gametophytic and sporophytic tissues but the female gametophyte at the time of fertilization consists of only a few cells (Fig. 3.3). Nevertheless, close examination of this gametophyte, or embryo sac, particularly at the ultrastructural level, reveals a highly specialized structure that facilitates fertilization and the initiation of embryogenesis. Typically, just prior to fertilization, it consists of seven cells in three functional associations: the egg apparatus at the micropylar end, consisting of an egg and two synergids, three antipodals at the opposite end and the binucleate central cell that encompasses the remainder of the embryo sac. The egg and the synergids are walled cells, but the walls are incom-

Figure 3.1. *The archegon-*
ium of a fern before (a) and
after (b) fertilization. In (a) the
neck is closed and is filled by
a binucleate neck canal cell
and a ventral canal cell. In (b)
the neck canal is open, the egg
has been fertilized and the first
division of the zygote has
taken place.

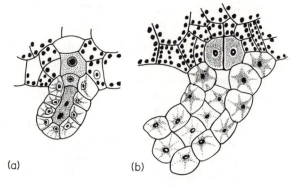

(a) (b)

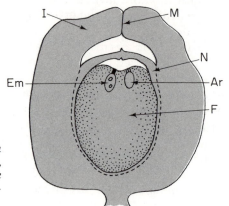

Figure 3.2. *Longitudinal section of a*
gymnosperm ovule (a cycad). Key: Ar,
archegonium; Em, two-nucleate stage
embryo; F, female gametophyte; I, in-
tegument; M, micropyle; N, nucellus.

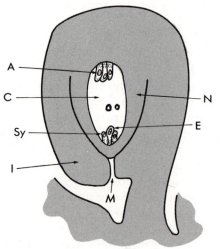

Figure 3.3. *Longitudinal section of an*
angiosperm ovule showing details of the
embryo sac shortly before fertilization.
Key: A, antipodal cells; C, central cell, E,
egg; I, integuments; M, micropyle; N,
nucellus; Sy, synergid.

pletely formed over the inner ends of the cells, and in these regions their plasma membranes lie close to that of the central cell. In angiosperms the pollen grains germinate on a receptive surface of the pistil (stigma), and pollen tubes must grow through the tissues of the pistil before entering the micropyles of the enclosed ovules. At the time of pollen tube entry, or even earlier, one of the synergids begins to degenerate. The pollen tube now penetrates this synergid by way of the filiform apparatus, an ingrowth of wall material somewhat resembling the wall projections of transfer cells. After some growth of the pollen tube within the degenerating synergid, contents of the tube, including two sperm cells, are discharged. Because of the degeneration of the synergid, one of the sperm cells is brought directly into contact with the plasma membrane of the egg, and fusion follows. Subsequently, the egg and sperm nuclei fuse and fertilization is complete. With some variations in the structure of the embryo sac and in certain details of the fertilization process, the events just described set the stage for embryogenesis in angiosperms.

It is a unique feature of the reproduction of angiosperms that a second fusion accompanies the fertilization that initiates embryogenesis, and this also sets in motion a developmental process of considerable significance. The second sperm cell released from the pollen tube fuses with the plasma membrane of the central cell, and its nucleus then fuses with the two nuclei of that cell, which have often already merged. The resulting nucleus is thus triploid, in contrast to the diploid zygote, and the central cell has now become the first cell of the endosperm, a tissue specialized for the nutritional support of the embryo. In some cases endosperm cells are more than triploid because more than two nuclei of the female gametophyte are involved in the fusion. In addition, it is not uncommon for the nuclear DNA content to be increased further through endopolyploidy. Thus, in contrast to the situation in the gymnosperms where the nutrients that support embryo development are accumulated in a massive female gametophyte prior to fertilization, double fertilization in the angiosperms causes the embryo and its supporting tissue to develop contemporaneously. Commonly, the early development of the endosperm is reminiscent of the gymnosperm female gametophyte in that a phase of free nuclear division precedes septation by the formation of walls. In other cases, however, the first one or several divisions are accompanied by cytokinesis, but free nuclear division usually follows in the several resulting compartments before the final septation. In the late stages of endosperm development the pattern of cell division may become very regular so that if a mutation causes a recognizable change in a cell, the derivatives of that cell give rise to a distinctive sector of the endosperm. It was the analysis of such sectors in maize kernels that resulted in the identification of genetic-controlling elements that are believed to play a role in differentiation (see Chapter 13).

The developmental fate of the endosperm varies in different groups of angiosperms. In the orchid family (Orchidaceae), for example, it degenerates after forming a few nuclei at the most, but this is an extreme situation. In many cases the endosperm is clearly functional but has been depleted and obliterated by the embryo before seed maturity. The reserves for germination are then contained within the embryo, for example, in the cotyledons of legumes. In other cases the endosperm persists to the maturation of the seed and, in addition to providing nutrients during embryogeny, is the primary source of metabolites for early seedling growth. In such cases cells in various parts of the endosperm may become specialized in relation to particular functions, as in the aleurone layer in cereal grains, which is the source of the hydrolytic enzymes required in germination. One of the most striking examples of specialization is seen in the widespread occurrence of endosperm haustoria, composed of one or a few cells, which invade surrounding ovular tissues and presumably take up nutrients from them. As the endosperm develops it becomes the repository for materials transported into the seed as well as the site of synthesis of metabolites produced from these materials. Some of these may be specific to the endosperm, such as the protein zein in the endosperm of maize. The widespread use of coconut milk and other endosperms to facilitate the growth of cells, tissues and even embryos in culture is evidence of the richness of their content of nutrients and growth-promoting factors.

The physical environment

It is now necessary to give consideration to the possible role of physical components of the embryonic environment. As the embryo develops it expands into regions previously occupied or simultaneously being occupied by other tissues, and the probability exists that at least in some cases the embryo experiences physical restraint on its expansion that might have morphogenetic significance. It is reasonable to expect that physical factors in the embryonic environment might be most easily analyzed in groups like the ferns in which the nutritional relationship between the embryo and its surrounding tissues appears to be relatively uncomplicated and the archegonium within which the embryo develops provides an obvious source of physical restraint. The first attempt to approach this problem experimentally in a fern, *Phlebodium* (*Polypodium*) *aureum*, was undertaken by Ward and Wetmore (1954). In this investigation various patterns of vertical and horizontal incisions were made in the prothallial tissues immediately surrounding the zygote (Fig.

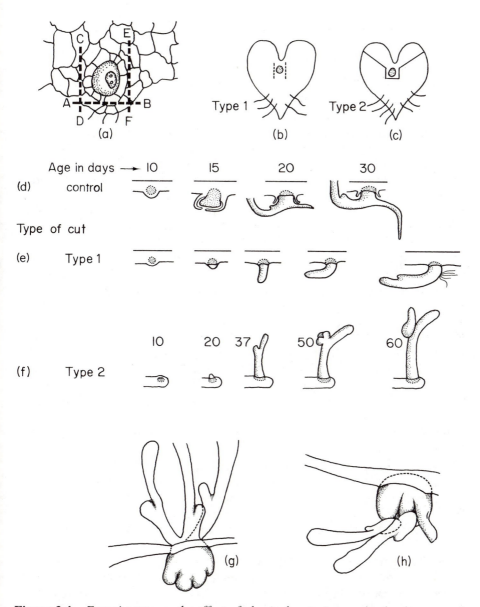

Figure 3.4. *Experiments on the effect of physical restraint on the development of the embryo of* Phlebodium (Polypodium) aureum. *(a) Longitudinal section of the zygote and surrounding gametophytic tissues showing the pattern of vertical and horizontal incisions. An incision along the line A–B removes the archegonial neck. (b,c) Gametophytes showing the positions of vertical incisions that are designated Types 1 and 2 below. In both types the archegonial neck is also removed. (d) Development of an unoperated embryo as the control for experimental treatments. (e,f) Development of embryos after Type 1 and Type 2 operations. (g,h) Development of embryos after removal of the archegonial neck only. (Adapted from Ward and Wetmore, 1954.)*

3.4). It was presumed that the incisions would reduce the restraint normally placed upon the zygote and early embryo by these tissues, but this was not verified by direct measurement. Following such operations the development of the embryo was markedly disturbed and the rate of growth was somewhat slower than in normal embryogeny. The embryos burst out of the surrounding tissues in fewer days after fertilization than in normal cases. At this time differentiation of leaf, shoot apex, and root had not occurred, and in many cases, the embryo consisted of a cylindrical mass of tissue that subsequently produced a leaflike organ at its tip, followed by a shoot apex in a position lateral to the leaf. Root development was considerably delayed. In those cases where the operation consisted only of the removal of the superficial part of the archegonium by a horizontal cut, the resulting outgrowth was tuberous and irregular rather than cylindrical. After considerable delay, leaflike appendages followed by laterally placed shoot apices arose in several places so that, in effect, one embryo produced several sporophytic buds. These experimental procedures clearly disturbed the normal pattern of embryogeny, but it is significant that in all cases the usual sporophytic organs differentiated. It is probable that the developmental abnormalities resulted from the reduction of physical restraint upon the embryo, but it cannot be overlooked that incisions in the gametophyte must also have impeded the normal nutritional and hormonal patterns and have caused the release of substances from damaged cells.

It might be expected that the role of external physical constraint could be investigated most effectively if the whole course of embryonic development could be made to occur in its absence. This was done experimentally by DeMaggio and Wetmore (1961) when the zygote of *Todea* was removed from the archegonium and grown in aseptic culture. Growth of the cultured embryos was slow, but the planes of early divisions were oriented normally, producing octet-stage embryos. Later divisions were less regular than in the normal embryo, and division soon ceased. The cells then expanded, causing the surface to protrude in various ways. Development was arrested at this stage, but some months later the embryos were found to have resumed development and to have produced flattened, two-dimensional outgrowths reminiscent of gametophytic prothalli. Although this experiment is difficult to interpret fully, it provided partial confirmation of the speculation that the difference in growth pattern of the fern sporophyte and gametophyte may result from the different environments in which the development of each begins. The sporophyte begins as a physically constrained zygote, whereas the gametophyte begins as a freely exposed spore.

The role of physical factors in the control of embryogeny in flowering plants is less easily approached. The effect of confinement upon later

Figure 3.5. *Embryo sacs of* Helianthus annuus *isolated from fixed (a) and fresh (b) material. The egg, one synergid, and one central cell nucleus are visible in each. Nomarski interference contrast microscopy was used to obtain these photographs. (a) ×265, (b) ×215. Scale bars = 30 μm. (Zhou and Yang, 1985.)*

stages of embryo development is evident in many cases in the curvature and other deformations that fit the embryo neatly into the seed and are notably lacking in embryos grown in culture. However, such effects of restriction in late stages of development are scarcely comparable to those suggested for the ferns that operate at very early stages, and they seem to be of little morphogenetic significance. It is precisely at the early stages of embryogeny in angiosperms that physical factors are least apparent. Only later as the endosperm develops does the embryo appear to come into close physical contact with other tissues. The possibility exists that the growing endosperm offers resistance to the enlargement of the embryo, but there are no measurements to test this suggestion.

It would, of course, be desirable to remove the zygote of an angiosperm from its ovular environment and culture it in isolation as was done for the fern *Todea*. While this has not yet been accomplished, recent reports of the isolation of intact, living embryo sacs by enzymatic degradation of ovules hold out considerable promise for the future (Zhou and Yang, 1985) (Fig. 3.5). It seems probable that zygotes could be removed from such embryo sacs and placed in culture or at least that the major part of the embryo sac could be dissected away leaving the zygote essentially free.

There is good reason to expect that an isolated zygote might be able to produce an embryo, because microspores and immature pollen grains of a number of species, isolated from the anther on a nutrient medium, have been able to give rise to normal-appearing embryos (Narayanaswamy and George, 1982). While the early divisions do not follow precisely the pattern of zygotic embryos, later stages are closely comparable. These embryos are haploid, of course, and the plants they produce are extremely valuable to plant breeders because doubling of the chromosome number results in immediate homozygosity without lengthy inbreeding. It has also been found that isolated somatic cells in culture can give rise to embryos but by a less direct pathway than that of either zygotic or pollen embryos (see Chapter 17). In view of these experiments it is difficult to assign a significant role to physical restraint in the early development of the angiosperm embryo.

The chemical environment

In attempting to investigate how the development of the embryo is controlled, the technique of embryo culture has proved to be extremely valuable. In this technique the embryo is removed from its natural environment at the desired stage of development and placed in a nutrient medium of known composition (Raghavan, 1976). Embryo culture has a long history, extending back into the past century. During the early period a number of workers removed fully developed or nearly mature embryos and found that these were able to germinate satisfactorily on nutrient media of very simple composition. In some cases the nutrients consisted of solutions of inorganic salts only. Although the culture of mature embryos has been of great significance to horticulturalists concerned with overcoming dormancy problems in some seeds, it tells little about the morphogenetic control of embryogeny. In order to approach problems in this area it is necessary to culture embryos at earlier stages of development. From the earliest attempts to excise and culture immature embryos, it has been apparent that progressively younger embryos have more exacting requirements than older, more mature embryos. These results generally have been interpreted as revealing a progressive transition from a heterotrophic mode of nutrition towards autotrophy.

This is illustrated in studies in which progressively younger embryos have been grown successfully in culture. Early work on embryo culture of *Datura* (Van Overbeek et al., 1942) showed that mature embryos, or even well-differentiated immature embryos, could be grown into seedlings on a medium consisting only of inorganic salts and a low concentration of sugar. Younger embryos failed to develop satisfactorily on this medium but could be cultured at the so-called torpedo stage – when the embryo is cylindrical with all organs formed but not fully enlarged – if the medium was supplemented by a mixture of organic substances including vitamins, amino acids, and other compounds believed to have growth-promoting activity. Still younger embryos, however, failed to develop even on this enriched medium. Embryos at the heart stage could be grown successfully only if the medium was further supplemented by coconut milk, a liquid endosperm selected because of its natural role in embryonic nutrition. A significant feature of the growth of these embryos was that for about ten days they developed in an embryonic way rather than undergoing the changes that lead to germination and development of seedlings. Rapid cell elongation associated with precocious germination had characterized the growth of larger embryos cultured without coconut milk. From these results it was concluded that coconut milk contains one or more growth factors, collectively designated "embryo factor," that promote the growth of young embryos without germination. Results of this type have been interpreted as revealing the progressive development in the embryo of the capacity to synthesize essential substances.

Other substances have been found that replace coconut milk in supporting the development of young embryos of *Datura* in culture, and that were, therefore, presumed to contain embryo factor. Among these are yeast extract, wheat germ, almond meal and sterile-filtered malt extract. Subsequently it was learned that a mixture of amino acids in the form of casein hydrolysate will essentially replace malt extract in the culture medium. In work with isolated *Hordeum* embryos (Ziebur et al., 1950), it was found that casein hydrolysate stimulates embryonic development as opposed to precocious germination. By testing the amino acid, inorganic phosphate, and sodium chloride components of commercial casein hydrolysate separately they found that much of the effect of casein hydrolysate was attributable to the increased osmolarity of the medium, although the amino acids also played a nutritional role.

The significance of high osmolarity in the culture of young embryos has intruded into the consideration of nutritional and hormonal aspects of embryo growth on a number of other occasions. It was found in *Datura* (Rietsema et al., 1953) that embryos excised at different developmental stages require different minimal sucrose concentrations for growth: Globular, preheart stages require 8 to 12 percent, heart stages require 4 percent, and later stages require progressively lower concentrations, until mature embryos can be grown in the complete absence of sugar. Interestingly enough, at a uniform concentration of 2-percent sucrose, embryos of different stages respond in exactly the same way to varied concentrations of mannitol, which suggests that the total effect of increased sucrose concentration is osmotic, the youngest embryos growing best in the medium with the highest osmolarity. Rijven (1952), working with *Capsella* embryos, showed essentially the same relationship and demonstrated also that very small embryos are isotonic with higher osmotic values than are more advanced embryos. Related to these observations is the more recent report (Smith, 1973) that in *Phaseolus vulgaris* (bean) the osmolarity of the liquid endosperm is very high around the late globular embryo and decreases in later stages. Rijven's experiments clearly showed that germination or expansion of embryos in culture occurred only after transfer to a medium with low osmotic value, thus strongly suggesting the role of high osmotic pressure in maintaining embryonic development.

Thus the dependence of continued embryonic development of immature excised embryos in culture upon complex substances had appeared to be clear from early studies, but subsequent analysis greatly reduced the clarity of the picture. A later study by Raghavan and Torrey (1963) on globular embryos of *Capsella* has brought the picture into much clearer focus. These workers found that relatively mature embryos could be cultured satisfactorily on simple media consisting of vitamins, inorganic salts, and a 2-percent concentration of sugar. Embryos as young as the heart stage grew satisfactorily, if somewhat slowly, on this medium, but younger embryos did not. To obtain growth of globular stages it was necessary to supplement this medium with indoleacetic acid, kinetin, and adenine sulphate, and they found a sharp optimum for each of these substances. There appeared to be no requirement for a high osmotic pressure of the medium. However, the surprising fact was that the requirement for the additives was in large part replaced by raising the sugar concentration of the basal medium to 12 or 18 percent or the salt concentration by a factor of ten. Thus it appeared that growth can be supported either by organic additions or by high osmotic concentrations in the surrounding medium, and the real basis of the maintenance of embryonic development in isolated embryos growing in culture remains obscure. Nevertheless, these experiments do support the contention that the embryo becomes progressively less dependent as embryogeny proceeds.

THE ROLE OF THE SUSPENSOR

The nearly universal occurrence of an early separation of the embryo into a group of meristematic cells, the embryo proper, and a region of vaculoated cells, either suspensor or foot, which functions only in the embryonic phase, suggests that there is a special function for this latter region (Fig. 3.6). Traditionally, as the name suspensor would imply, this role has been regarded as simply attaching the embryo proper and orienting it in relation to its food supply. The structural evidence reviewed in the previous chapter, however, is not consistent with this interpretation. Neither is the cytological evidence, which suggests high metabolic activity, or the actual demonstration of such activity in some cases (Walbot et al., 1972). Rather, it seems highly probable that this embryonic organ has a more direct role in the nutrition of the embryo. The vacuolating cells of the suspensor may be a preferred path of entry for nutrients into the more meristematic regions of the embryo. The elaborate haustorial development of the suspensor in some species provides

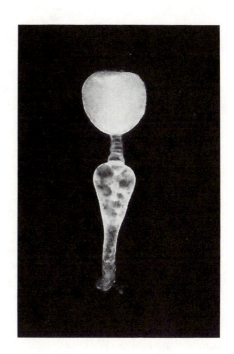

Figure 3.6. *Isolated live embryo of* Stellaria media *(chickweed) at an early heart stage. The distinction between suspensor and embryo proper is clearly visible.* ×150.

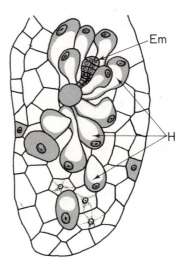

Figure 3.7. *Haustoria developed from the suspensor of* Asperula. *Key:* Em, *main body of the embryo;* H, *haustorial suspensor.* (F. E. Lloyd, Mem. Torrey Bot. Club *8:1, 1902.*)

support for this view (Fig. 3.7), as does the formation of internal wall projections that have the effect of increasing substantially the surface area of the plasma membrane in suspensor cells (Gunning and Pate, 1969). There are instances in which the haustorial suspensor actually penetrates the sporophytic tissues of the ovule or even tissues beyond the ovule, thus bypassing completely the endosperm nutritional mechanism (Maheshwari, 1950).

The significant role of the suspensor, particularly in early stages, is also demonstrated by embryo culture. Although earlier workers often appear to have ignored the suspensor, not realizing its importance, more recent workers have achieved success in culturing pre-heart-stage embryos by leaving the suspensor attached (Raghavan and Torrey, 1963). Even at the heart stage, which in *Phaseolus coccineus* (scarlet runner bean) does not require the suspensor for successful growth in culture, it has been shown that the presence of an intact suspensor greatly enhances both the survival and the development of the explants (Yeung and Sussex, 1979). If the suspensor is detached from the embryo proper but placed in contact with it on the surface of the culture medium, a substantial portion of the beneficial effect is retained. This suggests that the influence of the suspensor in this case is exerted by means of a diffusible material. Walthall and Brady (1986) have further emphasized the importance of the suspensor cells by demonstrating that at least one of its roles in *P. vulgaris* is the promotion of protein synthesis in heart-stage embryos. This effect is retained in part even if the suspensor is detached but included in the culture. With regard to the evidence in this and the previous experiment for the production of a diffusible substance, it is significant that much of the effect of an excised suspensor can be replaced by the inclusion of a low concentration of gibberellic acid in the culture medium.

The role of the suspensor in embryo development offers another example of progressive change in embryogeny. The increasing independence of the whole embryo is paralleled by a decreasing dependence of the embryo proper upon its suspensor. This is illustrated very pointedly in the experiments with cultured embryos of scarlet runner beans in which it was shown that the beneficial effect of the suspensor in the early heart stages has completely disappeared by the end of the heart stage (Table 3.1) (Yeung and Sussex, 1979). At this time the embryo proper has become self-sufficient for gibberellin (Pharis and King, 1985). It is also at about this time in most species investigated that the suspensor begins to degenerate, indicating that its functional life has been completed.

The growing awareness of the significant absorbing and biosynthetic capacity of the suspensor in the angiosperms has raised serious ques-

Table 3.1. *The Effect of the Suspensor on the Formation of Plantlets from* Phaseolus coccineus *Embryos Cultured in Vitro. The Results Were Recorded After Eight Weeks in Culture. Adapted from Yeung and Sussex (1979).*

Stage	Treatment	No. of Plantlets/ No. of Cultured Embryos	Percent of Plantlets
Heart (4.5 mm seed)	Embryo proper with suspensor attached	84/95	88
	Embryo proper only	37/89	42
	Embryo proper with detached suspensor in direct contact	37/51	73
	Embryo proper with suspensor 1 cm away	10/30	33
Late Heart (6.5 mm seed)	Embryo proper with suspensor attached	17/18	94
	Embryo proper only	17/18	94
Early Cotyledon (7 mm seed)	Embryo proper with suspensor attached	19/19	100
	Embryo proper only	18/18	100

tions about the role of the endosperm as the "nurse tissue" of the embryo throughout its development. There is growing evidence that at least the early, critical stages of embryogenesis are accomplished before the endosperm is sufficiently developed to provide much or any support. This is illustrated in *Papaver nudicaule* (Iceland poppy), in which endosperm digestion does not begin before the ninth or tenth day after pollination when the embryo has reached the late globular stage (Olson, 1981). During mid-embryogenesis stages the endosperm is certainly the major source of nutrients in most cases and may continue to be to the end of embryogeny and even to the germination phase. However, there are many instances in which the endosperm is absorbed before the completion of embryogeny so that in its final phases the embryo must again draw directly upon other tissues of the ovule.

THE COMPLETION OF EMBRYOGENY

In ferns and other lower vascular plants, the formation of the apical meristems in an embryo is followed, without pause, by their production of the postembryonic plant. On the other hand, in the seed plants generally, the morphogenetic activity of the apical meristems is arrested at some point after their formation so that there is a clear hiatus between embryogeny and the initiation of postembryonic development known as germination. This interruption does not immediately terminate embryogeny, and processes leading to the maturation of the embryo occur prior to complete developmental arrest. Prominent among these events is the synthesis of reserve nutrients, including abundant embryo-specific storage proteins (Higgins, 1984) that will be utilized during germination. Since it is characteristic of each species to produce a small number of these proteins in relatively large amounts and to initiate their synthesis at different times, this maturational phase of embryo development provides a clear example of a sequence of developmental events. As embryogeny draws to a close, the synthesis of these proteins ceases and the embryo becomes dehydrated and enters a period of metabolic quiescence. In some species germination can now occur simply with the addition of water, but in others special mechanisms of dormancy are superimposed so that germination can occur only upon reception of an appropriate signal. Of particular interest in this connection is the occasional occurrence of vivipary, or precocious embryo germination within the fruit. In *Rhizophora mangle* (mangrove), perhaps the best known viviparous angiosperm, there is a direct and uninterrupted transition from embryogeny to postembryonic growth as in the lower vascular plants and neither dehydration nor metabolic quiescence occurs (Sussex, 1975).

In several other species there is evidence that a hormone, abscisic acid, which accumulates in the embryo, is involved in the control of events in late embryonic development. Inclusion of abscisic acid in the culture medium suppresses germination of excised immature embryos, and high levels of the hormone enhance the synthesis and accumulation of some embryo-specific proteins (Finkelstein et al., 1985). Thus embryos that are capable of germinating when their meristems have been formed do not do so but continue embryonic development and the synthesis of specific proteins.

Sequential gene expression

The previous paragraphs have recorded a number of examples of the occurrence of sequential events in embryogeny. These events, and particularly those concerned with the timed synthesis of specific proteins, are suggestive of an underlying program of sequential gene activation. There is, in fact, direct evidence that this interpretation has validity. For example, in the embryos of *Brassica napus* (rapeseed) the two major storage proteins accumulate at different rates during different time periods. It has been possible to demonstrate that the messenger RNA for

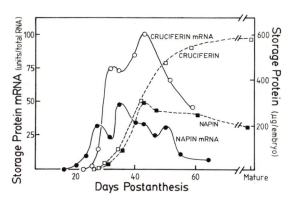

Figure 3.8. *The pattern of accumulation of the storage proteins Cruciferin and Napin and their respective mRNAs in the developing embryo of* Brassica napus *(rapeseed). (Adapted from Finkelstein et al., 1985.)*

each of these proteins appears in the embryo with a time course that parallels that of the protein but is slightly in advance of it (Finklestein et al., 1985) (Fig. 3.8). This is strong evidence for the activation of genes specific for those proteins.

Even more direct evidence for the involvement of gene activation in the sequence of developmental events is provided by the discovery in maize and *Arabidopsis* of a number of mutations that arrest embryo development at particular stages of the embryogenic sequence (Sheridan and Neuffer, 1981; Meinke, 1985) (Fig. 3.9). A reasonable interpretation of these mutants is that they represent mutations in genes required at specific stages of development, which, being defective, do not function at the proper time. In some of the mutants it has been shown that the embryos can be rescued by culturing them in an enriched nutrient medium, but in most cases the physiological basis of the defect is unknown. Particularly striking examples of the involvement of specific genes at particular stages of development are provided by viviparous mutants in both maize and *Arabidopsis* (Fig. 3.10). In these the normal phase of developmental arrest is not initiated because of the low level of abscisic acid and the embryos germinate precociously. It has been shown that the mutations must be present in the genome of the embryo and that it is the developmental program of the embryo itself that is affected (Karssen et al., 1983). In maize some viviparous mutants also contain low levels of ABA, but in others it appears that the embryo does not respond to normal levels of the hormone.

GENERAL COMMENT

We have seen in the previous chapter that embryogeny begins with a single cell and results in the formation of a bipolar embryo containing the meristems, which will give rise to the postembryonic plant. In this

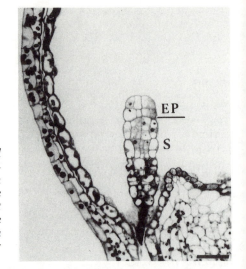

Figure 3.9. *Developmentally arrested mutant embryo of* Arabidopsis thaliana. *The embryo proper (EP) has been arrested at the eight-celled stage and the suspensor (S) has proliferated without undergoing normal differentiation. Scale bar = 20 μm. ×400. (M.P.F. Marsden and D. W. Meinke, Am J. Botany 72:1801, 1985.)*

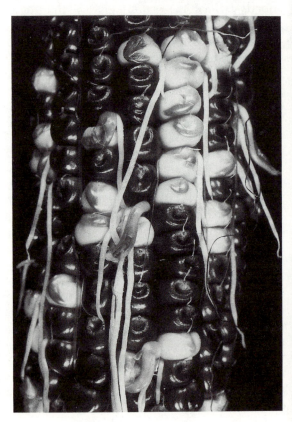

Figure 3.10. *Portion of a maize ear segregating for viviparous (pale) and wildtype kernels. The embryos of the viviparous kernels have already germinated and the roots and coleoptiles are extending from them. ×1.6.*

process new structural features, both tissues and organs, appear in an orderly sequence. Such a sequence is strongly suggestive of an underlying developmental program, components of which come into play in sequential order. In this chapter it has become apparent that there is a parallel series of physiological changes during embryogeny. Although the biochemical details are far from clear, a progressive acquisition of metabolic competence is evident from numerous embryo culture studies. This pattern is suggestive of the progressive elaboration of the genetically controlled enzymatic machinery that leads ultimately to autotrophy as a result of sequential gene activation. The occurrence of mutants that arrest embryonic development at specific stages and appear to do so through deficiencies in the synthetic machinery offers strong support for this interpretation. However, the most direct evidence for timed gene expression comes from the final stages of embryogeny, during which specific storage proteins are synthesized in quantities that facilitate analysis. Here the activation of the genes responsible for these proteins is revealed by the appearance of their specific messenger RNAs shortly before the proteins themselves are detected. While there is no such direct evidence relating to the earlier stages of embryogeny, it is reasonable to apply a similar interpretation to the observed sequential events.

In different groups of plants the region in which the embryo develops is highly variable, but for each group this environment has recognizable and characteristic structural, physical, and chemical features. The question to be considered is what role these features play in the regulation of embryogeny. In the seed plants at least, the chemical environment in which the embryo is developing is highly complex; consideration of this fact led to the development of the concept of embryo factors that act as regulators of development. Experimental evidence has now accumulated that numerous factors, both chemical and physical, can substitute for those complex factors that had been held to be highly specific. Furthermore, the embryo can dispense with many of the factors of its environment even though its pattern of development may then be somewhat atypical. The special chemical and physical features of the embryonic environment act perhaps not so much to regulate development as to provide the proper milieu in which the zygote may express its innate genetic capacity. The conditions necessary for complete expression of its inherent developmental potentialities must be provided by the embryonic environment – and they normally are provided in the embryo sac or archegonium. In this environment the zygote must be regarded as a unique cell in terms of what it does. However, a growing body of experimental evidence, alluded to in this chapter and extensively reviewed in Chapter 17, persuades us that a wide range of cells, and perhaps any living cell in the plant body, can reveal the same potentiality under appropriately permissive conditions. Thus, although neither the zygote nor its environment is unique, the regular combination of the two in the life cycle results in a remarkably effective mechanism for the production of new plants.

REFERENCES

DeMaggio, A. E., and R. H. Wetmore. 1961. Morphogenetic studies on the fern *Todea barbara*. III. Experimental embryology. *Am. J. Botany* 48:551–65.

Finkelstein, R. R., K. M. Tenbarge, J. E. Shumway, and M. L. Crouch. 1985. Role of ABA in maturation of rapeseed embryos. *Plant Physiol.* 78:630–6.

Gunning, B.E.S., and J. S. Pate. 1969. Transfer cells: Plant cells with wall ingrowths, specialized in relation to short distance transport of solutes – their occurrence, structure and development. *Protoplasma* 68:107–33.

Higgins, T.J.V. 1984. Synthesis and regulation of major proteins in seeds. *Ann. Rev. Plant Physiol.* 35:191–221.

Karssen, C. M., D.L.C. Brinkhorst-van der Swan, A. E. Breekland, and M. Koornneef. 1983. Induction of dormancy during seed development by endogenous abscisic acid: Studies on abscisic acid deficient genotypes of *Arabidopsis thaliana* (L.) Heynh. *Planta* 157:158–65.

Maheshwari, P. 1950. *An Introduction to the Embryology of Angiosperms*. New York: McGraw-Hill.

Meinke, D. W. 1985. Embryo-lethal mutants of *Arabidopsis thaliana:* Analysis of mutants with a wide range of lethal phases. *Theor. Appl. Genet.* 69:543–52.

Narayanaswamy, S., and L. George. 1982. Anther culture. In *Experimental Embryology of Vascular Plants*, ed. B. M. Johri, 70–103. Berlin: Springer-Verlag.

Olson, A. R. 1981. Embryo and endosperm development in ovules of *Papaver nudicaule* after in vitro placental pollination. *Can. J. Botany* 59:1738–48.

Pharis, R. P., and R. W. King. 1985. Gibberellins and reproductive development in seed plants. *Ann. Rev. Plant Physiol.* 36:517–68.

Raghavan, V. 1976. *Experimental Embryogenesis in Vascular Plants*. New York: Academic Press.

Raghavan, V., and J. G. Torrey. 1963. Growth and morphogenesis of globular and older embryos of *Capsella* in culture. *Am. J. Botany* 50:540–51.

Rietsema, J., S. Satina, and A. F. Blakeslee. 1953. The effect of sucrose on the growth of *Datura stramonium* embryos *in vitro*. Am. J. Botany 40:538–45.

Rijven, A.H.G.C. 1952. *In vitro* studies on the embryo of *Capsella bursa-pastoris*. *Acta Bot. Neerl.* 1:157–200.

Sheridan, W. F., and M. G. Neuffer. 1981. Maize mutants altered in embryo development. In *Levels of Genetic Control in Development*, eds. S. Subtelny and V. K. Abbott. New York: Liss.

Smith, J. G. 1973. Embryo development in *Phaseolus vulgaris*. II. Analysis of selected inorganic ions, ammonia, organic acids, amino acids and sugars in the endosperm liquid. *Plant Physiol.* 51:454–8.

Sussex, I. 1975. Growth and metabolism of the embryo and attached seedling of the viviparous mangrove, *Rhizophora mangle*. *Am. J. Botany* 62:948–53.

Van Overbeek, J., M. E. Conklin, and A. F. Blakeslee. 1942. Cultivation *in vitro* of small *Datura* embryos. *Am. J. Botany* 29:472–7.

Walbot, V., T. Brady, M. Clutter, and I. Sussex. 1972. Macromolecular synthesis during plant embryogeny: Rates of RNA synthesis in *Phaseolus coccineus* embryos and suspensors. *Devel. Biol.* 29:104–11.

Walthall, E. D., and T. Brady. 1986. The effect of the suspensor and gibberellic acid on *Phaseolus vulgaris* embryo protein synthesis. *Cell Differ.* 18:37–44.

Ward, M., and R. H. Wetmore. 1954. Experimental control of development in the embryo of the fern *Phlebodium aureum. Am. J. Botany* 41:428–34.

Yeung, E. C. and I. M. Sussex. 1979. Embryogeny of *Phaseolus coccineus:* The suspensor and the growth of the embryo-proper *in vitro. Z. Pflanzenphysiol.* 91:423–33.

Zhou, C., and H. Y. Yang. 1985. Observations on enzymatically isolated, living and fixed embryo sacs in several angiosperm species. *Planta* 165:225–31.

Ziebur, N. K., R. A. Brink, L. H. Graf, and M. A. Stahmann. 1950. The effect of casein hydrolysate on the growth *in vitro* of immature *Hordeum* embryos. *Am. J. Botany* 37:144–8.

CHAPTER 4

The structure of the shoot apex

It is an interesting fact that in plant science the study of development is not equated with embryology. Although the study of embryos has made significant contributions, it is clear that the framework of developmental study in the higher plants has been provided by postembryonic stages. A very important aspect of embryonic differentiation is the establishment of shoot and root apical meristems at approximately opposite poles of the embryonic body. These meristems, whose origins differ somewhat in the various groups of vascular plants, contribute relatively little to the actual development of the embryo, but they are the centers of postembryonic development, and by their continued activity they give rise to the shoot and root systems. The shoot- and root-building activity of these meristems does not represent a mere unfolding of embryonic rudiments; rather, it is a true epigenetic formation of organs and tissues that were not present in the embryo. Thus, all aspects of development – growth and differentiation, histogenesis and organogenesis – may be investigated in relation to the activity of apical meristems, and the size and accessibility of these formative regions, in comparison to the enclosed embryo, has made them favorable sites for both descriptive and experimental studies of plant development.

In the total development of the primary plant body via its meristems, it is obvious that many processes are taking place simultaneously. There can be little doubt that these processes are interrelated and that the interaction among them holds many important keys to the understanding of the plant body, its organization, and its integrated development. Thus it is difficult to give serious consideration to the functional organization of the shoot apical meristem without simultaneously examining the process of leaf primordium initiation, and it is almost impossible to consider leaf inception apart from the functional organization of the shoot meristem. It is, therefore, difficult to fit an analysis of these contemporaneous events into the linear sequence of exposition. As this treatment progresses, although particular aspects of development will be dealt with individually, an attempt will be made to compensate for the arbitrary isolation of the parts of an integrated whole by discussing the nature of the interactions among them whenever possible.

Ultimately it is the functional organization of the shoot apex that the student of development seeks to interpret, the mechanism by means of which this formative region remains capable of continued growth while giving rise to organs and tissues that mature. Although it is improbable that structure alone can provide the basis for such an interpretation, it is evident that a functional interpretation must be based upon, or at least be consistent with, the structure of the shoot apex. Accordingly, the structural organization will be examined at the outset.

MORPHOLOGY

The best way in which to acquire an overall appreciation of the positional relationships of the various components of a shoot tip is to remove from it, under a stereoscopic microscope, successively younger leaf primordia in the sequence of their origin. This will ultimately reveal the terminal meristem (Figs. 4.1, 4.2). Situated at the distal extremity of the axis and surrounded by the youngest leaf primordia, the terminal meristem may have a variety of geometric forms ranging from conical or dome-shaped to flat or even slightly depressed. The diameter of this initiating region at the level of the insertion of the youngest leaf primordium may be as great as 3,500 μm in some species of cycads and considerably less than 50 μm in some species of flowering plants. In most species that have been examined, however, apical diameter falls within the 100- to 250-μm range. Furthermore, within the same species, or even within the same plant, there may be considerable variation in dimensions and shape depending upon the age of the plant, the season of the year, and the involvement of the meristem in the formation of a leaf primordium. In most cases, even where the mature shoot is dorsiventral, the symmetry of the terminal meristem is radial, but some dorsiventral shoots, as in some species of ferns and *Selaginella*, are initiated by a meristem that is elliptical in shape.

HISTOLOGY

Much can be learned from the external examination of apical topography. It is clear, however, that the activity of the terminal meristem involves cellular processes, and it is of paramount importance, therefore, that the cellular organization of the meristem be studied. With the exception of a few investigations in which cellular changes have been examined in living meristems, this study involves the preparation of stained sections from fixed material. The examination of such sections has led

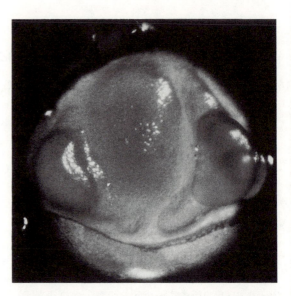

Figure 4.1. *The living shoot apex of* Lupinus albus *exposed by dissection. The youngest leaf primordium (P_1) is at the top in this photograph. P_2 is at the lower left and P_3 is at the lower right. The first lobes of leaflets are visible on P_3. ×170. (Courtesy D. DesBrisay and J. Waddington.)*

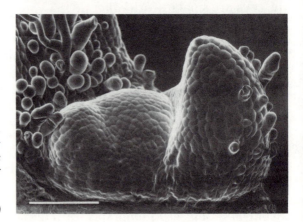

Figure 4.2. *Scanning electron micrograph of the shoot apex of tobacco. P_1 is at the lower left and P_2 at the lower right. Scale bar=100 μm. ×180. (R. S. Poethig and I. M. Sussex,* Planta *165:158, 1985.)*

to the recognition of an astonishing diversity of structural patterns in terminal meristems. A major difficulty in the interpretation of these patterns is that, whereas function must be understood in terms of the dynamic meristem, the structure can only be observed in a number of static states. This limitation, however, has not interfered with the development of an elaborate typology.

There have been several attempts to produce a workable classification of apical meristem types; perhaps the most often cited is that of Popham (1951), in which seven principal types were designated. On the other hand, one might argue that it is unwise to emphasize pattern types until sufficient functional information is available to indicate which of the structural variations have real significance. In this chapter the approach will be to describe the apices of several species that give some indication of the range of patterns encountered in the vascular plants rather than to define types. In addition to the patterns actually described, other forms that illustrate special features will be given brief mention.

Angiosperms – tunica and corpus

It is probable that the shoot apices of flowering plants are more familiar to most biologists than are those of other groups, and it seems appropriate to begin with a representative of this group. The apex of *Nicotiana tabacum* (tobacco), which has been used in a number of experimental investigations that will be discussed in later chapters, will serve here as a useful example (Fig. 4.3). When this apex is examined in median longitudinal section, the region above the youngest leaf primordium presents an overall appearance of cellular homogeneity (Fig. 4.3). The cells are small, nearly isodiametric, thin-walled, and characterized by a high nucleocytoplasmic ratio and by relatively inconspicuous vacuolation. Closer examination reveals that there is a definite stratification of the more superficial regions of the apex, which reflects the orientation of planes of cell division in different layers (Fig. 4.4). This is most conspicuous in the two outermost layers of cells in which divisions are ordinarily restricted to the anticlinal plane – that is, at right angles to the surface. In the more deeply seated regions of the apex, the planes of cell division are less regularly oriented, and in the interior they seem to be essentially random. The stratification becomes obscured at the margins as leaf primordia are initiated. Furthermore, the cytological homogeneity of the meristem gives way below to increasing heterogeneity related to the initial differentiation of tissues.

The apical pattern of stratification illustrated by *Nicotiana* is characteristic of the angiosperms and is also found in a few other plants. A terminology for the description of such apices, which was devised by Schmidt (1924), is widely used because of its simplicity and general applicability. The name *tunica* has been given to the one or more superficial layers of the apex that, above the level of the youngest leaf primordium, show only anticlinally oriented cell divisions. To the remainder of the meristem beneath the tunica the term *corpus* has been applied. The number of tunica layers observed in various angiosperms ranges from one to five, with the greatest number of species having a two-layered tunica. In *Nicotiana* (Fig. 4.4), the tunica is considered to be two-layered. It is interesting also to note that, among the gymnosperms,

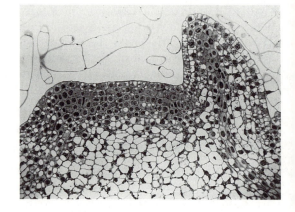

Figure 4.3. *Median longitudinal section of a tobacco shoot apex with leaf primordia at about the same stage as in Fig. 4.2. ×180. (R. S. Poethig and I. M. Sussex,* Planta *165:158, 1985.)*

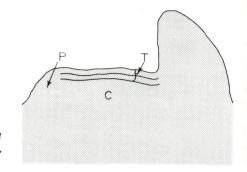

Figure 4.4. *Diagrammatic illustration of the apex shown in Fig. 4.3. Key:* C, *corpus;* P, *leaf primordium;* T, *tunica.*

Ephedra and *Gnetum,* as well as several species of conifers, have been found to possess one or more surface layers that correspond precisely to the tunica of the angiosperms, and the tunica–corpus terminology is generally applied in these cases.

The tunica–corpus concept has been used widely in the description of angiosperm shoot apices, but as more and more of these have been described, problems have arisen concerning the proper application of the terms. This, of course, is to be expected when terminology based upon a static pattern is applied to a dynamic system. A major difficulty concerns the degree of exclusiveness required in the use of the term tunica. Some workers have preferred to follow Schmidt's rather exclusive definition, which restricts the term tunica to those layers in which only anticlinal divisions occur. Others have chosen to adopt his actual procedure of admitting an occasional and local periclinal division in a layer designated tunica. If the very rigid definition of tunica is applied, some angiosperms, particularly among the grasses, lack even a single tunica layer, although the apical construction is clearly one of stratification.

There are also flowering plants in which the entire meristem is stratified, but even the permissive definition allows only two or three of the surface layers to be designated tunica. An example of this is seen in *Heracleum* (Majumdar, 1942) in which the meristem consists of eight or nine distinct layers of which only the outermost three constitute tunica.

There are well-documented cases in which the number of tunica layers varies in the same plant and presumably in the same apex (Gifford and Corson, 1971). In *Brassica campestris* (Chakravarti, 1953) the number of tunica layers increased from one in the embryo to three or four in the reproductive shoot. There are also indications that the degree of stratification in apices may fluctuate in relation to leaf initiation, reaching a maximum just before the inception of a primordium. In different instances this fluctuation has been interpreted as resulting from changes in the tunica layers or from changes in the division pattern in the outer region of the corpus. Finally in some perennial plants the number of tunica layers fluctuates seasonally, as for example in *Daphne pseudomezereum*, where it is highest during periods of active growth (Hara, 1962).

As more and more species have been investigated, it has also become increasingly evident that neither the tunica nor the corpus is cytologically homogeneous. In a number of species a group of cells at the summit of the apex, including both tunica and corpus, have been recognized as being somewhat larger and less densely stained than those around them. These cells, because of their position in the apex, have often been designated *tunica* and *corpus initials*.

The tunica–corpus concept has value in that it provides a framework for descriptive studies of angiosperm shoot apices within which comparisons may be made. The question concerning the limitation of the use of the term tunica poses some problems, but it seems that these are best met by a flexible approach rather than by absolute rigidity. Each investigator may then exercise his own judgment in applying the term and has the responsibility of making clear how he has applied it. The biological value of the concept is considerably less evident. Attempts to find correlations between the kind of tunica–corpus organization, ordinarily the number of tunica layers, and phylogenetic position or taxonomic relationships have not yielded consistent results. The significance of tunica and corpus is that they reflect patterns of cell divisions within the meristem and thus provide information on an important activity of this region. It has been suggested that the different planes of division in superficial and internal regions of the apex may be related to the geometry of surface and volume growth, but this relationship is by no means clear. On the other hand the predominance of anticlinal divisions in the superficial layers of the meristem does have significant consequences in the derivation of tissues, as will be discussed in the next chapter.

Gymnosperms – apical zonation

A striking aspect of meristem structure is the differences that are found in apices of different species. This is very well illustrated by comparison of the terminal meristem of *Nicotiana* with that of the ancient gymnosperm *Ginkgo biloba* as described by Foster (1938) (Figs. 4.5, 4.6). In *Ginkgo*, although some evidence of stratification may be detected, the structural pattern has a radial organization that centers around a cluster of enlarged, rather highly vacuolated cells that appear to divide infrequently and have been designated the *central mother cells*. The meristem is bounded distally by a *surface layer* in which anticlinal divisions predominate. Periclinal divisions also occur throughout the surface layer, but, together with oblique divisions, are considered to be more frequent in the cells at the summit of the apex. These summit cells have been designated the *apical initial group*. Their inner derivatives enlarge and become part of the immediately subjacent zone of central mother cells. Around the zone of central mother cells in a ringlike arrangement is a region of small and densely cytoplasmic cells called the *peripheral subsurface layers*. This region is augmented by periclinal divisions in the overlying surface layer and by division of cells at the margin of the zone of central mother cells. Underlying the zone of central mother cells is a *rib meristem* – that is, a series of rows or files of small, vacuolated cells. The arrangement of these cells suggests an orientation of cell divisions at right angles to the shoot axis. Foster (1938) recognized that between the central mother cells and the cells that surround them laterally and basally there is a *transition zone* in which the cytological characteristics of the central mother cells are gradually replaced by those of the surrounding zones.

Such a pattern, which directs attention to cytological heterogeneity in the meristem, is referred to as a cytohistological zonation, and it has been found, with various modifications, to be typical of gymnosperms. The apices of cycads have been interpreted as having this kind of organization, although these remarkable meristems bear little superficial resemblance to *Ginkgo*. In a cycad such as *Microcycas* (Foster, 1943) there appears to be a surface layer over the meristem as in *Ginkgo*, and there is also a large, funnel-shaped group of central mother cells. Between these two there is a region in which short vertical tiers of cells indicate frequent periclinal division. This region is not present in *Ginkgo*. Foster has designated this region, together with the surface layer, the *initiation zone*. The cells of the central mother cell zone also show a

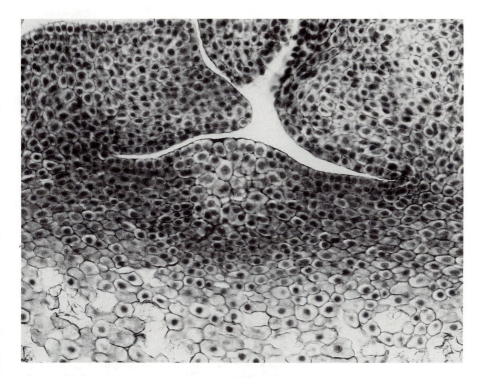

Figure 4.5. *Median longitudinal section of the shoot apex of* Ginkgo biloba. *×200.*

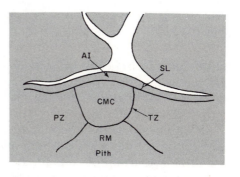

Figure 4.6. *Diagrammatic illustration of the apex shown in Fig. 4.5. Key:* AI, *apical initial group of cells;* CMC, *central mother cells;* PZ, *peripheral subsurface layers of cells;* RM, *rib meristem;* SL, *surface layer of cells;* TZ, *transition zone.*

tiered arrangement, which apparently represents the pattern acquired in the initiation zone and retained as cells enlarge and pass into the lower zone. The other regions of the apex are much as in *Ginkgo* except that the peripheral zone is considered to consist of two parts, an outer region derived from the initiation zone and an inner region derived from the zone of central mother cells (Fig. 4.7). In cycad seedlings and in

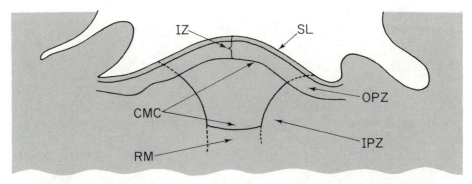

Figure 4.7. *Zonation in the shoot apex of* Microcycas calocoma. *Key:* CMC, *central mother cells;* IPZ, *inner peripheral zone;* IZ, *initiation zone;* OPZ, *outer peripheral zone;* RM, *rib meristem;* SL, *surface layer. (Modified from Foster, 1943.)*

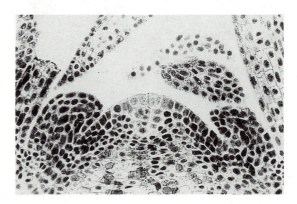

Figure 4.8. *Median longitudinal section of the shoot apex of* Pinus palustris *showing the radial zonation pattern.* ×150.

adult plants of species with apices considerably smaller than that of *Microcycas*, apical structure resembles that of *Ginkgo* very closely, suggesting that the unusual structure of the apex of *Microcycas* is related to its exceptionally large size.

Although differing in some details, descriptions of the structural organization of apices of a number of conifers seem to agree in identifying a distinctive group of cells at the summit of the apex. These cells are often somewhat enlarged and are also distinguished by a low affinity for histological stains (Fig. 4.8). The meristem as in *Ginkgo* usually has a recognizable surface layer and in a few cases, as has been noted, this qualifies to be designated a tunica. The cells at the summit of the apex in the surface layer have sometimes been called *apical initials,* and those beneath them have been called *subapical initials* or, if they are particularly conspicuous, central mother cells. Together they constitute a cen-

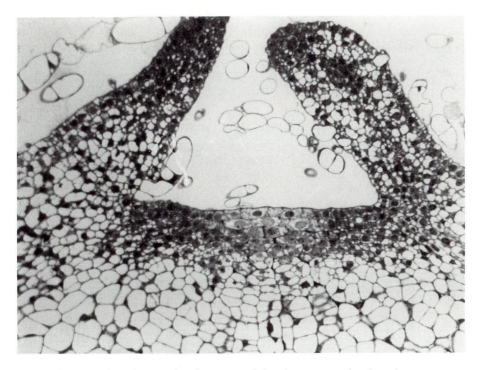

Figure 4.9. *Median longitudinal section of the shoot apex of* Helianthus annuus.
×*340. (Davis, Rennie, and Steeves, 1979.)*

tral zone surrounded by a peripheral zone of small, densely staining
cells, beneath which lies the pith rib meristem.

Angiosperms – apical zonation

Following Foster's recognition of cytohistological zonation in *Ginkgo*
and the extension of this interpretation to other gymnosperms, it soon
was found that apices of certain angiosperms could be more fully described in these terms than according to the tunica–corpus concept alone.
For example, Majumdar (1942) called attention to this kind of pattern
in the shoot apex of *Heracleum* and specifically compared it to the structure found by Foster in *Ginkgo*. A striking example of this type of cytohistological zonation in an angiosperm apex is provided by *Helianthus
annuus* (sunflower) (Davis, Rennie, and Steeves, 1979). This apex, like
those of many angiosperms, has a two-layered tunica overlying a corpus
(Figs. 4.9, 4.10). However, the most conspicuous structural feature of the
apex is a distinctive radial zonation suggestive of that in *Ginkgo*. A cen-

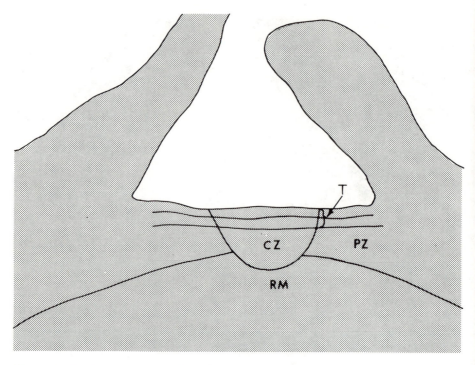

Figure 4.10. *Diagrammatic illustration of the apex shown in Fig. 4.9. Key:* CZ, *central zone;* PZ, *peripheral zone;* RM, *rib meristem;* T, *tunica.*

tral cluster of cells, including both tunica and corpus, stands out conspicuously because of their large size and weak affinity for histological stains in both nucleus and cytoplasm. The cells of this central region appear to divide infrequently. They are surrounded by a peripheral region composed of small, densely staining cells in which divisions appear to be more frequent. Immediately below these zones is a pith rib meristem in which divisions also appear to be frequent. Patterns of this type associated with a tunica–corpus organization have been described in many species of flowering plants (Gifford and Corson, 1971), although frequently the cytological characteristics of the central zone are much less distinctive than in *Helianthus* (cf. Fig. 4.3).

Angiosperm apices with radial zonation thus have little to distinguish them from those of gymnosperms that have a distinct tunica. In fact it is becoming generally recognized that a common pattern underlies the structural organization of shoot apices in the seed plants. This pattern consists of a *central zone* surrounded by a *peripheral zone* and underlain by a pith-producing *rib meristem*. Upon this basic plan a tunica–corpus pattern may or may not be superimposed.

Lower vascular plants

When attention is turned to the shoot apices of lower vascular plants a rather different picture emerges from what has been seen in the seed plants. In many of these plants the most prominent feature of apical organization is a superficial layer of axially elongate cells. A fern such as *Osmunda cinnamomea* (cinnamon fern) will serve to illustrate this type of pattern (Steeves, 1963) (Fig. 4.11). In *Osmunda* the superficial cells are rectangular in shape and are lightly stained by ordinary histological dyes. The center of the surface layer is occupied by a somewhat larger cell that is pyramidal in shape and has been called the *apical cell*. This cell divides only anticlinally and its derivatives augment the surface layer. The other cells of the surface layer divide both anticlinally and periclinally, but the periclinal divisions are unequal, so small inner derivatives are cut off and the distinctive shape of the surface cells is retained. However, at the margins of the apical mound these cells become segmented and lose their distinctive appearance. The apical cell in *Osmunda* is not very distinct morphologically, and during periods of active growth it can be distinguished only with difficulty because its segmentation pattern is irregular. There is some doubt as to whether a single cell functions continuously as the apical initial. Immediately beneath the surface layer, differences are apparent among the cells in different regions, and these have been interpreted in two different ways. On the one hand, they have been considered to represent initial stages of differentiation of the tissues of the mature stem, so only the surface layer constitutes the true initiating region. In a more recent study of fern shoot apices (McAlpin and White, 1974) it was proposed that the initiating region consists of the distinctive cells of the surface layer together with a group of centrally placed cells beneath them. If this interpretation proves to be correct, the fern shoot apex may be more comparable to those of the seed plants than has generally been thought to be the case.

An apical organization in which there is a layer of distinctive superficial cells in the initiating region seems to be general among the lower vascular plants. In some cases the apical cell is enlarged and very conspicuous as in *Equisetum* (Fig. 4.12) and many ferns. In other cases, a group of initial cells replaces the single apical cell, and considerable variation may be found in the same plant, or perhaps even in the same apex at different times. *Osmunda*, just described, is of this second type. However, Bierhorst (1977) has reinvestigated the shoot apices of a number of fern species and has concluded that a single apical cell is a nearly constant feature. Shoot apices of *Lycopodium* also can be included in this group even though the cells of the superficial layer are not so conspicuous in size and vacuolation as in other lower vascular plants.

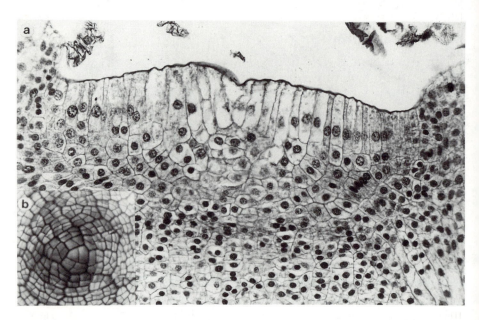

Figure 4.11. *Shoot apex of* Osmunda. *(a) Median longitudinal section of* O. cinnamomea *shoot apex in which the apical cell is visible in the center of the surface layer of enlarged cells. (b) Surface cellular configuration of a shoot apex of* O. regalis *showing the relationship of the apical cell to its derivatives. (a) ×170, (b) ×100. ([a] Steeves, 1963. [b] Bierhorst, 1977.)*

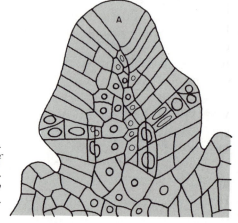

Figure 4.12. *Diagrammatic illustration of a median longitudinal section of the shoot apex of* Equisetum arvense. *Key: A, apical cell. ×120. (S. J. Golub and R. H. Wetmore,* Am. J. Botany. *35:755, 1948.)*

GENERAL COMMENT

Having now sampled the diversity of shoot apical structure and drawn from it such generalizations as present knowledge permits, we are prepared to explore the function of the shoot apex by means of analytical and experimental methods. But before proceeding it may be appropriate to conclude this chapter with a brief account of one rather widely applied interpretation that, although histological in nature, has definite functional implications.

Throughout the history of the study of shoot apices, most investigators have identified, or at least designated, certain cells at the summit of the apex by the term *apical initials*, and this has appeared several times in descriptions given earlier in this chapter. In many lower vascular plants a distinctive apical cell may be recognized, which by its regular segmentation appears to function as an initial cell for the entire shoot, while retaining its individual identity. In this sense it functions as the ultimate, although of course not the immediate or direct, source of all of the cells of the shoot. In other lower plants several such initial cells seem to fulfill this role. Undoubtedly the conspicuous presence of one or more initials in the apices of lower plants was an important factor in the widespread acceptance of the existance of apical initials even where they are not cytologically distinctive, as in the seed plants. The concept of a special group of cells, presumably in a particular physiological state, that imparts to the shoot meristem its property of continued meristematic activity is an attractive one, and one that has been inherent in most descriptive studies of the shoot apex.

The presence of apical initials usually has been associated, if not directly, at least by implication, with another feature of the organization of the apex, which is often designated the *promeristem* (Sussex and Steeves, 1967). The promeristem is defined as including the apical initials and their most recent derivatives, which have not as yet undergone any of the changes associated with tissue differentiation. The recent derivatives of the initials are presumed to remain in more or less the same physiological state as the initials, but because of their position they, unlike the initials, are not totipotent. Thus, although it has important functional implications, the promeristem is a histological concept based upon the absence of true histodifferentiation. It has definite structural features that vary in different plants, but these features, unlike those of immediately subjacent regions, are not related to the differentiation of the tissues of the shoot. A major problem in the application of this interpretation, however, is the extreme difficulty in recognizing the initial stages of tissue differentiation and consequently the boundaries of the promeristem, if indeed these boundaries can be thought of as being sharply defined. Rarely in studies of apical zonation has there been any attempt to designate which of the features observed represent the beginnings of the cytological changes associated with tissue differentiation. In several ferns it has been suggested that tissue differentiation can be detected immediately beneath the surface layer of enlarged cells, and the surface layer itself has been designated as the promeristem. In seed plants the promeristem would include the central zone, possibly all or part of the peripheral zone but almost certainly not the rib meristem, which appears to represent an early stage in the differentiation of the pith.

The concept of a promeristem composed of cells that have not undergone tissue differentiation, surrounding one or more initial cells that confer upon the apex its ability for continued growth, seems to be in accord with the structural patterns already described; at least it does not conflict with them. It is, in fact, no more than a working hypothesis based upon observed structure, and it is in no sense a true functional interpretation explaining how the shoot apex performs its all-important role in shoot growth. Moreover, it has been tested in recent years by a growing body of analytical and experimental information. It remains now to examine some of this evidence against the background of the traditional interpretation.

REFERENCES

Bierhorst, D. W. 1977. On the stem apex, leaf initiation and early leaf ontogeny in filicalean ferns. *Am. J. Botany* 64:125–52.

Chakravarti, S. C. 1953. Organization of shoot apex during the ontogeny of *Brassica campestris* L. *Nature* 171:223–4.

Davis, E. L., P. Rennie, and T. A. Steeves. 1979. Further analytical and experimental studies on the shoot apex of *Helianthus annuus:* Variable activity in the central zone. *Can. J. Botany* 57:971–80.

Foster, A. S. 1938. Structure and growth of the shoot apex in *Ginkgo biloba. Bull. Torrey Botan. Club* 65:531–56.

—— 1943. Zonal structure and growth of the shoot apex in *Microcycas calocoma* (Miq.) A.DC. *Am. J. Botany* 30:56–73.

Gifford, E. M., Jr., and G. E. Corson, Jr. 1971. The shoot apex in seed plants. *Bot. Rev.* 37:143–229.

Hara, N. 1962. Structure and seasonal activity of the vegetative shoot apex of *Daphne pseudo-mezereum. Bot. Gaz.* 124:30–42.

Majumdar, G. P. The organization of the shoot in *Heracleum* in the light of development. *Ann. Botany (London) (NS)* 6:49–82.

McAlpin, B. W., and R. A. White. 1974. Shoot organization in the Filicales: The promeristem. *Am. J. Botany* 61:562–79.

Popham, R. A. 1951. Principal types of vegetative shoot apex organization in vascular plants. *Ohio J. Sci.* 51:249–70.

Schmidt, A. 1924. Histologische Studien an phanerogamen Vegetationspunkten. *Botan. Arch.* 8:345–404.

Steeves, T. A. 1963. Morphogenetic studies of *Osmunda cinnamomea* L. The shoot apex. *J. Indian Bot. Soc.* 42A:225–36.

Sussex, I. M., and T. A. Steeves. 1967. Apical initials and the concept of promeristem. *Phytomorphology* 17:387–91.

5

Analytical studies of the shoot apex

If the study of structural patterns in the shoot apices of vascular plants does not lead to an understanding of this region in functional terms, other methods of investigation must be employed to attain such an understanding. A considerable body of research has attempted to analyze more precisely the activities of the shoot tip and its component parts and has produced some new information and brought some of the more challenging unsolved problems into sharper focus. This chapter will discuss some of the significant contributions that analytical studies have made.

GROWTH OF THE SHOOT APEX

One essential function of the shoot meristem is that of producing cells. If the region is treated as a population of dividing cells, some interesting quantitative estimates of cell production can be derived. For example in *Pisum sativum* (garden pea) Lyndon (1968) has determined the increase in cell number in the shoot apex during a *plastochron*, the interval between the initiation of two successive leaves. Immediately after the initiation of a leaf primordium the apex contains 900 to 1,000 cells. While the next primordium is being formed, approximately 1,600 new cells are added so that there is nearly a threefold increase in cell number. Since the duration of a plastochron can be determined, in this case forty-eight hours, it is possible to know the average rate at which the cells are dividing. This average rate is often expressed in terms of a *mean cell generation time,* the average time required for all of the cells to double, that is to divide once. In the shoot apex of the pea the mean cell generation time has been calculated to be approximately twenty-eight hours. Since plants are known to grow at different rates it is not surprising that the mean cell generation time varies widely in different species. It often falls in the range of one to three days, but it may be much longer as in *Elaeis* (oil palm), where it is estimated to be nearly fifty days (Lyndon, 1973).

DISTRIBUTION OF CELL DIVISION IN THE SHOOT APEX

While it is useful to know something about the rate of cell multiplication in the shoot apex as a whole, the structural studies reported in the previous chapter indicate that the meristem is not composed of homogeneous cells. It has in fact been suggested that the organizational patterns that have been noted are related to the distribution of cell divisions in the apex. It is, therefore, worthwhile to consider whether cell divisions are more or less uniformly distributed throughout the terminal region of the shoot or whether they tend to be localized in certain cell-producing regions, with the remaining regions of the meristem having other functions.

The presence of a cluster of enlarged and somewhat vacuolated cells in the center of the meristem in the shoot apices of a number of vascular plants has suggested to some workers in the past that this region may be characterized by a rate of cell division somewhat slower than that in surrounding regions (Philipson, 1954). Despite the presumed low rate of division in this region, however, it includes the cells, variously designated as tunica and corpus initials or apical and subapical initials, which have been regarded as the ultimate source of all the cells of the primary body of the shoot. Early observations on this region did not usually include any quantitative estimates of rates of cell division in various zones of the apex, but were based largely upon visual inspection.

Fortunately there have been in more recent years a number of careful analyses of the division frequency in the various zones of the shoot apex. The number of species investigated is small in comparison with the total number of vascular plant species, yet the picture that emerges is consistent enough to permit cautious generalization. It has been established that there is at the summit of the apex a group of cells that clearly divide less frequently than do those around and beneath them and that appear to correspond to the central zone described in the previous chapter. In some cases they have very distinctive cytological characteristics, but in others they are much less recognizable structurally.

The simplest demonstration of the difference in activity is achieved by counting mitotic figures in the various zones of the apex and expressing the results as the percentage of nuclei in division at a given time, the *mitotic index*. When this was done for the shoot apex of sunflower (Davis, Rennie, and Steeves, 1979), for example, it was found that the mitotic index of the peripheral zone (4.6) was approximately 3.5 times greater than that of the central zone (1.3), and comparable differences are reported for other species. This method has been criticized because it assumes that the duration of mitosis is the same in all zones of the apex, and if it is not, one cannot legitimately compare the mitotic in-

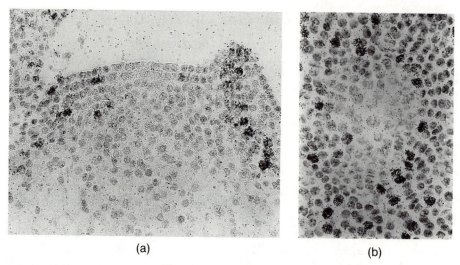

(a)   (b)

Figure 5.1. *Median longitudinal (a) and transverse (b) sections of the shoot apex of* Helianthus annuus *after feeding with tritiated thymidine and autoradiography of sections. Nuclei that had synthesized DNA during the feeding period may be identified by the accumulation of black silver grains over them. Note the absence of silver grains over nuclei in the central zone. The transverse section was cut slightly below the surface of the apex. ×290. ([a] Steeves et al., 1969. [b] Davis, Rennie, and Steeves, 1979.)*

dices of different zones. Fortunately, there is evidence that indicates the duration is usually the same throughout the meristem (Lyndon, 1976). There are methods, however, that avoid this problem. Treatment of the apex with colchicine, which blocks mitosis at metaphase, causes mitotic figures to accumulate over time, and the rate of accumulation is equivalent to the rate of cell division. Alternatively, if the apex is provided with tritiated thymidine, a precursor of DNA, it is possible to use autoradiography to localize the nuclei in which DNA synthesis and presumable mitosis are taking place (Fig. 5.1). This method substantiates the existence at the summit of the apex of a group of cells that divide less frequently than do those in the surrounding zones. However, these latter methods are not without difficulties either. Both require rather drastic manipulation of the shoot apex, often a partial dissection with removal of young leaves, so that the results obtained may not be entirely representative of the natural state of the apex. In the sunflower it has been shown that the removal of young leaves near the apex results in a stimulation of mitotic activity in the apex that is more pronounced in the central zone than in the peripheral regions (Davis, Rennie, and Steeves, 1979). Nonetheless, the consistency of the results obtained by

Table 5.1. *Duration of the Cell Cycle
(Hours) at the Summit (Central Zone)
and on the Flanks (Peripheral Zone) of
Vegetative Shoot Apical Meristems.
Adapted from Lyndon (1976).*

	Region of Apex	
	Summit	Flanks
Rudbeckia	>40	30
Pisum	70	28
Datura	76	36
Trifolium	108	69
Solanum	117	74
Chrysanthemum	140	70
Coleus	237	125
Sinapis	288	157

all methods inspires confidence that the phenomenon is real, even though the quantitative values may not be entirely accurate.

It is important to emphasize that the central zone cells are only relatively inactive mitotically; that is, they do divide, but less frequently than do cells in the surrounding and underlying zones. This is perhaps best expressed in terms of the duration of the cell cycle, the sequence of events that includes synthesis of DNA, mitosis, and periods that intervene both before and after mitosis. The duration of the cell cycle can be calculated from the mitotic frequency, which, as has been stated, is obtained from the accumulation of mitotic figures with colchicine treatment. It may also be obtained by the autoradiographic method, in which a particular group of nuclei is labeled while they are in the DNA-synthesis phase of the cell cycle and is then followed through the time required for two successive mitoses. Table 5.1 shows the length of the cell cycle in the central and peripheral zones of a number of species that have been analyzed. It is evident that, although the cycle times vary widely in different species, they are consistently two to three times longer in the central zone than in the peripheral or flanking regions of the apex. Again, in spite of possible inaccuracies because of the experimental manipulation, a consistent picture emerges. In the central zone, where the cell cycle is extended, the mitotic index is low when compared to values for the peripheral zone, as would be expected if the cells are dividing less frequently.

It is reassuring to discover that the results of more careful quantitative analysis have confirmed the conclusions derived from earlier obser-

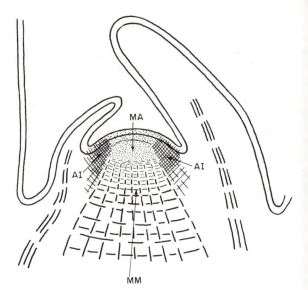

Figure 5.2. *Diagram of the shoot apex of* Cheiranthus cheiri *interpreted according to the* méristème d'attente *concept. Key:* AI, *anneau initial;* MA, *méristème d'attente;* MM, *méristème medullaire.* *(R. Buvat,* Ann. Biol. *31:595, 1955.)*

vations regarding the zonation of the shoot apex. However, the signifi-cance of this pattern of mitotic activity is far from clear at the present time. To most observers it is entirely consistent with the long-held view that the cells at the summit of the apex are indeed the ultimate initials of the developing shoot. This interpretation, however, is not universally accepted and has been challenged.

The méristème d'attente concept

The relative paucity of mitotic figures observed in the summital cells of the apex in histological preparations has led to a radically different con-cept of apical organization and function that was first proposed by Bu-vat and his coworkers in the early 1950s in France (Fig. 5.2) (Buvat, 1952). According to this theory, the summit of the apex in a vegetative shoot is occupied by a distinctive zone called the *méristème d'attente* that plays no histogenetic or organogenetic role in the development of the shoot – this being reflected in the absence or near absence of mitotic activity in the zone. The méristème d'attente is surrounded laterally by the *anneau initial* and is subtended by the *méristème médullaire,* both of which are characterized by the occurrence of frequent mitotic figures. To the anneau initial is ascribed the major role in the development of the shoot in that it gives rise to leaf primordia and the associated stem tissues. The méristème médullaire gives rise to the cells of the pith. This organization that prevails throughout vegetative growth of the shoot, changes drastically with the onset of floral or inflorescence development. Abundant mitoses may be detected in the méristème d'attente at this time, and the previously inactive region gives rise to most or all of the reproductive structures. According to the theory, this is the function of the méristème d'attente for which it "waits" throughout vegetative development.

The hypothesis was based originally upon the examination of apices of several species of dicotyledons, but it soon was extended to monocotyledons and to several gymnosperms. The distinctive meristems of the lower vascular plants, including *Equisetum arvense* and several ferns, have been reported to conform to this model.

The interpretation of shoot apical organization proposed by the French School gained very limited support elsewhere. Critics were quick to point out that mitoses could be observed in the supposedly quiescent region and supported these observations with autoradiographic evidence for the synthesis of DNA in the central zone (Clowes, 1961). They also demonstrated, however, as has been pointed out, that mitotic activity is lower in the region that corresponds to the méristème d'attente than in the equivalent of the anneau initial. It is important to realize that the demonstration of some mitotic activity at the summit of the apex does not in itself invalidate the theory, since the French workers themselves have reported the same observations and have taken it into account in formulating the theory. In its original statement, the méristème d'attente theory pictured the shoot apex as consisting of a dormant cap or dome of cells continually carried forward by proliferation in the anneau initial and the méristème médullaire. As the theory has been developed, the occurrence of a limited amount of mitotic activity in the méristème d'attente has been recognized and along with this some interaction among the zones of the apex. The inception of a leaf primordium in a particular region of the anneau initial reduces the extent of this zone in that part of the apex, and this is followed by the restoration of the anneau initial to its original dimensions. This process, which has been referred to as *regeneration*, involves increased mitotic activity in a localized region of the anneau, and this activity may encroach upon the adjacent region of the méristème d'attente. Thus, in some species mitotic figures may be observed in the méristème d'attente from time to time in localized regions. This is especially significant in small apices (Catesson, 1953).

It thus appears that the difference between the méristème d'attente theory and the more traditional concept of apical initials is one of interpretation since there is substantial agreement on the facts. The former theory holds that during vegetative growth the méristème d'attente is essentially passive and it is the anneau initial that is the true generating center of the shoot. Cells of the méristème d'attente are activated during reproductive development and participate in the formation of flowers or inflorescences. There is, of course, the problem that in many cases, particularly among the gymnosperms and lower vascular plants, this region never gives rise to reproductive structures. In such cases the inactive cells presumably are never activated and have no function in the life of the plant.

The question of apical initials

A contrasting view is that the summit cells of the apex are the true initials that serve as the ultimate, although of course not the direct, source of all of the cells of the developing shoot. Proponents of this interpretation (Wardlaw, 1957) point out that a small group of cells at the summit of the apex would need to divide very infrequently in order to provide a constant replenishment for the organogenetic and histogenetic regions that subtend them and that actually produce the organs and tissues of the shoot. For example Stewart and Dermen (1970) have calculated that in *Ligustrum* (privet) an initial cell in the apex could function as the ultimate source of the derivative cells that can be traced to it if it divided no more frequently than once in twelve days, because of the subsequent divisions of the derivatives. A decision in favor of one or the other of these conflicting interpretations clearly requires evidence in addition to that provided by the data on mitotic frequency.

The question of functional apical initial cells is debated with particular intensity in the case of lower vascular plants in which a structurally distinctive apical cell is often situated in the center of the apex. Some workers, following the méristème d'attente concept, have concluded that the apical cell and the cells immediately surrounding it divide so infrequently that their role is essentially a passive one except possibly in very young plants. The shoot is said to be initiated by the flanking regions of the meristem. In some cases it is reported that the apical cell is highly polyploid as a result of endoreduplication without subsequent mitosis (D'Amato, 1975). On the other hand, careful analysis of cellular patterns in the meristem provides evidence that the apical cell produces a regular sequence of daughter cells, which are then further segmented (Bierhorst, 1977). Thus, the traditional role of the apical cell is reaffirmed. Observed mitotic frequency in the apical cell is held to be adequate to sustain its role as the ultimate initial cell of the meristem. The high level of polyploidy reported in some cases has not been found in other species, and it is suggested that it may be a phenomenon associated with determinate shoots no longer growing actively (Gifford, 1983).

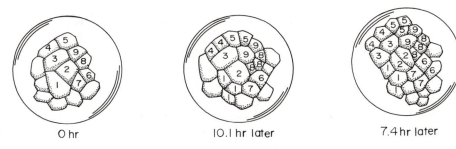

O hr   10.1 hr later   7.4 hr later

Figure 5.3. *Record of cellular changes in the surface layer of the shoot apex of* Asparagus officinalis. *The drawings are tracings of time-lapse photographs. Nine cells that could be identified at zero time are numbered, and these numbers are retained for the derivatives resulting from their divisions. Thus, in the first picture there is one cell designated 5. It had given rise to two cells 10.1 hours later, each labeled 5, and 7.4 hours later one of these had divided again, so there are then three cells labeled 5. (Drawn from Ball, 1960.)*

Observations on living apices

Although rather substantial agreement seems to have been reached regarding the histological evidence for the occurrence of significantly different cell-division frequencies in the several zones of the shoot apex, a different approach to the question of cell proliferation in shoot apices, pioneered by Ball (1960), has led to very different conclusions. This approach is the direct observation of cell division patterns in living shoot apices. In these studies sequential observations of cells in the surface layer of the meristem of several species of angiosperms were made and recorded by drawings or by photography. Changing cellular configurations and the appearance of new walls at or near the summit of the apex provided convincing evidence that cell division does occur in this region of the apex. Ball's photographic records indicate that the frequency of division is high in the summit cells of the apex. For example, in one apex of *Asparagus* he found that in approximately 17.5 hours, nine cells originally placed under observation all divided once and three derivatives also divided, producing a final total of 21 cells (Fig. 5.3). Other apices showed less rapid division, but in no case was there any evidence for an axial group of nondividing or slowly dividing cells.

Although the direct observation of living meristems is an attractive approach to the question of cell division frequency, the experiments thus far reported have serious limitations. The only cells in which divisions were actually recorded were those at the summit of the apex, and no comparison was made with cells in more lateral positions. Thus the relative rates of division in different parts of the meristem, the key issue, were not revealed. Equally serious is the probable influence of the treatment of the shoot apices required to make the observations. The shoot tips were excised and grown in sterile culture, and many young leaves and primordia were removed in order to expose the apex for observation. As noted earlier, experiments with sunflower have shown that such drastic manipulation can result in an increase of mitotic activity in the apex, especially in the central zone. Thus these observations, although intriguing, cannot be accepted as invalidating a large body of evidence for a lower mitotic frequency in the summital cells.

Observations of another type on living shoot apices have also been brought to bear upon the question of the role of the summital cells (Soma and Ball, 1964; Ball, 1972). In these studies the shoot apices of intact plants of several species were exposed by bending back the surrounding leaves, and the central cells at the surface were marked either by applying small carbon particles to the summit or by making small needle punctures that killed individual cells. After a period of further growth, the positions of these markers were located and it was observed that many of them had been displaced from the summit of the apex. This was interpreted to mean that the summital cells do divide actively and contribute to the subjacent region. It was also considered to provide evidence that there are no permanent or long-term apical initials but rather that the summit of the apex is occupied by a shifting population of dividing cells. However, as in the previous reports, there is no indication of relative rates of cell proliferation in central and peripheral regions. In some cases the displacement was, in fact, very sluggish. For example in one experiment with lupin, after twenty-one days more than 20 percent of the apices marked with carbon particles showed the marker still located at the center. These observations do support an initiating role for the central region of the apex, but they do not provide conclusive evidence for the absence of permanent or long-term apical initials.

CYTOHISTOLOGICAL STUDIES

The zonation patterns in shoot apices described in the previous chapter show an apparent relationship to the distribution of mitotic activity. It should be recalled, however, that these patterns can be recognized because of the distinctive cytological characteristics of the cells that compose the various zones. Some of these features are suggestive of differences in metabolic activity, and it has been suggested by some workers that they are related to differences in mitotic activity. It is, therefore, of interest to consider what specific information is available about the cytohistological basis of apical zonation.

Workers of the French school, in support of the concept of the méristème d'attente, have carried out detailed cytological studies on a variety of angiosperms and gymnosperms in which they have shown important localized differences within the meristem (Nougarède, 1967). The summit cells of the apex are said to be characterized by large vacuoles, filamentous mitochondria, small nucleoli, and the presence of differentiated plastids, whereas those of the anneau initial possess small vacuoles, granular mitochondria, large nucleoli, and undifferentiated plastids. These cytological features have been associated with the differences in mitotic activity of the cells in the two regions. Cytochemical tests carried out on shoot apices have shown localized differences in the cellular content of RNA, which is low in the méristème d'attente and high in the more peripheral regions of the meristem. Incorporation studies using radioactive precursors have indicated that the synthesis of RNA is low in the axial cells of the meristem as compared to the surrounding regions.

Other workers, while they have usually found cytological differences related to zonation and particularly a contrast between the central and peripheral regions of the apex, have not reported such a consistent pattern. A common finding is that the cells of the central zone are larger than those of the peripheral zone and that their nuclei are larger and less densely stained with the Feulgen reaction, which specifically identifies DNA. This is correlated with a less active incorporation of DNA precursors as noted earlier, but cytophotometric measurements indicate that the DNA content per nucleus is the same as in other diploid cells (Steeves et al., 1969). In the larger nuclei the DNA is simply less concentrated (Fig. 5.4). Another commonly reported feature of the central zone is a lower content of RNA, which appears to correlate with the generally fainter staining of this region. The synthesis of RNA is also reported to be less active in the central zone than in the peripheral zone, but it does occur, indicating that the cells are metabolically active but less so. A similar pattern of protein content is also reported. Unfortunately, many of the descriptions of RNA and protein distribution in the apex have been based on visual inspection of staining patterns and have not included quantitative measurements. The fainter staining may be related primarily to the more highly vacuolated state of the central zone cells. It is important to note that the vacuolation of these cells is not like that of mature or maturing cells in which one or a few large vacuoles occupy much of the cell volume. Rather, the cytoplasm contains a large number of small vacuoles so that it may have what has been described as a "frothy" appearance.

Clearly, however, it is not possible to generalize such descriptions of apical zonation because there are differences in different species. For example, in *Brachychiton* West and Gunckel (1968) found that both RNA

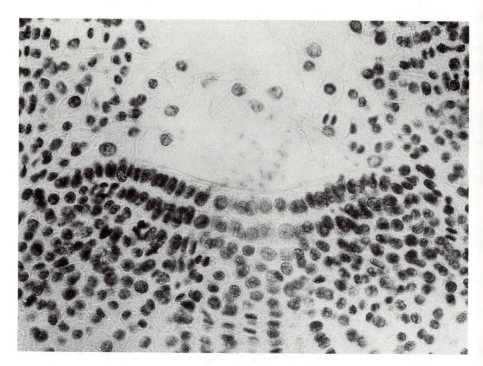

Figure 5.4. *Median longitudinal section of a shoot apex of* Helianthus annuus *treated by the Feulgen reaction, which stains only DNA. The large nuclei in the central zone are less intensely stained than are those in the peripheral zone. ×450. (Steeves et al., 1969.)*

and protein were more concentrated (on a per cell basis) in the peripheral than in the central zone. RNA synthesis was also higher in the peripheral zone, but the cells of the central zone were surprisingly high in activity. By contrast, in *Pisum* Lyndon (1970) has reported no difference in RNA or protein content, again on a per-cell basis, in the central and peripheral zones, and RNA synthesis was equally high in both zones. Yet in *Pisum*, as has been noted, the central cells have a distinctly lower division frequency than do the peripheral cells.

Many of the investigations described above were undertaken with the objective of identifying a basis for metabolic differences among the zones of the shoot apex. A few workers have chosen to make direct observations on the metabolic status of the meristem by investigating the localization of enzyme activity. Wetmore, Gifford, and Green (1959) found that in the apices of several species histochemical tests for oxidases and dehydrogenases give indication of greater amounts of both in the mitochondria of the centrally located cells of the meristem than in periph-

eral and subjacent tissues. Vanden Born (1963) has carried out extensive observations on unfixed apices of *Picea glauca* (white spruce) in
which attempts were made to localize a number of enzymes in the meristem and subjacent tissues. The most interesting pattern was that of
peroxidase, which was abundant in the peripheral regions of the meristem and the immediately subjacent tissues as well as in differentiating vascular tissue and in other regions in which mitotic activity appears to be high. It was significantly absent from cells at the summit of
the apex and thus from the zone where mitotic activity is lowest. A similar pattern was observed by Riding and Gifford (1973) in *Pinus radiata*.
It is difficult to interpret results of this type on the basis of a few species,
but they do indicate that there are differences in metabolic activity among
the zones of the shoot apex.

Ultrastructure of meristem zones

Attempts to achieve a greater refinement in the characterization of the
cytological basis of zonation by turning to the ultrastructural level have,
by and large, proved disappointing. They have not in general revealed
any consistent fine structural differences among the zones of the apex.
Plastids are sometimes less well differentiated in the central zone than
in the peripheral regions, but in other cases this is not true. For example, cells in the central zone of the shoot apex of *Helianthus* contain
conspicuous starch grains in the plastids (Fig. 5.5). Similarly, mitochondria in the central zone have been reported to be large in some cases
(Gifford and Stewart, 1967). A commonly noted feature is the apparent
lower density of ribosomes in the central region, which correlates with
the generally lower RNA content noted earlier. However, this may depend upon the type of quantitative analysis used, whether on the basis
of unit area, unit volume or individual cell, and it is not clear which is
the physiologically significant analysis (Lin and Gifford, 1976).

In the absence of distinctive qualitative differences among the zones,
careful quantitative analysis assumes particular importance. In a study
of this type Lyndon and Robertson (1976) found no significant differences in cellular components in the various regions of the shoot apex of
Pisum, and there were no changes during the course of a plastochron.
The only significant ultrastructural differences were those clearly associated with early stages of histodifferentiation. In a quantitative study
of ultrastructure of the zoned shoot apex of the cactus *Echinocereus engelmannii*, Mauseth (1981) determined for each zone the proportion of
the cell volume occupied by each of the following cell components: nucleus, nucleolus, vacuole, mitochondria, chloroplasts, dictyosomes, and
residual cytoplasm. Each zone differed from all other zones in some

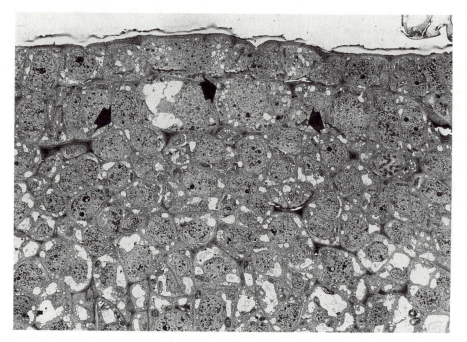

Figure 5.5. *Electron micrograph of a longitudinal section of the shoot apex of* Helianthus annuus. *Arrows indicate starch-containing plastids.* ×1,075. (V. K. Sawhney, P. J. Rennie, and T. A. Steeves, *Can. J. Botany 59:2009, 1981.)*

aspect of its ultrastructure, but none of these quantitative differences, although often statistically significant, could be considered a major distinction. As in *Pisum,* all of these differences were dwarfed by the contrast between any cell of the meristem and any cell of the adjacent differentiating tissues.

When Foster (1938) first initiated the description of shoot apices in terms of a radial zonation, this was based upon the presumption that the zones have distinctive cytological and corresponding functional characteristics. The studies just described have attempted to document these characteristics. Although a substantial body of detailed information has been assembled, it is evident that many inconsistencies remain to be clarified before general conclusions can be drawn. Much of this information does support the recognition of a group of centrally placed cells that differ cytologically and biochemically from those adjacent to them. These are the cells that also have been shown to have a lower mitotic frequency than others in the apex. It remains to be determined, however, what, if any relationship exists between the cytological and biochemical attributes of these cells and their relatively low division rate.

DEVELOPMENTAL CHANGES IN ZONATION

Up to this point the organization and function of the vegetative shoot apex have been considered without regard to the development of a plant. It is important not to overlook the fact that the organizational patterns can, and frequently do, change ontogenetically. In particular there is often an early phase during which a characteristic zonation pattern is established and a final phase, usually associated with reproduction, in which the characteristic pattern may be lost or drastically altered.

The establishment of zonation in the apex of *Coleus* during germination and early seedling growth has been described by Saint-Côme (1966). In the mature seed the shoot apex of the embryo consists of three layers of cells and shows no zonation. The distinction between central and peripheral zones is first established by the occurrence of more frequent divisions in the latter resulting in a difference in cell size and is heightened from about the twelfth day after germination by characteristic differences in the intensity of RNA specific staining. Full adult zonation is not attained until five pairs of leaves have been initiated.

In conifers the establishment of zonation in the shoot apex apparently occurs at different times in different species. In *Pinus radiata* Riding and Gifford (1973) have reported the progressive elaboration of a zonate pattern that was not complete until eighty-four days after germination. In the dormant embryo there was an apparent zonation, but this was based on differences in the distribution of storage products rather than on cell sizes and configurations. On the other hand, in *P. lambertiana* there is a structural zonation in the embryo but this is elaborated into the typical adult pattern during germination and early seedling development (Fosket and Miksche, 1966).

Instructive examples of the establishment of a zonation pattern are provided by Mauseth's (1978) observations on germination and early seedling growth of a number of species of Cactaceae. The adult plants of all of these species have a distinct radial zonation comparable to those that were described earlier. Mauseth showed that in certain species, at the time of germination, the shoot apex has a tunica–corpus organization but no evidence of a radial zonation superimposed upon it. Upon germination, usually within a few days, zonation begins to be evident because cells in the peripheral regions of both tunica and corpus become smaller than those placed centrally by virtue of more frequent cell divisions. Usually slightly later the pith rib meristem appears. It is significant that there is considerable variation in the stage at which these changes occur. In some species they are not completed until more than 30 leaves have been produced, while in others the zonation is present at germination, indicating that the changes must have occurred during embryogeny. These observations suggest that, while zonation is a characteristic feature of shoot apical organization, it cannot be considered essential for the normal functioning of the meristem.

On the other hand, Niklas and Mauseth (1981) have calculated the relative contributions of the various zones to the volume of the adult shoot apex and have shown that, in spite of substantial changes in the size and shape of the apex, once the adult pattern has been established, the relative proportions of the zones remain remarkably constant. It was also demonstrated that the quantitative ultrastructure of the apical zones is strikingly invariable. Thus, although apparently not absolutely necessary, a stable zonation pattern is certainly characteristic of each species.

The cytological features that characterize the vegetative meristem are usually modified extensively with the initiation of reproductive development, often after a very brief period of induction in the case of photoperiodically sensitive species (see Chapter 10). This transition results in the development of uniformity throughout the terminal meristem as cells of the central zone acquire the characteristics previously found only in the peripheral zone. This, of course, is associated with the acceleration of mitotic activity in this region. Work on photoperiodically sensitive plants has revealed an interesting aspect of these changes that may go a long way toward explaining some of the differences in staining patterns reported by earlier workers. It now appears that if such a plant is maintained in a light regime that is unfavorable for the initiation of flowering, the shoot apex may nonetheless undergo some of the cytological changes associated with flowering while still remaining vegetative in function (Nougarède, Gifford, and Rondet, 1965). Thus, some of the zonation that has been attributed to vegetative apices, or lack of it, may in fact be applicable to this transitional stage, and great caution must be exercised in the interpretation of such results.

CLONAL ANALYSIS

There is another way in which the activity of cells in various regions of the shoot apex could be analyzed – that of marking individual cells and tracing their progeny through successive cell divisions into the mature regions of the plant. Since the progeny of a particular cell can be considered a clone, this method is often called *clonal analysis*, and it has been used very effectively in studies of animal development. It has also aided in establishing the developmental destinies of portions of certain plant embryos. In fact, this method has been employed in plant developmental studies for many years, but its significance has seldom been fully appreciated.

The chief requirement of this method is the marking in an identifiable

way of individual cells in the shoot apex so that if they divide, the subsequent fate of their derivatives can be traced. This kind of analysis can
be achieved in plants known as *chimeras* (Tilney-Bassett, 1986). In a
chimera tissues composed of cells with distinctive markers form a kind
of mosaic in the plant body. The markers are permanent cellular characteristics and include such features as chromosome number, anthocyanin pigmentation, or chloroplast pigmentation. They arise as a result
of somatic mutation, either in the nucleus or in organelles, changes in
chromosome number, or alterations in chromosome structure, and although they are known to occur spontaneously, they can also be induced artificially by exposure to radiation or to chemical agents. The
original artificially produced chimeras developed from mixtures of stock
and scion tissues in the region of union in grafts between varieties or
species whose cells had identifiable features. In chimeras tissues with
specific markers are mixed in various ways (Fig. 5.6). In a *periclinal chimera* parallel layers of cells with different characteristics are found, as
for example in a shoot that is entirely diploid except for a tetraploid
epidermis. In a *sectorial chimera* tissues with a particular characteristic
occupy a sector of an organ, and in a *periclinal chimera* a periclinal
chimera occupies only a sector of the organ (Fig. 5.7).

When chimeras are relatively persistent it is inferred that they result
from correspondingly stable changes in the apical meristem, and, in
fact, in the case of periclinal polyploid chimeras such changes can often
be seen in the meristem. In *Datura* Satina, Blakeslee, and Avery (1940)
were able to identify three layers in the shoot apical meristem that responded independently to the polyploidy-inducing substance colchicine
(Fig. 5.8). Each of the meristem layers could be related to a layer or
layers of tissue in the resulting mature shoot by a comparison of ploidy
levels. In a number of angiosperms the existence of three such layers
has been reported, and these correspond to two tunica layers (the commonest pattern), in which divisions in the meristem are restricted to
the anticlinal plane, and the corpus, in which the plane of division is
not restricted.

In conifers that have been investigated in this way it is reported
(Stewart, 1978) that two distinct apical layers appear to be involved.
This agrees with the known structure of the shoot apex (see Chapter 4)
in which one surface layer is usually seen. When features other than
ploidy level are involved, they cannot usually be identified in the meristem itself, but the pattern of distribution in the mature regions indicates a derivation similar to that of polyploid chimeras. The close relationship between layers at the summit of the apex and tissues of the
mature body of the shoot provides strong evidence that the summit cells
do in fact have an initiating role.

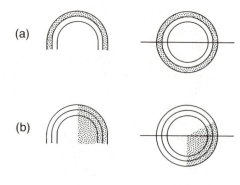

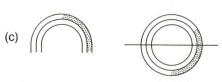

Figure 5.6. *Diagrammatic representation of the shoot apex of chimeric plants: (a) periclinal, (b) sectorial, and (c) mericlinal. On the right, cross sections are shown, and on the left, longitudinal sections in the plane indicated by the line.*

Figure 5.7. *A mericlinal chimera of* Helianthus annuus. *The albino sector extends from the lower part of the stem into the inflorescence. 1/10 nat. size. (Photograph provided by D. Jegla.)*

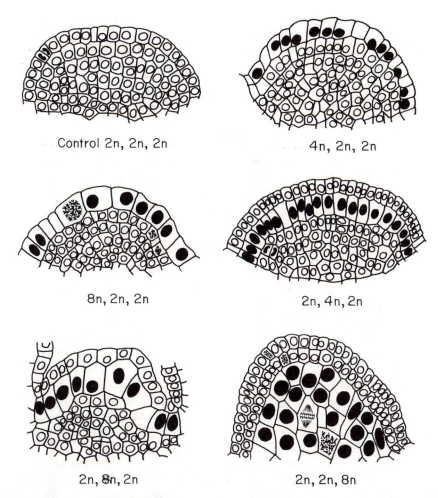

Figure 5.8. *Periclinal polyploid chimeras seen in median longitudinal sections o₁ the shoot apex. The ploidy level of each of the three apical layers is shown for each apex. Diploid nuclei are shown as open circles and polyploid nuclei as filled circles. (Redrawn from Satina, Blakeslee, and Avery, 1940.)*

Control 2n, 2n, 2n

4n, 2n, 2n

8n, 2n, 2n

2n, 4n, 2n

2n, 8n, 2n

2n, 2n, 8n

Chimeral plants have also facilitated the study of a different but related problem, that of the existence of permanent or long-term apical initial cells. Supporters of a passive role for the summit cells (méristème d'attente) of course deny the existence of any such cells. For entirely different reasons, some observers of living shoot apices (Soma and Ball, 1964; Ball, 1972) have argued that the apical meristem is a dynamic or shifting population of actively dividing cells, with no cell having any long-term role as an initial. Mericlinal chimeras, that is, periclinal chimeras that occupy a sector of the shoot only, provide convincing evidence bearing upon this question. These sectors arise in periclinal chimeras following a periclinal division in a cell of one layer and the insertion of one of the derivatives into an adjacent layer. The sectors that include the altered tissue ordinarily occupy one-third, one-half, or two-thirds of the shoot, and they persist for an extended period of growth (Fig. 5.6). In cranberry Dermen (1945) was able to trace altered sectors along 1.3 m of the stem through approximately 100 nodes. Similar sectors were followed through 104 nodes in juniper, 26 nodes in privet, and 50 nodes in poinsettia (Stewart and Dermen, 1970). Observations such as these support the conclusion that each initiating layer in the apex contains two or three initial cells that function over an extended period. The distinctive features of one of these would be expressed in its derivatives in one-third or one-half of the shoot only, the remainder being unaffected.

The chimeral patterns vary in permanence, however, and may disappear after a period of growth, indicating that the apical initials are not to be regarded as permanent initials but can be displaced by shifting configurations in the apex. In a study of albino mericlinal chimeras in *Juniperus* Ruth, Klekowski, and Stein (1985) have observed that, although some sectors disappeared, many others actually widened, ultimately giving rise to completely albino shoots. This was interpreted as the result of a displacement by the albino initials of the other initials in the same layer.

There are also narrower mericlinal sectors that persist for shorter periods, for example three to five nodes in privet. These presumably result from the insertion into an adjacent layer of a recent derivative of an initial cell rather than the initial itself. Because these limited sectors define the extent of tissue production by recent derivatives of initials, they permit calculation of the frequency of division in the initials under a particular set of growing conditions. In privet, as noted earlier, division no more frequently than once in 12 days would be sufficient to allow the apical initials to perform their role as the ultimate source of the shoot. Thus, the evidence derived from clonal analysis not only demonstrates the initiating role of the summital cells of the apex but it also supports the traditional view that among these cells are to be found true apical initials.

GENERAL COMMENT

This chapter has assembled a considerable body of information on the distribution of mitotic activity in the shoot and on patterns of cytohistological zonation in the apex. Although it now appears that zonation is widespread, if not universal, at least in the seed plants, the specific characteristics of the zones appear to be varied, and this makes it difficult to interpret the functional significance of zonation. One correlation has been noted that appears to be reasonably consistent. The cytohistological distinction between the central and peripheral regions of the meristem in general parallels the pattern of mitotic frequency in which the summit of the apex is occupied by cells that divide less frequently than do those surrounding them. The present state of our knowledge, however, does not permit general conclusions about the nature of the relationship between mitotic frequency and cytological characteristics.

The traditional view of shoot apical organization and function, which places one or a group of initial cells in a summital position, would conclude that the central group of slowly dividing and sometimes cytologically distinctive cells are the initials or at least include them. This view, however, has been challenged. Supporters of the méristème d'attente concept, while acknowledging that these cells do divide, assign to them only a passive role in vegetative development and ascribe the initiating function to peripheral and subjacent regions of the meristem. On the other hand some studies on living shoot apices purport to show a high division frequency in the summit of the apex and seem to rule out the existence of any permanent or long-term initial cells. Although there are conflicting observations that contribute to the divergent views of the shoot apex, one is inclined to conclude that the major remaining differences are matters of interpretation. Data that to one observer provide conclusive evidence, to another prove nothing because the methods are held to be invalid.

One method of analysis, which is certainly not new but has stimulated renewed interest in recent years, is providing direct support for the traditional view of apical organization. The study of chimeral plants – clonal analysis – makes possible the tracing of cell lineages, in some cases directly from the meristem, and thus provides direct information about division patterns in the meristem. This method has demonstrated that in seed plants, particularly the angiosperms, the layered structure of the shoot apex has developmental significance in that the layers function with a high degree of independence in the generation of the tissues of the shoot. Of even greater significance is the evidence provided by mericlinal chimeras that each layer is ultimately initiated by a small group of initial cells that functions for a protracted period, longer in some species than in others. It is difficult to interpret the continued generation of a stem sector with altered cells through the development of more than 100 nodes in any other terms. When the mutational event occurs in a derivative cell rather than in an initial, this is readily deduced from the abbreviated generation of altered cells and the narrowness of the mutant sector. It has also been possible to calculate the frequency of divisions in the initial cells and to demonstrate that a low mitotic frequency is consistent with the role of these cells.

Supporters of the concept of apical initials have recognized the probability that the initials are not permanent in the sense of functioning throughout the life of a meristem. As will be shown in the next chapter, they can be replaced in the event of injury, and there is no reason to suppose that they are not replaced in the course of normal development by derivatives having equivalent potentialities. The only requirement is that, in the initiating region, some cells be retained to continue the initiation process while derivatives begin to undergo processes of differentiation while still proliferating. This understanding of apical initials is verified by the results of clonal analysis. It is somewhat at variance with the idea often implied in statements about apical initials, that they are permanently identifiable cells, presumably with unique properties. It is entirely consistent with the concept of initials advanced by Newman (1965) in which he emphasized this interpretation by naming these cells a "continuing meristematic residue." Even in the lower vascular plants in which an apical cell can be distinguished morphologically, it is doubtful in many cases (see Chapter 4) that a unique individual cell functions permanently as the initial.

This understanding of the nature and role of initial cells leads to the recognition of their striking similarity to certain cells that play an important role in animal development and are designated *stem cells* (Barlow, 1978). Stem cells, which are responsible for the continued growth of structures, such as teeth, hair, skin, intestinal epithelium, and blood, are described as populations of relatively unspecialized cells that produce derivatives that differentiate while perpetuating themselves for extended periods. Like shoot apical initials, stem cells generally have a lower mitotic frequency than is found among their derivatives and, although they persist for long periods, they can be replaced by cells from among their recent derivatives. On the other hand, these remarkable similarities must not be allowed to obscure the distinctive aspects of plant development (see Chapter 1). The initial cells of a shoot apex participate in the formation of a major portion of the plant body, including the development of organs, processes that at least in the higher animals are restricted to the embryonic phase. Moreover, most stem cell activity in animals is associated with the replacement of tissues within the body and is balanced by cell loss so that there is no overall growth. In plants there is a continued accumulation of cells, and overall growth continues throughout the life of the organism.

At the conclusion of the previous chapter it was suggested that the shoot apex could be interpreted structurally as consisting of a promeristem, containing apical initials and surrounded laterally and basally by differentiating derivatives. It now appears that the same interpretation may be applicable functionally. The apical initials are associated with their recent derivatives, which have not yet begun to differentiate and are therefore presumably equivalent in potency to the initials and interchangeable with them. The difficulties of relating this organization to the zonation patterns that can be recognized histologically have been discussed, particularly the distinction between cytological features associated with initial differentiation and those that reflect the organization of the promeristem. It is hoped that the refinement of cytological techniques may shed some light upon this question.

There is also another method of investigating the functional organization of the shoot apex, and this is through the medium of experimental manipulation of the apical region. It is appropriate to proceed in the next chapter to a consideration of some of this experimental work.

REFERENCES

Ball, E. 1960. Cell divisions in living shoot apices. *Phytomorphology* 10:377–96.
———. 1972. The surface histogen of living shoot apices. In *The Dynamics of Meristem Cell Populations,* ed. M. W. Miller and C. C. Kuehnert, 75–97. New York: Plenum Press.

Barlow, P. W. 1978. The concept of the stem cell in the context of plant growth and development. In *Stem Cells and Tissue Homeostasis,* ed. B. I. Lord, C. S. Potten, and R. J. Cole, 87–113. Cambridge: Cambridge University Press.

Bierhorst, D. W. 1977. On the stem apex, leaf initiation and early leaf ontogeny in filicalean ferns. *Am. J. Botany* 64:125–52.

Buvat, R. 1952. Structure, évolution et functionnement du méristème apical de quelques dicotylédones. *Ann. Sci. Nat. Bot.* Ser. 11. 13:199–300.

Catesson, A. M. 1953. Structure, évolution et functionnement du point végétatif d'une monocotylédone: *Luzula pedmontana* Boiss. et Reut. (Joncacées). *Ann. Sci. Nat. Bot.* Ser. 11. 14:253–91.

Clowes, F.A.L. 1961. *Apical Meristems.* Oxford: Blackwell.

D'Amato, F. 1975. Recent findings on the organization of apical meristems with single apical cells. *Giornale Bot. Ital.* 109:321–34.

Davis, E. L., P. Rennie, and T. A. Steeves. 1979. Further analytical and experimental studies on the shoot apex of *Helianthus annuus:* Variable activity in the central zone. *Can. J. Botany* 57:971–80.

Dermen, H. 1945. The mechanism of colchicine-induced cytohistological changes in cranberry. *Am. J. Botany* 32:387–94.

Fosket, D. E., and J. P. Miksche. 1966. A histochemical study of the seedling shoot apical meristem of *Pinus lambertiana. Am. J. Botany* 53:694–702.

Foster, A. S. 1938. Structure and growth of the shoot apex in *Ginkgo biloba. Bull. Torrey Bot. Club* 65:531–56.

Gifford, E. M., Jr. 1983. Concept of apical cells in bryophytes and pteridophytes. *Ann. Rev. Plant Physiol.* 34:414–40.

Gifford, E. M., Jr., and K. D. Stewart. 1967. Ultrastructure of the shoot apex of *Chenopodium album* and certain other seed plants. *J. Cell. Biol.* 33:131–42.

Lin, J., and E. M. Gifford, Jr. 1976. The distribution of ribosomes in the vegetative and floral apices of *Adonis aestivalis*. *Can. J. Botany* 54:2478–83.

Lyndon, R. F. 1968. Changes in volume and cell number in the different regions of the shoot apex of *Pisum* during a single plastochron. *Ann. Botany* 32:371–90.

———. 1970. DNA, RNA and protein in the pea shoot apex in relation to leaf initiation. *J. Exp. Botany* 21:286–91.

———. 1973. The cell cycle in the shoot apex. In *The Cell Cycle in Development and Differentiation*, edited by M. Balls and F. S. Billett, 167–83. Cambridge: Cambridge University Press.

———. 1976. The shoot apex. In *Cell Division in Higher Plants*, edited by M. M. Yeoman, 285–314. New York: Academic Press.

Lyndon, R. F., and E. S. Robertson. 1976. The quantitative ultrastructure of the pea shoot apex in relation to leaf initiation. *Protoplasma* 87:387–402.

Mauseth, J. D. 1978. An investigation of the morphogenetic mechanisms which control the development of zonation in seedling shoot apical meristems. *Am. J. Botany* 65:158–67.

———. 1981. A morphometric study of the ultrastructure of *Echinocereus engelmannii* (Cactaceae) II. The mature, zonate shoot apical meristem. *Am. J. Botany* 68:96–100.

Newman, I. V. 1965. Pattern in the meristems of vascular plants. III. Pursuing the patterns in the apical meristem where no cell is a permanent cell. *J. Linn. Soc. (Bot.)* 59:185–214.

Niklas, K. J., and J. D. Mauseth. 1981. Relationships among shoot apical meristem ontogenetic features in *Trichocereus pachanoi* and *Melocactus matanzanus* (Cactaceae). *Am. J. Botany* 68:101–6.

Nougarède, A. 1967. Experimental cytology of the shoot apical cells during vegetative growth and flowering. *Internat. Rev. Cytol.* 21:203–351.

Nougarède, A., E. M. Gifford, Jr., and P. Rondet. 1965. Cytological studies of the apical meristem of *Amaranthus retroflexus* under various photoperiodic regimes. *Bot. Gaz.* 126:281–98.

Philipson, W. R. 1954. Organization of the shoot apex in dicotyledons. *Phytomorphology* 4:70–75.

Riding, R. T., and E. M. Gifford, Jr. 1973. Histochemical changes occurring at the seedling shoot apex of *Pinus radiata*. *Can. J. Botany* 51:501–12.

Ruth, J., E. J. Klekowski, and O. L. Stein. 1985. Impermanent initials of the shoot apex and diplontic selection in a juniper chimera. *Am. J. Botany* 72:1127–35.

Saint-Côme, R. 1966. Application des techniques histoautoradiographiques et des méthodes statistiques à l'étude du functionnement apical chez le *Coleus blumei* Benth. *Rev. Gén. Botan.* 73:241–323.

Satina, S., A. F. Blakeslee, and A. G. Avery. 1940. Demonstration of the three germ layers in the shoot apex of *Datura* by means of induced polyploidy in periclinal chimeras. *Am. J. Botany* 27:895–905.

Soma, K., and E. Ball. 1964. Studies of the surface growth of the shoot apex of *Lupinus albus*. *Brookhaven Symp. Biol.* 16:13–45.

Steeves, T. A., M. A. Hicks, J. M. Naylor, and P. Rennie. 1969. Analytical studies on the shoot apex of *Helianthus annuus*. *Can. J. Botany* 47:1367–75.

Stewart, R. N. 1978. Ontogeny of the primary body in chimeral forms of higher plants. In *The Clonal Basis of Development*, ed. S. Subtelny and I. M. Sussex, 131–60. New York: Academic Press.

Stewart, R. N., and H. Dermen. 1970. Determination of number and mitotic activity of shoot apical initial cells by analysis of mericlinal chimeras. *Am. J. Botany* 57:816–26.

Tilney-Bassett, R.A.E. 1986. *Plant Chimeras.* London: Arnold.

Vanden Born, W. H. 1963. Histochemical studies of enzyme distribution in shoot tips of white spruce (*Picea glauca* [Moench] Voss.). *Can. J. Botany* 41:1509–27.

Wardlaw, C. W. 1957. The reactivity of the apical meristem as ascertained by cytological and other techniques. *New Phytol.* 56:221–9.

West, W. C., and J. E. Gunckel. 1968. Histochemical studies of the shoot of *Brachychiton.* II. RNA and protein. *Phytomorphology* 18:283–93.

Wetmore, R. H., E. M. Gifford, Jr., and M. C. Green. 1959. Development of vegetative and floral buds. In *Photoperiodism and Related Phenomena in Plants and Animals,* edited by R. B. Withrow, *Am. Assn. Adv. Sci. Publ.* 55:255–73.

CHAPTER 6

Experimental investigations on the shoot apex

Most of the work reviewed thus far has consisted of observations of various types of shoot apices that are for the most part normal or at least intact. It is also possible to investigate the organization of the meristem by subjecting it to experimental treatments and analyzing its reaction to these manipulations. The most widely used and the most successful of the experimental procedures applied to shoot apices has been microsurgery, in which delicate instruments are used to make punctures, incisions, and excisions in various regions of the meristem. Experiments of this sort have also been carried out in sterile culture so that the surgically isolated region of the meristem can be grown in isolation from the remainder of the plant. By these methods it is possible to obtain information about the roles played by various portions of the meristem and about interactions between parts of the meristem and between the meristem and other parts of the plant. One limitation of the surgical method is that it cannot be assumed that the effects of operations are just the isolation or removal of particular regions of meristematic tissue. Wounding may itself produce its own responses that are difficult to evaluate in terms of the normal apex, for example, cell proliferation. Thus, methods involving surgical manipulations tell a great deal about the potentialities of portions of a meristem, information that is extremely valuable, but experiments must be controlled carefully and the results screened rigorously if the information is to be applied validly to the interpretation of the organization of the intact meristem.

APICAL AUTONOMY

One of the problems confronting students of plant development is the question of the relationship between the meristem and the mature regions of the plant it has produced. Clearly, the meristem is a center of cell production, and it provides the structural units of which the mature body is constructed. It is not, however, immediately obvious whether the meristem functions in a completely plastic manner, governed in all of its activities by the patterns of more mature regions of the plant,

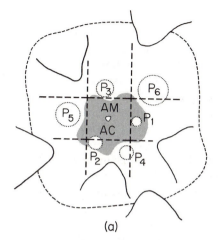

(a)

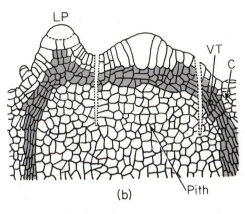

(b)

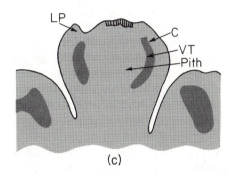

(c)

Figure 6.1. *Surgical isolation of the apical meristem of* Dryopteris dilatata. *(a) A view of the shoot tip showing the position of the four incisions that iso-*lated the terminal meristem from the surrounding leaf primordia and differentiating tissues. (b) Longitudinal section of a surgically isolated meristem showing the tissues that have been severed by the incisions, and the intact pith connection below the meristem. (c) A longitudinal section of the shoot tip five weeks after an operation isolating the meristem. The meristem has continued to grow, producing new leaf primordia, and cortical, vascular, and pith tissues have been differentiated internally. Key: AC, apical cell; AM, apical meristem; C, cortex; LP, leaf primordium; $P_1–P_6$, the six youngest leaf primordia around the meristem; VT, vascular tissue. (a) ×45, (b) ×80, (c) ×10. (Adapted from C. W. Wardlaw, Phil. Trans. Roy. Soc., London, Series B 232:343, 1947.)*

whether it is largely autonomous, perhaps to a considerable extent controlling the destinies of its own cellular products, or whether its true role lies somewhere between these two extremes. Some of the experimental work on shoot apices has provided evidence favoring the view that the meristem is largely autonomous in its activity.

In both ferns and angiosperms (Wardlaw, 1950; Ball, 1952b) the center of the apical meristem has been isolated inside the youngest leaf primordia by three or four deep, vertical incisions that leave it supported on a plug of maturing pith tissue and separated laterally from all leaf primordia (Fig. 6.1). Its normal vascular supply is, of course, severed by the incisions. Under these conditions the meristem continues its growth, although often at a reduced rate at the outset, initiates a sequence of new leaf primordia, and gives rise to normal tissues of the shoot. Vascular differentiation, furthermore, proceeds basipetally through the original immature pith tissue and often establishes a connection with the original shoot vascular system around the bases of the incisions. The seemingly normal development of the meristem under these conditions makes it difficult to believe that it has any dependence upon mature regions of the plant for anything other than basic nutrients.

With the development of sterile culture techniques it was possible to effect an even more dramatic isolation of the shoot meristem. In an early investigation of this type Ball (1946) isolated pieces from the shoot tip approximately 0.5 mm^3 in volume, including the shoot meristem and several of the youngest leaf primordia. These were explanted to a culture medium containing inorganic salts, sugar, agar, and water. Such explants regularly gave rise to complete shoots and, through rooting, to whole plants that could subsequently be transplanted to a greenhouse. Results of this sort, though they are suggestive of apical autonomy, do not test the potentialities of the meristem alone because several leaf primordia and considerable subjacent tissue were included in the explant. Subsequently Ball (1960) was able to grow meristems of *Lupinus* isolated without any leaf primordia into shoots 5 to 10 cm in length and bearing seven to nine leaves. This result was achieved only by the incorporation of coconut milk and gibberellic acid into the medium, and shoot growth was ultimately terminated by the loss of meristematic activity in the meristem.

In contrast to this result, similar explants of several species of vascular cryptogams have been grown into entire rooted plants on a medium consisting of sugar and inorganic salts only (Wetmore, 1954) (Fig. 6.2). These early results, which indicated a difference in culture requirements of isolated meristems of angiosperms and vascular cryptogams, were suggestive of a fundamental difference in meristem autonomy between these two groups of plants. However, more recent studies have revealed differences in the culture requirements of different angiosperms, many of which can be grown on media approaching the simplicity of those used for vascular cryptogams. Smith and Murashige (1970) obtained whole plants from explants consisting only of the terminal meristem without leaf primordia in five species of flowering plants. The medium required for this development was remarkably simple, containing in addition to inorganic salts and sugar only myo-inositol, thiamine hydrochloride, and indoleacetic acid. It was established that indoleacetic acid was essential for the complete development of the meristem.

The pattern of development of the isolated meristems was followed in

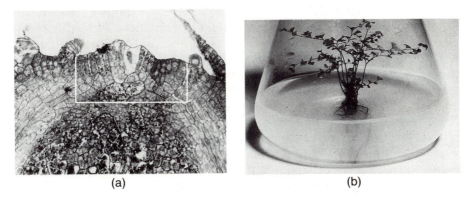

(a)   (b)

Figure 6.2. Adiantum. *(a) Longitudinal section of the shoot tip showing the size of the meristem piece that was excised and grown in sterile culture. (b) A plant grown in culture from an explant like that shown in (a). (a) ×125, (b) ⅔ nat. size.*

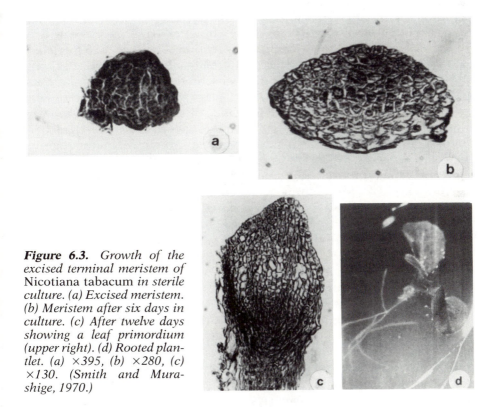

Figure 6.3. *Growth of the excised terminal meristem of* Nicotiana tabacum *in sterile culture. (a) Excised meristem. (b) Meristem after six days in culture. (c) After twelve days showing a leaf primordium (upper right). (d) Rooted plantlet. (a) ×395, (b) ×280, (c) ×130. (Smith and Murashige, 1970.)*

detail only in tobacco (Fig. 6.3), and it differed strikingly from that of larger explants cultured earlier in which a leafy shoot formed first and roots were initiated subsequently. During the first six days in culture the meristem underwent considerable cell division and cell enlargement, particularly near the cut surface, but it retained its essential organization. In six to nine days a root apex was initiated within the tissue opposite the shoot apex, forming a bipolar axis. By the twelfth day a leaf primordium was initiated and the shoot–root axis contained a well-developed provascular system. Subsequent growth led to the development of a plantlet that ultimately could be transplanted to soil.

In comparable experiments Ball (1980) was able to grow explants that consisted of an isolated shoot meristem or the meristem and the three youngest leaf primordia of *Trachymene coerulea* on a culture medium entirely lacking hormones or plant growth regulators and containing sucrose, inositol, thiamine hydrochloride, and a chelating agent as the only organic constituents. Almost all of the explants grew into leafy shoots. Most produced adventitious roots and, when transferred into soil in a greenhouse, grew into mature flowering individuals. If the meristem explants were made smaller by either excising only the central region or bisecting the meristem, regeneration of a functional meristem occurred in about half the cultures, but the results were somewhat complicated by the extensive development of somatic embryos from the bases of the explants.

The possibility that explants that include different portions of the shoot tip might have different nutrient requirements has been investigated by Shabde and Murashige (1977), who studied growth regulator requirements of the excised shoot meristem of *Dianthus caryophyllus* (carnation). On a medium that contained, in addition to inorganic salts, only sucrose, inositol, and thiamine hydrochloride isolated meristems failed to develop. Neither did meristems isolated with either one or two pairs of leaf primordia, but over half of those isolated with two pairs of leaf primordia and the youngest pair of expanding leaves developed into complete rooted plants. Addition of IAA to the culture medium resulted in the development of a small proportion of those explants that included one or two pairs of leaf primordia but not of explants that consisted of the meristem only. A small number of these isolated meristem explants would develop in a medium that contained kinetin, but when the medium contained both IAA and kinetin, 65 percent of the isolated meristem explants developed into complete plants. From these results Shabde and Murashige concluded that hormonal substances required for meristem growth and leaf formation are synthesized in young, expanding leaves and are transported into the meristem.

The long-debated difficulty in culturing angiosperm apices seems to have resulted from a failure to find the precise nutrients and growth factors required for normal development, and it now appears that these differ in different species. The meristems of all plants are clearly dependent on the rest of the plant to supply the basic nutrients and water required for growth. The experiments described above indicate that the meristems of vascular cryptogams and of some flowering plants may require nothing further to be supplied from other parts of the plant but that the meristems of some flowering plants are also dependent for auxin and others for both auxin and cytokinin that are synthesized in young leaves and transported to the meristem.

One may reasonably ask whether an organ that is dependent upon exogenous sources of such specific substances as auxins and cytokinins can be regarded as autonomous in any ordinary sense of the word. If by the term *apical autonomy* is meant the capacity for continued growth and production of organized structure in the absence of exogenously supplied developmental stimuli, then the above experiments can be considered to have revealed a high degree of apical autonomy in both vascular cryptogams and flowering plants. However, the meristem cannot be entirely independent in its function. There are well-known phenomena in plants in which the shoot meristem responds dramatically to a precise exogenous stimulus. Processes like the initiation of flowering, shoot tip abortion, and other similar developmental responses clearly indicate that the meristem may be extremely sensitive to stimuli proceeding from more mature regions of the plant, even though we do not yet know the exact nature of these stimuli. One aspect that must be kept in mind when dealing with the concept of apical autonomy is the distinction between those exogenous factors that simply permit the meristem to regulate its own development and those that specifically induce new patterns of development.

ORGANIZATION AND INTEGRATION IN THE SHOOT APEX

The structural and analytical evidence presented in earlier chapters suggests that there may be localizations of function within the shoot meristem. A variety of surgical methods may be used to investigate this possibility, the simplest being the separation of the meristem into two portions by a vertical incision. In a series of experiments that marked the beginning of the modern era of experimental investigation of shoot apices, Pilkington (1929) carried out operations of this type on young plants of *Vicia faba* and *Lupinus albus* (Fig. 6.4). One might anticipate several different results of such an operation, each of which would lead to a different interpretation of the organization of the meristem. The apex might simply cease its growth or the two halves might graft to-

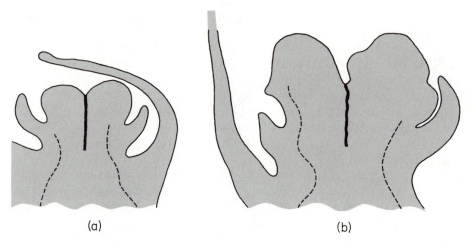

(a)   (b)

Figure 6.4. *The shoot tip of* Vicia faba *sectioned several days after the meristem had been bisected by a vertical incision. (a) Seven days after the operation the two halves of the meristem are enlarged. (b) Thirteen days after the operation new leaf primordia have emerged on the two regenerating apices. (Pilkington, 1929.)*

gether, reestablishing a single apex, suggesting that the apical meristem must be intact if it is to function at all. Alternatively, the half apices might continue their development separately. In this event, each half meristem might give rise to what would correspond to one half of a normal shoot in its symmetry and in the tissues and organs that it contains. On the other hand, each half meristem might become reorganized into a complete meristem capable of giving rise to a whole shoot with normal symmetry and placement of organs and tissues. Pilkington's experiments gave clear-cut evidence that this last alternative – the development of two complete shoots – is the correct one, and this finding has been confirmed by a number of subsequent workers.

In a reexamination of this phenomenon in *Lupinus* Ball (1955) has described the reestablishment of apical organization in the half meristems. In each half a new center was established on the original apical flank at some distance from the incision and not involving the original center. Surgical bisection has also been performed on excised shoot apices in culture with similar results. Ball (1980), for example, obtained regeneration of entire shoot apices from cultured halves of the apical meristem of *Trachymene cerulea* excised without leaf primordia. Thus it appears that a portion of the meristem, if separated, is capable of reestablishing a new meristem comparable in organization and function to the one from which it was separated.

The results of apical bisection raise the further question of whether

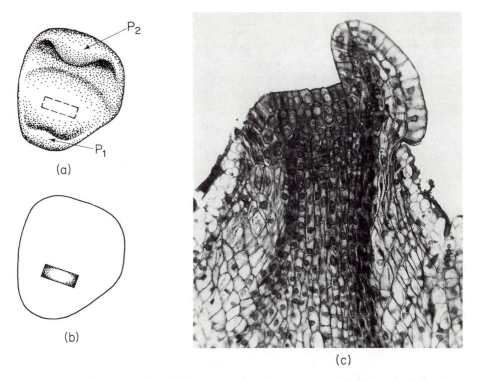

Figure 6.5. *Regeneration of the shoot apical meristem of* Solanum tuberosum. *(a) A diagram of the shoot apex showing the two youngest leaf primordia,* P_1 *and* P_2, *and the position of the apical panel that will be tested in the experiment. (b) The shoot apex after excision of all meristematic tissue except that in the apical panel. (c) A regenerating apical panel that has formed a leaf primordium and contains procambium. (c) ×150. (Sussex, 1964.)*

smaller subdivisions of the meristem might be able to reestablish whole meristems. When the apex of *Lupinus* was separated into four segments (Ball, 1948), each of the quadrants regenerated a complete meristem and produced an entire shoot. When the apex was divided into six segments (Ball, 1952a), some of these regularly regenerated, but always fewer than the total number possible. The fact remains, however, that segments as small as one-sixth of the original meristem were capable of reestablishing a complete meristem. When the meristem was divided into eight portions, none showed regenerative capacity. However, Sussex (1952) was able to obtain complete regeneration from a panel on the flank of the meristem of potato, which represented only about one-twentieth of the original meristem (Fig. 6.5). This regeneration occurred only if the rest of the meristem was excised from the plant, leaving the panel, with about 12 cells in its surface layer, alone at the summit of the shoot. Such small pieces of meristem pass through an initial period during which they enlarge considerably before initiating any leaf primordia. It must be concluded that the minimal size of a piece of meristem that can regenerate completely is very small and has not yet been defined. In fact one might ask whether, if the individual cells of the meristem could be isolated, they would reestablish whole meristems.

These experiments show that small portions of the terminal meristem have the capacity to regenerate entire apices. Some experiments have been carried out to test the possibility that there may be localization of regenerative capacity within the meristem by the isolation of specific regions. Loiseau (1959) destroyed large areas of the apical surface, leaving relatively small groups of cells either in the central (méristème d'attente) or in the peripheral (anneau initial) regions. The resulting development indicated an equivalent ability to regenerate in the two regions. The sterile culture technique would seem to offer excellent opportunities to test the regenerative capacity of specific portions of a meristem. Both Sussex and Rosenthal (1973) and Ball (1980) have used this method to demonstrate that the center of the meristem, devoid of all or most of the mitotically more active peripheral regions, can regenerate an entire shoot apex.

In the study of developmental potentialities of precise regions in the animal embryo it has been profitable to use the technique of grafting in which a particular organ rudiment or tissue mass is excised and implanted in another location. The possibility that comparable experiments might be carried out in the analysis of the shoot meristem of plants was explored by Ball (1950) in *Lupinus*. He cut wedge-shaped pieces from the summit of the apex and attempted to reimplant them in the original position as a preliminary test and to exchange such pieces between comparable apices. Unfortunately, this promising approach was not successful. Even in the few instances in which the grafted piece remained alive for some time, no growth was detected. There is reason to hope, however, that this technique can be perfected. Working with axillary buds of seedlings of *Pisum*, Gulline and Walker (1957) were able to obtain rejoining of severed meristems as little as 50 μm in height by maintaining a very humid atmosphere around the plants during and following the operation. Partial grafting of leaf primordia of the fern *Osmunda cinnamomea* has also been reported (Hicks, Steeves, and Sweeny, 1967) on excised shoot apices growing on a liquid medium in sterile culture.

Although it appears that any portion of the shoot meristem, even a relatively small portion, has the capacity to regenerate an entire apex, the fact remains that in the intact apex this potentiality is not ex-

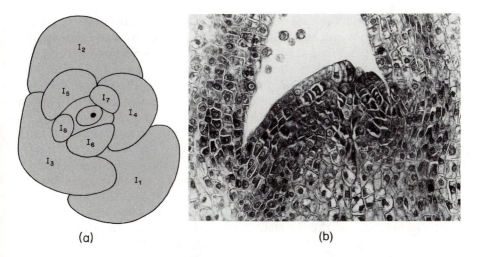

Figure 6.6. *Apical puncture in* Solanum tuberosum. *(a) diagram of a shoot apex in which eight leaves (I₁–I₈) had been initiated while the puncture remained in a central position in the meristem. (b) A longitudinal section of an apex showing the puncture. (b) ×135. (Sussex, 1964.)*

pressed. Clearly, there is a mechanism that serves to maintain the functional integrity of the shoot meristem. Two of the possible mechanisms that might be suggested are hormonal integration and competition for nutrients among various parts of the apex. It has been shown, particularly in the case of the isolation of small portions of meristem (Sussex, 1953), that shallow incisions not penetrating beyond the upper four or five layers of the apex are considerably less effective in permitting the regeneration of isolated panels than are deep incisions extending into the maturing subapical region. Such results are suggestive of a competitive nutritional basis of the integration mechanism. On the other hand, many plant physiologists would find it impossible to consider distribution of nutrients without invoking a hormonal mediation of this process.

Integration of the meristem has also been explored by surgical experiments of another type in which the distal cells of the apex are destroyed by puncturing with a needle. In *Lupinus albus* and *Vicia faba* Pilkington (1929) reported that puncturing the summit of the apex was invariably followed by regeneration of only one new apex from the flank region of the original meristem. Sussex (1964) observed a similar response to needle punctures of varying diameters in the shoot apex of potato (Fig. 6.6), but a different result was apparently obtained in *Impatiens* by Loiseau (1959). After punctures had been made in the summit there was a general expansion of the apex followed by reorganization into two, three, or four new shoot apices. When the same experiment was performed on the fern *Osmunda cinnamomea*, two to six new apices regenerated from the uninjured portions of the meristem. In this case the original puncture was small, including only the apical cell and a few adjacent derivatives, but subsequent necrosis produced an effect corresponding to an extensive puncture. In the fern *Dryopteris* Wardlaw (1949) observed that similar punctures led to the regeneration of one or more buds from the peripheral parts of the original meristem.

Though experiments such as these do not indicate the mechanism by which integration in the meristem is achieved, they do suggest that the central group of cells may be especially important in this phenomenon. In all of these cases, the destruction of these cells has been followed by a reorganization of the meristem such that one or more new growth centers have been established from uninjured portions of the meristem. Equivalent injuries in peripheral regions (Loiseau, 1959) have no such effect and, in fact, are soon displaced by further growth of the apex along its original axis. In plants that branch terminally by separation of the meristem into two equal or unequal parts the establishment of new growth centers is accompanied by a loss of apical meristem characteristics by cells of the original center. This phenomenon, which occurs commonly in the lower vascular plants and more rarely in seed plants, will be examined more fully in Chapter 8.

Nonuniform distribution of mitotic activity and cytohistological zonation have been described in Chapter 5 as indications of organization within the shoot apex. Unfortunately, there is only a limited amount of experimental evidence that can contribute to the interpretation of these patterns. Surgical division of the apex clearly disrupts the central zone and subsequent regeneration of organized apices occurs in the peripheral regions of the meristem. In two cases the reorganization of functional meristems has been followed in sterile culture using ^{3}H-thymidine incorporation to identify localized differences in DNA synthesis and presumably mitotic activity. In *Nicotiana* Sussex and Rosenthal (1973) divided the original meristem into quarters, and in *Helianthus* (Davis and Steeves, 1977) the apices were bisected. In neither case was there a localized central region of low mitotic activity in the early stages of regeneration, but in both cases such a zone was established as the new apices began to function normally. In *Helianthus* lower activity in the central zone was detected within three to five days of the operation and its appearance was thought to be correlated with the beginning of leaf initiation. At this time there was no indication of the other cytological features associated with zonation in the normal apex. These developed later and the time of appearance was variable. These experimental observations suggest that, while obviously not essential for the reorganization of a meristem, a central group of slowly dividing cells is important for the maintenance of normal function. Other cytological features often associated with this basic pattern appear to be of less significance.

GENERAL COMMENT

The experiments that have been described in this chapter provide useful information about the functional organization of the shoot apex. Although the apex has been shown to be complex in its structural organization and not to be a homogeneous group of dividing cells, subdivision of the apex into small units reveals the capacity of portions of the meristem to regenerate entire apices. Moreover this capacity is equally present in both central and peripheral regions of the heterogeneous apex. However, in the normal, unoperated apex all of the cells function together as an integrated system, indicating that there must be a regulating mechanism that suppresses the tendency for regeneration of the various parts and maintains the organization of the apex as a whole. The high degree of apical autonomy revealed by experimentation, moreover, indicates that both the ability to regenerate and the organization that controls it are largely contained within the apex itself. Unfortunately, although experimental investigations have demonstrated these important characteristics of the shoot apex, they have not led to an understanding of the mechanism of integration in the meristem.

The hope that it would be possible to relate functional organization to structural patterns has not yet been realized. About all that can be said in this regard is that the structural complexity of the meristem – within which can be discerned zonation patterns, localized differences in cell division frequency, and different orientations of cell divisions in surface and interior layers – reflects a delicate balance maintained by refined controls. It is the nature of these controls that awaits elucidation. Ten years ago it seemed probable that refined techniques of biochemical analysis that were beginning to be applied to the shoot apex would reveal a more fundamental organization than that shown by structural patterns and one more closely reflecting the functional integration of the apex. Although some useful information has been obtained by these methods, in general they have not been sufficiently sensitive to reveal subtle differences in the localization of hormones or other substances that could play a role in functional integration. Furthermore, it must be recognized that such localizations – and the visible structural patterns themselves – may be the result rather than the cause of meristem organization and thus, in the long run, may contribute little to the interpretation of apical integration.

The continuing failure to expose the fundamental nature of apical integration may in fact be the result of inadequate concepts and the consequent restriction of methods used to explore the phenomenon. For example, the emphasis upon chemical regulation has tended to ignore a wide range of possible physical factors, such as pressure, tension, or the electrical properties of cells and tissues, that should be more intensively investigated. The near impossibility of following cell lineages accurately in experimental systems has long been accepted as inevitable, but the prospect of using chimeral plants in experimental studies holds real promise for overcoming this obstacle. Moreover, as a wider variety of plant species is investigated, there is a growing recognition that some features of apical organization are related to the growth characteristics of particular plants, others depend upon the stage of development, and still others appear to represent aspects of tissue differentiation. This recognition is beginning to focus attention upon more fundamental features of apical organization that probably reflect functional integration of the meristem. Thus there are hopeful indications that real progress may occur in the next decade.

REFERENCES

Ball, E. 1946. Development in sterile culture of stem tips and subjacent regions of *Tropaeolum majus* L. and of *Lupinus albus*. L. *Am. J. Botany* 33:301–18.
———. 1948. Differentiation in the primary shoots of *Lupinus albus* L. and of *Tropaeolum majus* L. *Symp. Soc. Exp. Biol.* 2:246–62.
———. 1950. Isolation, removal and attempted transplants of the central portion of the shoot apex of *Lupinus albus* L. *Am. J. Botany* 37:117–36.
———. 1952a. Experimental division of the shoot apex of *Lupinus albus* L. *Growth* 16:151–74.
———. 1952b. Morphogenesis of shoots after isolation of the shoot apex of *Lupinus albus*. *Am. J. Botany* 39:167–91.
———. 1955. On certain gradients in the shoot tip of *Lupinus albus*. *Am. J. Botany* 42:509–21.
———. 1960. Sterile culture of the shoot apex of *Lupinus albus*. *Growth* 24:91–110.
———. 1980. Regeneration from isolated portions of the shoot apex of *Trachymene coerulea* R.C. Grah. *Ann. Botany* 45:103–12.
Davis, E. L., and T. A. Steeves. 1977. Experimental studies on the shoot apex of *Helianthus annuus:* The effect of surgical bisection on quiescent cells in the apex. *Can. J. Botany* 55:606–14.
Gulline, H. F. and R. Walker. 1957. The regeneration of severed pea apices. *Austral. J. Botany* 5:129–36.
Hicks, G. S., T. A. Steeves, and P. R. Sweeny. 1967. A method for grafting fern leaf primordia *in vitro*. *Can. J. Botany* 45:2232–6.

Loiseau, J. E. 1959. Observation et expérimentation sur la phyllotaxie et le fonctionnement du sommet végétatif chez quelques Balsaminacées. *Ann. Sci. Nat. Bot.* Ser. 11, 20:1–214.

Pilkington, M. 1929. The regeneration of the stem apex. *New Phytol.* 28:37–53.

Shabde, M. and T. Murashige. 1977. Hormonal requirements of excised *Dianthus caryophyllus* L. shoot apical meristem *in vitro*. *Am. J. Botany* 64:443–8.

Smith, R. H. and T. Murashige. 1970. *In vitro* development of the isolated shoot apical meristem of angiosperms. *Am. J. Botany* 57:562–8.

Sussex, I. M. 1952. Regeneration of the potato shoot apex. *Nature* 170:755–7.

———. 1953. Regeneration of the potato shoot apex. *Nature* 171:224–5.

———. 1964. The permanence of meristems: Developmental organizers or reactors to exogenous stimuli? *Brookhaven Symp. Biol.* 16:1–12.

Sussex, I. M. and D. Rosenthal. 1973. Differential ^{3}H-thymidine labeling of nuclei in the shoot apical meristem of *Nicotiana*. *Bot. Gaz.* 134:295–301.

Wardlaw, C. W. 1949. Further experimental observations on the shoot apex of *Dryopteris aristata* Druce. *Phil. Trans. Roy. Soc. London* Ser. B 233:415–51.

———. 1950. The comparative investigation of apices of vascular plants by experimental methods. *Phil. Trans. Roy. Soc. London* Ser. B 234:583–604.

Wetmore, R. H. 1954. The use of *in vitro* cultures in the investigation of growth and differentiation in vascular plants. *Brookhaven Symp. Biol.* 6:22–40.

CHAPTER

7

Organogenesis in the shoot: leaf origin and position

Any attempt to study the apical meristem of the shoot, even a simple dissection to expose it for observation, leads immediately to a consideration of the initiation and development of lateral appendages, particularly leaf primordia. It is evident that the formation of leaf primordia is a major activity of the shoot meristem and that the early development of these primordia in such close proximity to the meristem must result in important developmental interactions between the two. This intimate relationship is reflected in the unity of the mature shoot system in which any attempt to isolate stem and leaf either structurally or functionally is artificial. The student of phylogeny finds an interpretation of this relationship, at least in the ferns and seed plants, in the evolution of stem and leaf from a primitively undifferentiated branching system. The object of developmental analysis, however, must be to understand how, in the individual living plant, structures that originate as outgrowths of the meristem acquire distinctive characteristics and interact with one another and with the meristem that produced them. This analysis is further complicated by the fact that the same meristem frequently gives rise to other appendages that develop as replicas of the original shoot.

Whereas the shoot is characterized by potentially unlimited or indeterminate growth and this feature is retained by lateral branches, the leaf is an organ of transient, although in some cases extensive, growth. Thus the leaf at its inception has a definable developmental destiny, which is to produce a lateral appendage mature in all its parts and devoid of further growth potentialities. This is equally true for the minute scales of a juniper or for the massive fronds of a tree fern, and it may be correlated with the fundamental role of the leaf as a photosynthetic organ with a limited functional life. Often the leaf is further distinguished by its dorsiventrality and elaborate structural specialization. In developmental terms, the problem is to account for the change in growth pattern of a particular group of cells of the apical meristem that results in the formation of an organ as distinctive as a leaf. Moreover, because this change is localized and occurs only in precise positions in the shoot

100

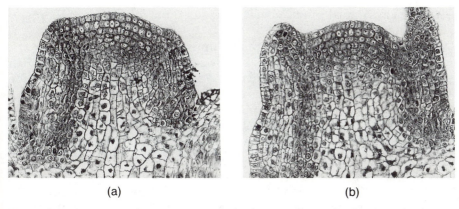

(a)   (b)

Figure 7.1. *Stages in the emergence of a leaf primordium of* Solanum tuberosum. *(a) A well-developed leaf buttress is formed on the left flank of the meristem. Below it procambium can be seen in the stem. On the right flank of the meristem can be seen the edge of an older leaf. (b) A slightly older stage of leaf emergence. The new leaf appears on the left side of the meristem as a protrusion at the summit of the buttress. ×170. (Sussex, 1955.)*

meristem, it is necessary to seek the mechanisms that regulate this activity of the meristem. In many ways the origin and development of a determinate organ like the leaf resemble the formation of appendages in some animals and may provide an opportunity to discover similarities in plant and animal development.

LEAF INITIATION AND EARLY DEVELOPMENT

There have been many descriptions of leaf initiation based upon conventional histological methods, largely the examination of median longitudinal sections of shoot apices. These have resulted in a reasonably consistent picture of the relationship between the emerging leaf primordium and the meristem that gives rise to it. This approach may be illustrated by considering a specific instance in which a detailed study has been carried out. *Solanum tuberosum* (potato) will serve as an example (Sussex, 1955). The first suggestion of leaf initiation is the swelling of the apical flank in the position that may be predicted to be the site of the next leaf primordium (Fig. 7.1a). This protrusion is the result of accelerated cell division and growth in at least three layers of the meristem and probably in deeper tissues as well. Because potato has a two-layered tunica, both tunica and corpus are involved in the initiation process. In the surface layer only anticlinal divisions occur, but in the underlying layers both anticlinal and periclinal divisions may be observed. The protruding flank of the apex that precedes the emergence of a distinct primordium has been designated the *foliar buttress* in many species of dicotyledons.

The emergence of the leaf primordium is accomplished by the localization of growth activity at the summit of the foliar buttress (Fig. 7.1b). It is not, however, from these cells only that the primordium arises, for as the mound enlarges it expands over the apical surface, incorporating additional tissues into itself. In potato meristematic activity at the leaf apex continues until four younger primordia have been initiated and a length of approximately 200 μm has been achieved. At this point apical growth ceases and the terminal leaf cells become vacuolated. There is great variation among dicotyledons in the extent of apical meristematic activity, but in the great majority thus far investigated this process ceases before the primordium has attained a length of 1 mm. There are, however, notable exceptions, such as *Nicotiana tabacum*, where leaf apical growth continues until the primordium reaches a length of 3 mm, and *Angelica archangelica* with a length of 15 mm. The important point is that after a definite period, growth at the apex ceases so that, in contrast to the shoot as a whole, the leaf is determinate. There is also increasing evidence that the apical region of the leaf primordium never functions as a true apical meristem in the sense of playing a special role in the formation of the primordium. This will be considered further in Chapter 9 in the context of the clonal analysis of leaf development.

Descriptions of leaf initiation and early development in different species indicate that there is considerable variation in the size relationship between a leaf primordium and the meristem that produces it. If the primordium is sufficiently large that its initiation involves a sizable portion of the meristem, its initial stage is evident as a lateral bulge of the meristem, that is, a leaf buttress. If, however, the primordium is small in relation to the meristem, as in the high-domed apices of grasses and some other monocotyledons or the broad conical apices of many ferns, the first emergence is ordinarily recognizable as a distinct leaf primordium and no buttress stage is designated. Published accounts of leaf initiation reveal the variations in size relationships between apical meristem and leaf primordium, but the accounts themselves vary in the recognition of a buttress stage. The foliar buttress, where it occurs, is the leaf in its earliest stage of emergence. On the other hand, many workers have considered buttress formation as a phase of apical expansion preceding leaf initiation and have described a sequence of maximal and minimal sizes of the meristem preceding and following successive leaf initiations. Indeed, the buttress is both of these things, a fact that emphasizes the difficulty of separating stem and leaf in the unified shoot system. This question is more than a matter of terminology, because the very early stages of development must be identified precisely and de-

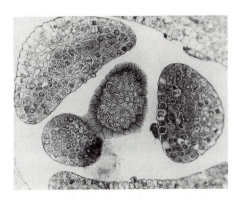

Figure 7.2. *A cross section of the shoot tip of* Solanum tuberosum *showing the shape changes of developing leaf primordia. The youngest primordium shown attached to the shoot apex at the lower left, is approximately circular in section, the two older leaves are increasingly bilateral as a result of tangential growth. ×150. (Sussex, 1955.)*

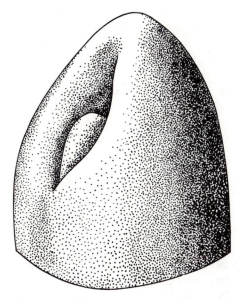

Figure 7.3. *Three-dimensional representation of the shoot tip of* Triticum *showing the hooded form and encircling base of a leaf primordium. The next younger primordium is just becoming visible. ×185. (Redrawn from Williams, 1975.)*

scribed clearly as a basis for experimental work on leaf initiation. The importance of this point will be evident in the interpretation of experimental studies that will follow.

In potato the newly emergent primordium is approximately circular in section. As growth continues a progressive change in shape occurs in the primordium, and it becomes somewhat flattened on the side facing the shoot meristem (Fig. 7.2). This change in shape results from more extensive growth throughout the primordium in the tangential dimension than in the radial. Thus the primordium becomes a bilaterally symmetrical, peglike outgrowth.

Whereas the initial manifestation of bilaterality in the primordium results from the growth pattern of the whole organ, the extension of this in later development to produce the dorsiventral leaf lamina is brought about by more localized growth along the leaf margins. Marginal growth is ordinarily treated as a distinct phase of leaf development, but because it is usually described as beginning before the end of apical growth, it is clear that apical and marginal phases overlap. The particulars of marginal meristematic activity will be described in a later chapter in connection with histogenesis in the leaf.

The initiation and early development of leaves of monocotyledons differ in some details from the pattern described for dicotyledons. In some monocotyledons there are periclinal divisions in the outermost apical layer so that this layer does not contribute only to the epidermis. The widespread monocotyledon characteristic of sheathing leaf bases is reflected in early primordial development. From the site of initiation, growth activity extends laterally around the flanks of the apex in both directions, usually encircling it completely. Growth at the apex of the primordium is even more restricted than in dicotyledons, and in some cases no recognizable apical growth phase has been identified. In grasses, such as *Zea* and *Triticum* (wheat), increase in length at the point of origin is accompanied by an upward growth of the margins so that the primordium acquires a hoodlike form surrounding the apex (Fig. 7.3). This lateral phase is also very limited in extent and subsequent growth is intercalary. In monocotyledons that develop a broad lamina, marginal growth resembles that found in dicotyledons.

In the lower vascular plants where the apical meristem consists of a single layer of cells, it might be expected that the relationship between leaf primordium and meristem would be different, at least in details if not in principle. Because much of the experimental work on leaf initiation has been carried out on several species of ferns, it will be advisable to consider early leaf development in such a plant. In *Dryopteris* (Wardlaw, 1949b) and in *Osmunda* (Steeves and Briggs, 1958) a group of cells at the periphery of the surface apical layer initiates the leaf by an acceleration of cell division activity (Fig. 7.4a). Wardlaw has described an initial stage in which the cells of this group enlarge; but no such enlargement has been noted in *Osmunda*. Accelerated growth soon leads to the protrusion of a primordium that is approximately circular in outline and, being small in relation to the total volume of the apex, is immediately distinct as a leaf primordium. At an early stage a centrally placed superficial cell of the primordium enlarges and becomes the distinctive apical initial cell of the leaf (Fig. 7.4b). In contrast to the situation in seed plants, apical growth is of long duration in fern leaves, and the apical cell remains visible and active for nearly three years in *Osmunda*. Undoubtedly the circinate, or coiled, pattern of the unexpanded

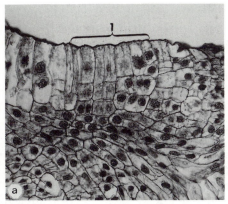
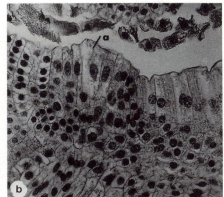
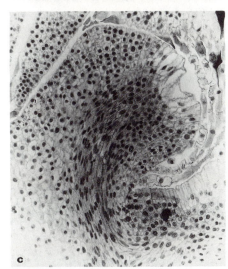

Figure 7.4. *Early stages of leaf development of* Osmunda cinnamomea. *(a) Longitudinal section of a very young leaf. Although it has not yet emerged as a mound on the surface of the shoot, active development in the form of cell division has taken place in the surface and subsurface layers. (b) A slightly older stage in which the leaf primordium is beginning to form a mound and its tip now contains an enlarged apical cell. (c) A later stage of leaf growth. The leaf curves toward the center of the shoot apex, and the larger size of cells on the outer side of the primordium is evident. Key: a, apical cell; 1, leaf primordium. (a) ×160, (b) ×170, (c) ×90. (Steeves and Briggs, 1958.)*

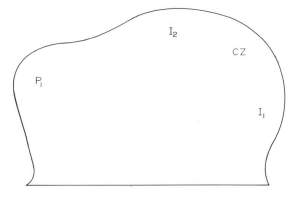

Figure 7.5. *Diagram of a longitudinal section of the shoot apex of* Pisum *cut in the plane of the leaves and showing the location of the youngest leaf primordium and two incipient primordia. Key: CZ, central zone; I_1 and I_2, sites of incipient leaf primordia; P_1, youngest visible leaf primordium. (Adapted from Lyndon, 1983.)*

frond axis is correlated with this long continuation of true apical growth. Ultimately, however, with one or two possible exceptions, the determinate nature of the organ is manifested in a cessation of apical growth and the differentiation of the apical cell. As in the seed plants, bilaterality is established at an early stage and is accentuated by the tendency of the primordium to curve toward the center of the meristem as a result of greater longitudinal growth on the abaxial side than on the adaxial (Fig. 7.4c).

ANALYSIS OF EARLY LEAF DEVELOPMENT

As in the case of the shoot apical meristem, descriptive accounts of leaf initiation and early development have been substantially extended and clarified by the use of analytical methods. In some cases conclusions based upon observations are brought into question by more rigorous analysis. It has long been accepted, for example, that an increased rate of cell division in a localized region of the shoot apex is an essential aspect of the emergence of a leaf primordium. Lyndon (1983) has undertaken to test this assumption in a quantitative analysis of cell division in the shoot apex of *Pisum* (pea). In this plant the leaf arrangement is distichous, that is, the leaves are in two opposite rows and occur at alternate nodes in each row. Thus a longitudinal section of the apex in the plane of the leaves (Fig. 7.5) includes, in addition to the youngest emergent primordium (P_1), the site of the next incipient primordium (I_1) opposite it, and above P_1 the site of yet a further incipient primordium (I_2). Lyndon determined the rate and orientation of cell division in these regions and was able to construct a sequence of events, not only in the emerging primordium, but in the incipient leaf sites during the two plastochrons preceding emergence. A plastochron, it will be recalled, is the interval of time between the emergence of two successive leaf primordia. He found that there is a modest increase in cell division rate approximately 1.5 plastochrons before the emergence of a primordium, which persists until the primordium has emerged, and then falls to the original rate. What appeared to be of greater importance was a shift in the orientation of cell divisions in the incipient primordial site near the end of the plastochron preceding emergence such that the proportion of periclinal divisions rose significantly. Lyndon concluded that a change in the plane of cell division is of greater significance than an increase in the rate of division in the initiation of a leaf primordium.

Evidence has been provided by Green and his associates (Green and Poethig, 1982) that changes in the orientation of cell division are associated with other profound alterations in cell structure that affect subsequent cell growth. They have studied in surface view the initiation of

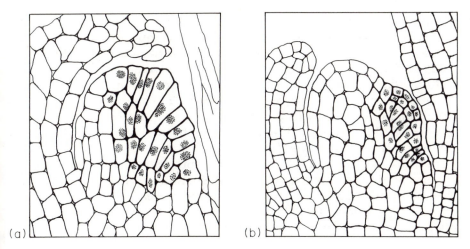

Figure 7.6. *Leaf initiation in* Triticum. *(a) Longitudinal section of the shoot apex of a gamma plantlet showing emergence of a leaf primordium in the absence of cell division. (b) Normal leaf initiation. ×175. (Foard, 1971.)*

the first leaf primordia of buds that regenerate on detached leaves of *Graptopetalum* (Crassulaceae) and have observed changes in the alignment of the anticlinal divisions in this layer at the sites of incipient leaf primordia. These changes are associated with altered patterns of orientation of cellulose microfibrils in the cell walls and by corresponding changes in the microtubule arrays within the cells. The altered patterns of cellular reinforcement are important in determining the direction of subsequent cell growth. Physical factors like this may be important in the directed growth necessary for the emergence of a leaf primordium.

The significance of directional growth in the emergence of a leaf primordium is strikingly illustrated by an experiment performed on seedlings of *Triticum* (Foard, 1971). In this species the first identifiable stage in leaf initiation is periclinal cell division in the outermost layer of the apex. In this experiment grains were gamma-irradiated with a dose sufficient to inhibit cell division in the shoot apex. After germination the seedlings (gamma plantlets) produced one additional outgrowth at the site of the next anticipated leaf by the directed enlargement of cells in the surface layer (Fig. 7.6a), which, in a normal plant, would have divided (Fig. 7.6b). This was accomplished without any concurrent cell division.

In the study of leaf initiation, a great deal of attention has been devoted to establishing the precise depth of tissues in the apical meristem from which the primordium arises. Although it is relatively easy to locate the site of early cell divisions associated with leaf initiation, it is much more difficult to determine which apical layers contribute tissues to the total development of the leaf. The only completely effective means of tracing meristem contributions to the leaf has been clonal analysis using periclinal chimeras in which meristem layers are marked by their cellular characteristics and may thus be traced into the developing leaf. The most useful of these have been the polyploid chimeras and the plastid chimeras discussed in Chapter 5. In a number of species of both monocotyledons and dicotyledons it has been shown that the leaf is composed of tissues derived from the three outermost layers of the meristem, thus including both tunica and corpus (Stewart, 1978). In *Zea* it appears that only two layers are involved, and in *Nicotiana* four layers may be implicated in some cases. Thus, in the cases in which reliable evidence is available it appears that leaf formation is not relegated simply to superficial regions of the meristem but involves its full depth.

QUESTIONS OF LEAF DEVELOPMENT

The initiation and early development of leaves as they have been reviewed here raise important, and probably complex, questions concerning the regulation of development in plants. The leaf appears to be an unusually favorable system in which to analyze developmental processes because it is an easily recognizable entity and one that is readily accessible to the experimenter. It is not surprising, therefore, that a great deal of experimental work has been done on this system. The body of experimental data resulting from these investigations presents a complex and sometimes confusing picture that must be examined closely if its meaning is to be grasped. In order to facilitate this examination it is advisable to state at the outset what seem to be the essential questions posed by the phenomenon of leaf development. These questions may be framed as follows:

1. Why does the peripheral region of the shoot meristem give rise to outgrowths of any sort?
2. By what mechanism is the placement of the outgrowths regulated in the meristem?
3. What is the nature of the influences that act upon the outgrowths in such a way as to cause them to be leaves?
4. What is the nature of the responses to these influences that result in leaf development?

In many ways, the first of these questions is the most fundamental, yet it is the aspect of leaf development about which the least information exists. One might postulate an inherent tendency of the meristem to grow out, at least in its peripheral regions. In fact Schüepp (1917) formulated a theory of leaf initiation for flowering plants based upon the idea that the anticlinally dividing surface layers should increase in area disproportionately to the irregularly dividing inner regions and thus be thrown into folds. The surgical work reviewed in Chapter 6 showed that, whereas the intact meristem functions as an integrated unit, small portions isolated laterally will grow out independently. The fact that such outgrowths give rise to shoots has no bearing upon the present question. Thus it is possible to visualize the initial phase of leaf development as resulting from an isolation of particular regions of the meristem from the correlative influences that otherwise prevent outgrowth. Unfortunately, there is no information whatsoever as to the nature of the correlative mechanism or the system by which it is selectively interrupted – or for that matter whether such a visualization has any validity.

LEAF POSITION – PHYLLOTAXY

Descriptive account

The importance of the second question is evident even from observations on mature shoots where, although patterns vary, leaves are always distributed along the stem in a definite and repetitive manner. The arrangement of leaves on an axis is called *phyllotaxy*. Leaves may occur singly at each node, in pairs, or in whorls of three or more. Where they are individually disposed along the stem, it has long been recognized that they form a helical pattern ascending the stem in order of decreasing age. This is referred to as the *generative spiral*. In plants with helical phyllotaxy it is also possible to recognize approximately vertical ranks of leaves, each of which is called an *orthostichy*. Customarily the phyllotatic pattern is described in the form of a fraction, the denominator of which is the number of orthostichies in the system and the numerator of which is the number of gyres of the helix between successive leaves on the same orthostichy.

In shoots in which there is little or no internodal elongation during maturation, phyllotaxy cannot be determined in this way because vertical ranks of leaves cannot be found. This can be seen very easily in a pine cone or in the head of a sunflower. This situation is of special interest to the student of development because it prevails generally in the unexpanded apical bud even though the mature shoot derived from the bud may have evident orthostichies. There is, however, another method of establishing the phyllotactic pattern in such cases; and this method may be described best by using an example. In *Osmunda cinnamomea*, if a transverse section passing through the apical meristem is drawn or

Figure 7.7. *A transverse section of the shoot tip of Os-munda cinnamomea with the contact parastichies drawn in. There are two sets of parasti-chies. One set, represented by dashed lines in the diagram, runs clockwise from older to younger leaves, and connects leaves differing in age by five plastochrons. There are five parastichy lines in this set. The other set, indicated by contin-uous lines, runs counter-clockwise and connects leaves that differ in age by eight plastochrons. There are eight parastichy lines in this set. Thus, the phyllotaxy is 5 + 8.* (T. A. Steeves, J. Indian Bot. Soc. 42A:225, 1963.)

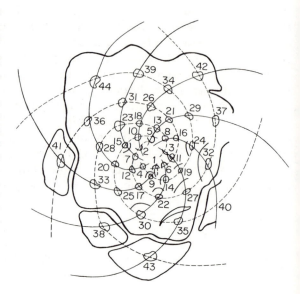

projected in such a way that the positions of the youngest primordia and the bases or traces of older primordia are accurately placed, a dif-ferent kind of pattern may be detected (Fig. 7.7). Again, there are no evident straight ranks, but it is possible to observe curved rows of leaves extending outward from the center. These curved rows, which have been shown to have the path of logarithmic spirals, are called *parastichies*.

The parastichies that have been of greatest interest in the study of development are those known as *contact parastichies* of which there are two sets, one in each direction. Each leaf is located at the intersection of two contact parastichies and the leaves most closely associated with it on the two parastichies are those closest to it in the bud and even in contact with it in cases in which close packing occurs. The leaf interval along parastichies is regular, but the two sets of parastichies in opposite directions have different characteristics. Those of one set are more nu-merous, have a larger leaf interval, and are shorter than those of the other set. Phyllotaxy can be expressed in terms of the parastichies by stating the number in each set (Church, 1904). Thus in *Osmunda* it is possible to recognize five long parastiches and eight short parastichies, and the phyllotaxy may be indicated as 5 plus 8 contact parastichies. It is also possible to translate phyllotaxy expressed in this way to a frac-tional form, by taking the number of long parastichies as the numerator of the fraction and the total number as the denominator.

The apparent anomaly that an apical bud that contains no straight ranks of leaves should give rise to a shoot with orthostichies may be

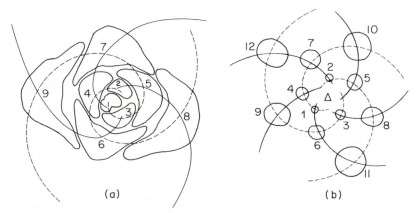

Figure 7.8. *Phyllotaxy of* Solanum tuberosum *(a) and of* Dryopteris dilatata *(b). In the potato shoot there are two parastichies in the clockwise direction (dashed lines) and three in the counterclockwise direction (continuous lines). The phyllotaxy is 2 plus 3. Leaves are large in relation to the size of the shoot meristem at the time of their initiation, and they grow rapidly. In the* Dryopteris *shoot there are three counterclockwise parastichies and five clockwise parastichies. The phyllotaxy is 3 + 5. Leaves are small in relation to the size of the shoot meristem at the time of their initiation, and their rate of lateral expansion is slow.*

explained by a consideration of events during elongation. In the apical bud, in addition to the contact parastichies, there are other less sharply curved rows of leaves that, although they are not straight, correspond to orthostichies in the particular system in terms of the leaf interval along them. When elongation of the stem occurs and the leaves are widely separated, the lateral displacement of successive leaves in the row is minimized and the vertical alignment is accentuated such that the eye is caught by an apparent straight line.

Apart from plants having whorled or opposite phyllotaxy, the vast majority of vascular plants whose phyllotaxy has been described falls into a series that has interesting mathematical properties. In this series, the numbers of contact parastichies in the two sets are consecutive numbers in the Fibonacci series 1 : 1 : 2 : 3 : 5 : 8 and so on, for example, 2 plus 3, 3 plus 5, 5 plus 8. The regularity of pattern revealed by the study of parastichies ultimately reflects the regularity of leaf initiation and particularly the fact that each new leaf primordium arises at a constant angular divergence from the next older one. If the *divergence angles* between successive leaves are measured in any plant, it will be found that, though fluctuations occur, the average is ordinarily an approximation of the so-called ideal angle, 137° 30′ 28″. Moreover it will be noted that if phyllotaxy is expressed in fractional form as just explained, successive fractions in the series, taken as fractions of 360°, converge on this angle. The important point is that, as far as can be determined by measurement, the angle of divergence remains constant, whereas phyllotactic patterns expressed in terms of parastichy numbers vary widely.

It is thus a matter of some interest to consider how the phyllotactic patterns can vary while the divergence angle remains constant. Examination of apical sections of a plant having a low phyllotaxy (2 plus 3) and of a plant having a higher phyllotaxy (3 plus 5) indicates how this difference may be explained (Fig. 7.8). The phyllotactic pattern is determined by the number of contact parastichies in each direction and thus, ultimately, by the position in the total leaf sequence of the two leaves that are the contacts for any particular primordium. In the 2 plus 3 pattern the contact leaves for primordium 1 are leaves 3 and 4, whereas in the 3 plus 5 pattern the corresponding contacts are leaves 4 and 6. Variations in contact relationships among the leaves of an apical bud would seem to have their explanation in the growth relationships of apical meristem and leaf primordia. This is illustrated in Fig. 7.9, in which the branches of different sizes of *Araucaria excelsa* show different phyllotactic patterns resulting not from variation in the angle of divergence but from different relative growth rates of apex and leaf primordia. Richards (Richards and Schwabe, 1969) has called attention to this fact in his use of the *plastochron ratio* in the interpretation of phyllotactic patterns. Briefly, this ratio is obtained by dividing the distance from the center of the apex to the center of one primordium by the distance from the apical center to the center of the next younger primordium; it thus gives a measure of the extent of radial growth by the apex during a plastochron, that is between the initiation of successive leaf primordia. The role of this variable in different phyllotactic patterns has been demonstrated by Richards both in theoretical diagrams and in actual measurements in shoots with changing phyllotaxy. The primordial contacts in the apex, and consequently the pattern of contact parastichies, also are influenced by the shape of the developing primordia following their initiation. Because neither plastochron ratio nor primordial shape has any influence upon angular divergence, this factor remains constant in varying phyllotactic patterns.

The regularity of the phyllotactic patterns and the properties of the ideal divergence angle have long intrigued mathematically oriented biologists, some of whom have sought explanations in terms of geomet-

Figure 7.9. *Transverse sections, drawn at the same magnification, of the apical regions of branches of different sizes in* Araucaria excelsa *representing different orders of branching. These show different parastichy patterns (a) 8 + 13, (b) 5 + 8, (c) 3 + 5, without significant differences in average angular divergence. (Adapted from Church, 1904.)*

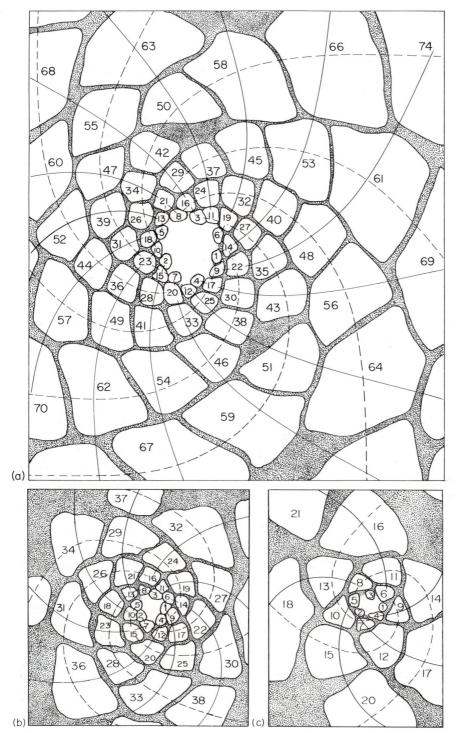

Figure 7.9. *(Caption on facing page)*

ric relationships. Others have attempted to produce models of phyllotactic systems based upon postulated physiological or physical mechanisms that might control the placement of leaf primordia. Most recently computer simulations have been used to test the ability of hypothetical interactions in the shoot apex to produce stable, repeating systems comparable to actual phyllotactic patterns. Erickson (1983) has recently summarized the principal points of many of the geometric analyses of phyllotaxy.

Experimental studies and interpretation

The foregoing analysis of phyllotaxy is a purely descriptive account of the regular placement of leaves on the axis and provides little information about the mechanism that underlies the pattern. For this kind of understanding it has been necessary to turn to an experimental approach, the most successful technique having been a surgical one. Experiments of this type first were performed by M. Snow and R. Snow (1931) on the shoot apex of *Lupinus* and led to the conclusion that the position at which a new leaf primordium is initiated is influenced by the preexisting leaf primordia adjacent to the site of initiation. If P_1, the youngest leaf primordium, was isolated from the apex by a tangential incision, the next leaf, I_1, arose in its expected position, but the following leaf, I_2, did not arise in its anticipated position (Fig. 7.10a, b). It was, in fact, abnormally close to P_1, or rather to the incision that separated P_1 from the apex. This was determined by a measurement of angular divergence, with increases in divergence of up to 29° being found. The phyllotactic pattern of *Lupinus* is 2 plus 3, with the result that the contacts for I_2 are P_1 and P_2, whereas those for I_1 are P_2 and P_3. Thus it may be argued that P_1 has an influence only upon positions adjacent to it. On the other hand, it could be argued that the position of I_1 was established before the operation was carried out. To test this possibility, the Snows performed another experiment in which I_1 was isolated by a tangential incision (Fig. 7.10c, d). In this case, I_2, which had been shown not to be fixed, arose in its normal position, and I_3 arose in a position abnormally close to I_1, with an increase of the divergence angle from I_2 of up to 67°. Thus some of the divergence angles were in excess of 180°, and reverses in the direction of the generative spiral were noted. These experiments established that preexisting leaf primordia influence the positions only of new primordia that arise in their immediate vicinity.

Wardlaw (1949a) has applied similar techniques to the study of phyllotaxy in the fern *Dryopteris* that have generally confirmed the observations of the Snows. These are of particular interest because in this plant the leaf primordia are widely spaced and could not possibly be

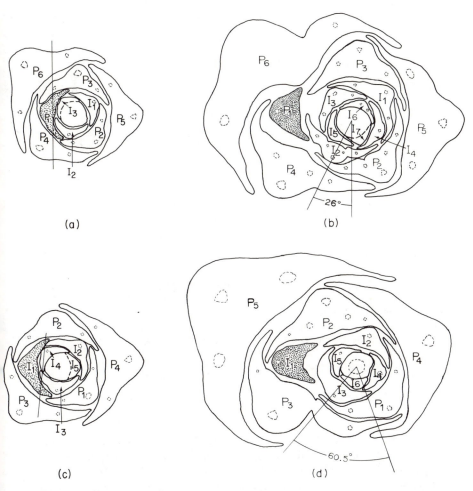

(a)

(b)

(c)

(d)

Figure 7.10. *The effect of isolating a leaf primordium of* Lupinus albus *(lupin) on the positions of subsequent primordia. (a) Diagram of a normal shoot apex showing the position of the tangential incision (the line in the diagram) that would isolate the P_1 primordium from the meristem. (b) Transverse section of a shoot apex in which P_1 was isolated from the meristem by a tangential incision. Several younger leaf primordia were formed after the operation. Primordium I_1 occupies its correct position, but the next primordium, I_2, is abnormally close to the isolated P_1 position, and the angular divergence between I_1 and I_2 has been increased by 26° from the normal 137.5°. (c) Diagram of a normal shoot apex showing the position of the tangential incision (the line in the diagram) that would isolate the I_1 position from the meristem. (d) Transverse section of a shoot apex in which the I_1 position was isolated from the meristem by a tangential incision. Several new leaf primordia were formed after the operation. Primordium I_2 occupies its normal position, but I_3 is abnormally close to the isolated I_1 leaf, and the angular divergence between I_2 and I_3 has increased by 60.5° to 198°. This has resulted in a reversal of the direction of the generative spiral, which up to this point had been acropetally clockwise. (M. Snow and R. Snow, 1931.)*

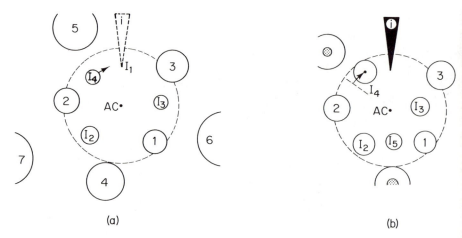

(a)   (b)

Figure 7.11. *The effect of destroying a presumptive leaf position in* Dryopteris
dilatata *on the positions that subsequent leaves occupy. (a) A diagram of the shoot
apex showing the predicted result of destroying the* I_1 *position by a knife cut* (i). *The
leaves* I_2 *and* I_3 *are predicted to occupy their normal positions, but leaf* I_4 *should
emerge closer to the* I_1 *operation site than in the normal apex. (b) An apex operated
on as described above. It is seen that the leaves* I_2 *and* I_3 *occupy their predicted
positions, as judged by their angular divergences and positions relative to their contact leaves, and that* I_4 *has emerged closer to the* I_1 *wound site. (Adapted from Wardlaw, 1949a.)*

influenced by physical contact. Incision of the I_1 position prevented the
formation of a leaf at this site and thus the effect of the absence of I_1
upon the emergence of later leaves could be assessed (Fig. 7.11). The
positions of I_2 and I_3 were unaffected by this operation, but I_4 arose out
of its expected position and in proximity to the incision at the I_1 site.
This result is actually an exact parallel of the Snows' because in *Dryopteris* the contacts of I_3 are P_1 and P_3, whereas those of I_4 are I_1 and P_2,
the phyllotactic pattern being 3 plus 5. This experiment and others in
which various combinations of leaf sites were incised supported the
conclusion that the position at which a leaf primordium arises is influenced by preexisting primordia adjacent to it. Wardlaw further observed that the abnormally placed primordia often grew faster than those
in normal positions and frequently outgrew older primordia. This suggested an inhibitory effect upon young primordia by older adjacent primordia. In order to investigate this relationship further, Wardlaw isolated various primordia laterally by pairs of radial incisions. In each
instance the isolated primordium outgrew other primordia that were
older and larger, thus enhancing the inhibition interpretation.

   In point of fact, the experiments in which the site of emergence of a

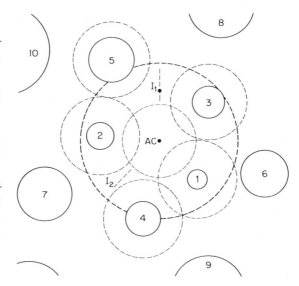

Figure 7.12. *A diagram of the shoot apex indicating the distribution of the hypothetical physiological fields that are associated with the terminal meristem of the shoot and the young emergent leaf primordia. The physiological fields are drawn as circles around the meristem and the leaves, and their lateral extent is arbitrarily fixed. The position of I_1 is the first to come to lie outside the existing fields as the shoot tip expands during growth, and I_2 is the next position to do so. (Wardlaw, 1949a.)*

new primordium was displaced towards the position of a preexisting primordium that had been surgically isolated or suppressed can be interpreted also in terms of regions of inhibition surrounding developing primordia. Further, the limitation of leaf initiation to the margins of the apical dome is strongly suggestive of the operation of inhibition in the central region. This concept has been developed into a generalized field theory of phyllotaxy by Schoute (1913), and more recently by Richards (1948), and by Wardlaw (1949a). Wardlaw has elaborated this theory in the interpretation of the specific phyllotactic pattern of *Dryopteris*, and his discussion may be used in illustrating the application of the theory. According to Wardlaw, as each primordium is initiated, it is surrounded by a physiological field within which the inception of new primordia is inhibited. Although the fields are presumed to be chemical, there is, as yet, no precise information on their nature. The fields are described as varying in intensity, decreasing with the age of the primordium, and each is presumed to have an intensity gradient decreasing from the center. A comparable field is seen as occupying the central dome of the shoot apex.

A simplified diagram in which the field theory is applied to a shoot apex is shown in Fig. 7.12, but such a diagram does not illustrate variations in field intensity, nor does it reveal three-dimensional relationships. However, even without these important features, the diagram does show that areas available for new leaf formation must appear in a regular pattern as the shoot apex grows. The physiological field theory seems to apply particularly well to the apex of *Dryopteris*, where it is necessary to account for the fact that leaf primordia do not occupy all of the space that would appear to be available to them. The principle, however, applies equally well to apices in which the leaf primordia are more closely packed.

The results obtained by the Snows in *Lupinus* could be given the same interpretation as are those from *Dryopteris*, but the Snows themselves have preferred a different explanation. They regard the major factor limiting primordium initiation as space and consider that a new leaf primordium arises in the first space that becomes available by reason of having attained sufficient width and distance from the summit of the apex. In this view the experiments just described, in which the isolation of particular leaves or leaf sites influenced the positions of subsequent leaves adjacent to them, may be interpreted as having limited the lateral expansion of the isolated primordia and thus having altered the space relationships in the region of new leaf origin. Although this theory of the first available space appears to be an adequate vehicle for the description of leaf positioning in apices in which the leaves are closely packed, it is less satisfactory for those cases, such as *Dryopteris*, in which leaf primordia are widely spaced and availability of space could hardly be a limiting factor. Because the field hypothesis is equally applicable to both types of apices and moreover includes a physiological rather than a geometrical mechanism for its operation, it does seem to be more acceptable than the available space theory, which has limited applicability.

Direct documentation of the nature of physiological fields within the confines of the shoot apex will be difficult to obtain and will require the perfection of sufficiently sensitive microanalytical methods. Nevertheless, there is evidence suggestive of the possible chemical nature of the mechanism controlling leaf placement. Schwabe (1971) found that the introduction of tri-iodobenzoic acid, a substance known to modify auxin transport, through the cut basal end of shoots of *Chrysanthemum*, caused a significant and long-lasting alteration of phyllotaxy in a number of cases. The normal spiral phyllotaxy was replaced by a distichous arrangement and the altered condition persisted for many weeks. The effect of more localized application of auxin antagonists was tested by Meicenheimer (1981). He treated individual youngest leaf primordia of *Epilobium hirsutum* with N-1-naphthylphthalamic acid, an inhibitor of auxin transport, or α-4-chlorophenoxyisobutyric acid, an auxin antagonist. With both substances the normal opposite phyllotaxy was replaced by a spiral pattern in which the divergence angle approached the ideal angle of the Fibonacci sequence. These and other experiments in which applied growth regulators have been shown to upset the developmental pattern of the shoot apex are suggestive of a role of natural

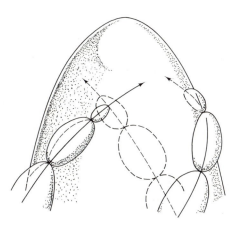

Figure 7.13. *A diagram of the shoot apex showing the hypothetical distribution of foliar helices. Three helices are shown with leaves being formed along each helix. (Plantefol, 1948.)*

regulators in the apex and, in this sense, support the field theory. However, they certainly do not offer conclusive proof.

Some years ago a completely different kind of interpretation of the phenomena of phyllotaxy was advanced by the French botanist Plantefol (1948). This theory, which was widely accepted and elaborated upon by other workers in France, interprets phyllotactic patterns in terms of a number of foliar helices that wind around the stem in one direction and terminate in leaf generating centers in the apex (Fig. 7.13). This theory of multiple foliar helices actually gave rise to the méristème d'attente concept of the shoot apex; the generative centers are thought to be located in the anneau initial. The stimulus to mitotic activity that initiates a leaf primordium is presumed to be transmitted acropetally along a foliar helix to the generative center in the meristem, but the activity of the several centers in an apex is held to be harmonized by an organizer located more apically in the meristem. In dicotyledons it is stated that there are generally two such foliar helices, one originating with each cotyledon, whereas in monocotyledons the number is variable but is usually greater than two. Because the continuity of the foliar helices is reflected in the apical bud in the contacts between adjacent leaves along the helix, contacts that may of course be removed by subsequent stem elongation, it is evident that the foliar helices have properties like those of contact parastichies long recognized by other workers apart from this theory. In fact, one of the major criticisms of the theory of multiple foliar helices is that it is difficult to decide which of the contact parastichies in any shoot ought to be interpreted as foliar helices. Furthermore, many workers find it difficult to understand what a generative center means structurally if it is, in fact, propagated helically along the axis of the plant. Clearly what is involved is the sequential and disjunct occurrence of centers of growth associated with leaf initiation, and the fundamental point of this theory is that the location of these centers is causally related to a pattern in the already mature regions of the shoot. This is difficult to reconcile with clear evidence just cited that the placement of leaf primordia can be interpreted in terms of phenomena occurring within the shoot apex itself. The sum total of evidence would seem to suggest that the phyllotactic patterns described as foliar helices are better interpreted as parastichies and given a geometric rather than a functional or causal significance.

The idea that leaf sites may be determined by stimuli proceeding acropetally from older regions of the shoot is not restricted to those workers who support the theory of multiple foliar helices. After observing what appear to be partially differentiated leaf traces present in the stem before the leaves with which they will be associated have been initiated, a number of workers (Sterling, 1945; Gunckel and Wetmore, 1946) have suggested that such precocious traces might in fact transmit the leaf-initiating stimulus. Although instances in which the trace clearly precedes the leaf primordium have not been described frequently, there can be little doubt that the phenomenon does occur in certain cases. It has recently been reaffirmed by Larson (1983), who has observed precocious procambial strands for presumptive leaves I_1 through I_6 in *Populus deltoides*. However, the presence of internal differentiation in the stem associated with presumptive leaf sites need not be inconsistent with the field theory if the physiological fields are envisioned as having depth as well as surface extension. The result of the operation of such a system would be the coordinated development of internal and external features regardless of the order of initiation.

A final question concerning phyllotaxy is the possible biological significance of the ideal Fibonacci angle, which is characteristic of most species having a helical arrangement. The fact that in actual apices the angle fluctuates widely, with only the average approximating the ideal, supports the contention that biological factors and not purely geometric considerations underlie the widespread occurrence of this angle. If various phyllotactic patterns are plotted out using the divergence angle of 137.5°, it will be seen that each primordium is placed between its contacts such that there is a constant ratio between angular distance from the primordium to the older and to the younger of the contacts. This system, moreover, can continue indefinitely as the apex grows without any change in this ratio. This suggests the possibility that the ratio of angular distance in some way reflects the relative intensities of the fields of the contact leaves and the decrease in intensity with age. Recently Richter and Schranner (1978) have proposed an inhibitory field theory along these lines, basing their calculations on the proportionality of inhibitory strengths of the fields of the three youngest primordia.

Other workers, such as Thornley (1975a and b), Young (1978) and Mitchison (1977), have used computer simulations to test the capacity of inhibitory fields to produce patterns corresponding to those found in nature. By varying factors like strength of the fields, rates of diffusion and degradation of inhibitors, and the threshold concentration at which a primordium can develop, they have been able to derive continuing stable systems that resemble actual phyllotactic patterns, including those that are characterized by the ideal divergence angle. Studies like these do help to dispel some of the mystery that has for too long surrounded the problem of leaf arrangement, but a real understanding of phyllotaxy will only be achieved when the nature of inhibitory fields, if they exist, has been elucidated. This will not be an easy task.

GENERAL COMMENT

The field theory has been presented in this chapter as the most plausible hypothesis to explain the placement of leaf primordia, at least in species with helical phyllotaxy. It may also have broader implications. In Chapter 6 it was shown that the terminal meristem of the shoot functions as an integrated unit in spite of the fact that any portion of it, even a very small one, is capable of forming a new apex if suitably isolated from surrounding tissues. The operation of a field of inhibition, possibly generated by the central cells of the apex, could provide a mechanism by which the meristem is integrated, that is, kept free from outgrowths. If there is an intensity gradient from center to margins in this central field, outgrowths might be anticipated at the periphery of the meristem where they do indeed occur. Similar inhibitory fields around the peripheral outgrowths, operating in conjunction with the central field could explain why the primordia arise in a regular pattern and not haphazardly.

The fields are probably most easily visualized in terms of the production of inhibitory substances, but the nature of such substances has not been explored, nor is there clear proof that they actually occur. The fields might, in fact, result from the withdrawal or diversion of nutrients or essential growth factors. Moreover, as pointed out earlier, these two alternatives may not be mutually exclusive. The next logical step in the development of this important concept is the direct demonstration of the existence of fields and the elucidation of their biochemical basis.

While this would seem to be a potentially fruitful approach, it must be borne in mind that other mechanisms are possible, and indeed some workers have interpreted the regular patterns of phyllotaxy in terms of physical determinants. Such factors as tensions or compressions in specific regions of the shoot apex have been proposed as controlling mechanisms in leaf placement. It is possible, in fact, to visualize fields around existing growth centers that have a physical rather than a biochemical basis. In fact, this concept has been developed in detail by Green (1986). It is also possible that fields that are both physical and biochemical are involved (Williams, 1975). Only further refined experimentation and analysis can resolve these questions.

The field theory offers a working interpretation for the regulation of outgrowths from the shoot meristem and their positioning. There remains the further problem of how such outgrowths can acquire highly distinctive characteristics that convert them into determinate and dorsiventral leaves rather than additional shoot apices. The problem is further complicated by the fact that the same apex often produces other appendages that do form shoot apices, that is, buds. It is to this question that the next chapter is devoted.

REFERENCES

Church, A. H. 1904. *On the Relation of Phyllotaxis to Mechanical Laws.* London: Williams and Norgate.

Erickson, R. O. 1983. The geometry of phyllotaxis. In *The Growth and Functioning of Leaves*, ed. J. E. Dale and F. L. Milthorpe, 53–88. Cambridge: Cambridge University Press.

Foard, D. E. 1971. The initial protrusion of a leaf primordium can form without concurrent periclinal cell divisions. *Can. J. Botany* 49:694–702.

Green, P. B. 1986. Plasticity in shoot development: A biophysical view. *Symp. Soc. Exp. Biol* 40:212–32.

Green, P. B., and R. S. Poethig. 1982. Biophysics of the extension and initiation of plant organs. In *Developmental Order: Its Origin and Regulation*, eds. S. Subtelny and P. B. Green, 485–509. New York: A. R. Liss.

Gunckel, J. E., and R. H. Wetmore. 1946. Studies of development in long shoots and short shoots of *Ginkgo biloba* L. II. Phyllotaxis and the organization of the primary vascular system; primary phloem and primary xylem. *Am. J. Botany* 33:532–43.

Larson, P. R. 1983. Primary vascularization and the siting of primordia. In *The Growth and Functioning of Leaves*, eds. J. E. Dale and F. L. Milthorpe, 25–52. Cambridge: Cambridge University Press.

Lyndon, R. F. 1983. The mechanism of leaf initiation. In *The Growth and Functioning of Leaves*, eds. J. E. Dale and F. L. Milthorpe, 3–24. Cambridge: Cambridge University Press.

Meicenheimer, R. D. 1981. Changes in *Epilobium* phyllotaxy induced by N-1-naphthylphthalamic acid and α-4-chlorophenoxyisobutyric acid. *Am. J. Botany* 68:1139–54.

Mitchison, G. J. 1977. Phyllotaxy and the Fibonacci series. *Science* 196:270–5.

Plantefol, L. 1948. *La Théorie des Hélices Foliares Multiples.* Paris: Masson et Cie.

Richards, F. J. 1948. The geometry of phyllotaxis and its origin. *Symp. Soc. Exp. Biol.* 2:217–45.

Richards, F. J., and W. W. Schwabe. 1969. Phyllotaxis: A problem of growth and form. In *Plant Physiology. A Treatise,* vol. 5A, ed. F. C. Steward. New York: Academic Press.

Richter, P. H., and R. Schranner. 1978. Leaf arrangement. Geometry, morphogenesis and classification. *Naturwissenschaften* 65:319–27.

Schoute, J. C. 1913. Beiträge zur Blattstellungslehre. *Réc. Trav. Bot. Néerl.* 10:153–235.

Schüepp, O. 1917. Untersuchungen über Wachstum und Formwechsel von Vegetationspunkten. *Jahrb. wiss. Botanik.* 57:17–79.

Schwabe, W. W. 1971. Chemical modification of phyllotaxis and its implications. *Symp. Soc. Exp. Biol.* 25:301–22.

Snow, M., and R. Snow. 1931. Experiments on phyllotaxis. I. The effect of isolating a primordium. *Phil. Trans. Roy. Soc. London* Ser. B 221:1–43.

Steeves, T. A., and W. R. Briggs. 1958. Morphogenetic studies on *Osmunda cinnamomea* L. – The origin and early development of vegetative fronds. *Phytomorphology* 8:60–72.

Sterling, C. 1945. Growth and vascular development in the shoot apex of *Sequoia sempervirens* (Lamb.) Endl. II. Vascular development in relation to phyllotaxis. *Am. J. Botany* 32:380–6.

Stewart, R. N. 1978. Ontogeny of the primary body in chimeral forms of higher plants. In *The Clonal Basis of Development,* eds. S. Subtelny and I. M. Sussex, 131–60. New York: Academic Press.

Sussex, I. M. 1955. Morphogenesis in *Solanum tuberosum* L.: Apical structure and developmental pattern of the juvenile shoot. *Phytomorphology* 5:253–73.

Thornley, J. H. M. 1975a. Phyllotaxis. I. A mechanistic model. *Ann. Botany* 39:491–507.

———. 1975b. Phyllotaxis. II. A description in terms of intersecting logarithmic spirals. *Ann. Botany.* 39:509–24.

Wardlaw, C. W. 1949a. Experiments on organogenesis in ferns. *Growth* (Suppl.) 13:93–131.

———. 1949b. Further experimental observations on the shoot apex of *Dryopteris aristata* Druce. *Phil. Trans. Roy. Soc. London* Ser. B 233:415–51.

Williams, R. F. 1975. *The Shoot Apex and Leaf Growth: A Study in Quantitative Biology.* Cambridge: Cambridge University Press.

Young, D. A. 1978. On the diffusion theory of phyllotaxis. *J. Theor. Biology* 71:421–32.

8

Organogenesis in the shoot: determination of leaves and branches

In the previous chapter four questions about leaf origin and development were posed. The first of these, which asked why any outgrowths of the shoot meristem occur at all, was left essentially unanswered. The second, which dealt with the location of outgrowth, was dealt with in that chapter. The last two, which questioned the nature of the influences that cause an outgrowth to become a leaf and the response of the outgrowth to these influences, could be phrased in another way: If outgrowths are initiated, why do some become leaves and others branch shoots? The reason for phrasing the question in this way is that in addition to the regular formation of leaves, it is characteristic of the shoots of all but a few vascular plants to give rise to a succession of branches such that the whole shoot becomes a ramifying system. Clearly the difference between a determinate and dorsiventral leaf and a branch that is a replica of the main axis is a striking one, and it is important to seek an explanation for this difference in the initiation or early development of both types of appendages. This chapter is devoted to a consideration of these questions.

LEAF DETERMINATION

In many ways, one of the most revealing approaches to the study of leaf development is that in which the partially developed organ is removed from the plant and allowed to continue its development on a culture medium of known composition in complete isolation from the parent organism. When this experiment was carried out on *Osmunda cinnamomea*, the rather surprising result was that immature leaves, including those as little as 1 mm in length, produced small mature leaves on a culture medium containing inorganic salts and sugars but no complex organic supplement (Steeves and Sussex, 1957) (Fig. 8.1). Subsequently similar results were obtained with a number of other species of ferns and with several species of flowering plants (Fig. 8.2). This experiment clearly demonstrated that early in its development the leaf becomes essentially self-controlling, or autonomous, and in the presence of a sup-

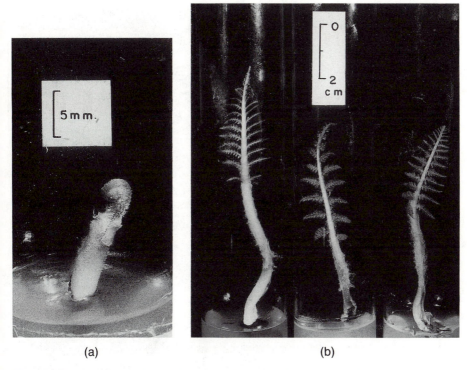

(a)   (b)

Figure 8.1. Osmunda *leaf cultures. (a) An excised leaf primordium growing on an agar culture medium. The leaf is just starting to uncoil. (b) Examples of leaves that have developed to maturity in sterile culture. (Steeves and Sussex, 1957.)*

ply of simple nutrients is able to complete its characteristic pattern of development. The resulting mature leaves are ordinarily much smaller than those that mature on the intact plant, and their morphological complexity may be reduced, but they are unquestionably determinate and dorsiventral leaves.

When, however, an attempt was made with *Osmunda* to discover whether the leaf primordium is autonomous from the time of its inception, a rather different result was obtained. If the ten youngest primordia in the shoot apex (P_1 to P_{10}) are explanted to culture medium, many give rise to shoots and ultimately to whole plants, after the initiation of roots, rather than to leaves (Steeves, 1961) (Fig. 8.3, Table 8.1). The proportion of primordia that give rise to shoots is very high in the case of P_1 and P_2 and diminishes progressively in older primordia. P_{10} invariably develops as a leaf. Thus it may be concluded that the ability to develop as a leaf without further control from the rest of the plant is acquired by the primordium early in its development but not necessar-

Figure 8.2. *Angiosperm leaf culture. An excised leaf of* Helianthus annuus *(sunflower) that has developed to maturity in sterile culture.* ×4. *(T. A. Steeves, H. P. Gabriel, and M. W. Steeves.* Science *126:350, 1957.)*

Figure 8.3. *Transformation of an excised leaf primordium into a shoot. A* P_3 *leaf primordium excised from the bud of* Osmunda cinnamomea *and grown in sterile culture has developed into a shoot on which several leaves have been initiated and from which roots have developed. Scale in millimeters. (Steeves, 1961.)*

Table 8.1. *Fates of Excised Primordia of* Osmunda
Cultured on Agar Medium Containing Mineral Salts and
2 Percent Sucrose. (Results from One Experiment.)
From Steeves (1966).

Primordium	Leaves	Shoots*	Doubtful or No Growth
P_1	2	7	11
P_2	2	12	6
P_3	4	10	6
P_4	4	11	5
P_5	8	11	1
P_6	12	8	0
P_7	16	4	0
P_8	17	1	1
P_9	19	0	1
P_{10}	20	0	0

* In many cases whole plants.

ily at its inception. The data in Table 8.1 indicate that in individual plants primordia as old as P_8 may still be undetermined, although in most plants determination occurs much earlier. However, Haight and Kuehnert (1969, 1971) have conducted similar experiments on *Osmunda* and have reached a different conclusion. They have suggested that determination occurs very soon after inception of a leaf primordium and that the shoots that arise from older primordia are in fact adventitious buds. Careful analysis of the development of these buds, however, has shown that it is the apical meristem of the explanted primordium that gives rise to the shoot, supporting the interpretation that the primordium was not irreversibly determined (Steeves, 1962).

In two species of angiosperms that have been investigated (sunflower and tobacco) the youngest primordium successfully explanted (P_2) has consistently developed as a leaf, indicating that, at least in these species, primordial destiny is established at a very early stage.

It is important to recognize the significance of the change effected in these minute primordial outgrowths as they acquire autonomy. Initially, in the ferns at least, the primordium does not differ in its potentialities from the shoot meristem that produced it. Quickly, however, it has imposed upon it a different and more specialized pattern of development. Moreover, what is imposed upon it at this stage is not a particular set of morphological characteristics but a program for future development that it is then able to carry out by itself to the formation of a mature organ. This process whereby a group of totipotent meristematic cells becomes established in a particular developmental pathway is often called *determination* in conformity with the usage of this term for similar phenomena in animal development. In fact the determination of a leaf, which arises in a definite site and has a limited period of development followed by complete maturation, has many similarities to the determination of organ primordia in animals. Although in the angiosperm species studied determination seems to occur at an earlier stage than in ferns, there is no reason to expect that the phenomenon is fundamentally different in the two groups.

Some of the early insights into the process of leaf determination were obtained by the use of surgical techniques in which, because the primordium is only partially isolated by incisions, it is possible to deduce something of the source of determining influences. In the fern, *Dryopteris*, Wardlaw (1949) isolated the I_1 position from the center of the meristem by a wide and deep tangential incision. The I_1 position thus isolated in most cases gave rise to a shoot bud rather than to a leaf, indicating that it was not at this stage determined. Later, in the same fern, Cutter (1956) was able to cause the three youngest visible leaf primordia (P_1 to P_3) to develop as shoots by isolating them as Wardlaw had done with the I_1 position or by isolating them from all lateral contacts on a plug of tissue by four deep vertical incisions. Here, as in the *Osmunda* culture studies, determination of the primordium as a leaf must occur sometime after its emergence. Similar experiments performed on angiosperms have revealed much less plasticity in the youngest foliar primordia. In all species thus far investigated the isolation of P_1 by an incision separating it from the apical meristem has been reported to result only in the development of a leaf. Even the isolation of the I_1 position in this way usually does not prevent the development of a leaf, although, as will be discussed later, it often does not have normal symmetry. In *Phaseolus* Pellegrini (1961) has reported that, in a few cases, I_1 produced a shoot after such isolation but only if all of the remainder of the apex was excised. Therefore, it would seem that leaf primordia are determined at a much earlier stage of development in angiosperms than in ferns, but it must be acknowledged that the information available is too scanty to permit generalizations.

In fact, surgical experiments carried out by Cutter (1958) on *Nuphar lutea*, in which P_1, I_1, or I_2 positions were isolated from the apical meristem by a tangential incision, could be interpreted as having resulted in the development of buds in certain cases, although the author did not adopt this interpretation. In most cases a leaf developed from the isolated P_1, I_1, or I_2 site, but in a few instances the panel developed as an expanding dome from which a shoot apex and a leaf primordium subsequently emerged. This does suggest that, as in ferns, the isolated sites

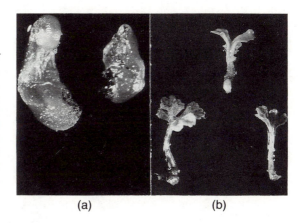

Figure 8.4. *Development of leaves from fragments of excised primordia. (a) Two immature leaves developed from the halves of a sagittally split sixth primordium (P_6). (b) Three leaves developed from the apex (top) and halves of the base of a seventh primordium (P_7). (a) ×8, (b) ×3. (Steeves, 1966.)*

(a)   (b)

may not have been irreversibly determined at the time of the operation. Furthermore, even though the process may occur more rapidly in angiosperms, there is no evidence to suggest that it is different in nature.

In the ferns that have been investigated the leaf primordium undergoes a certain amount of characteristic development prior to determination. This might suggest that it is determined only after it has achieved a particular level of morphological organization or that such organization is what constitutes determination. However, this possibility has been ruled out, at least in the case of *Osmunda*, by the demonstration that pieces of leaf primordia in many cases give rise to entire leaves when grown on nutrient culture medium (Fig. 8.4) (Steeves, 1966). The leaf fragments were obtained by subdividing primordia P_4 to P_{10} in various ways so that none contained the total structural configuration of the intact primordium. In a few cases three leaves have been obtained in this way from a single primordium, from the apex and two halves of the base, respectively. It begins to appear that the phenomenon of determination may be expressed in the primordium at the cellular level rather than at higher levels of organization, although this is certainly an extrapolation from existing data.

In a comparable experiment Sachs (1969) showed that leaves of an angiosperm (*Pisum*) behave in a similar fashion. When he operated surgically upon the youngest visible primordium (young P_1), he found that it was able to regenerate the missing portion and form a new leaf. Bisected primordia often gave rise to two entire leaves (Fig. 8.5). In progressively older primordia (up to P_3), however, excision of parts was not followed by complete regeneration. Rather, certain parts were missing even though the rest of the leaf was normal (Fig. 8.6). This suggests that as the leaf becomes more differentiated, the potentialities of any portion of it are restricted.

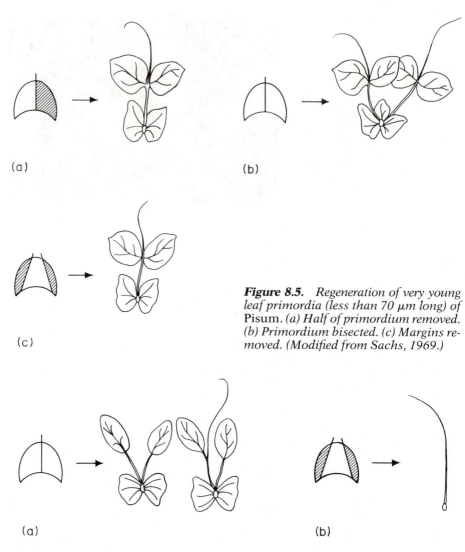

(a)

(b)

Figure 8.5. *Regeneration of very young leaf primordia (less than 70 μm long) of* Pisum. *(a) Half of primordium removed. (b) Primordium bisected. (c) Margins removed. (Modified from Sachs, 1969.)*

(c)

(a)

(b)

Figure 8.6. *Regeneration of leaf primordia 70 μm long of* Pisum. *(a) Primordium bisected. (b) Margins removed. (Modified from Sachs, 1969.)*

In the *Osmunda* experiments described above, it is evident that this restriction of potentialities must occur at a later stage, just as determination itself appears to be delayed in ferns. However, it does occur, although the time of its appearance has not been established clearly. When Caponetti (1972) excised the meristematic apices of relatively advanced leaf primordia of *Osmunda*, he found that they had apparently lost the potentiality to produce the basal region of the frond that had previously been initiated while still retaining the capacity to form those portions not yet initiated. Although the distinctive apical growth of fern leaves makes comparison with angiosperms difficult, this does suggest a similar kind of progressive restriction.

In considering the phenomenon of leaf determination, it also is necessary to seek the possible sources of the stimulus that produces this important change in a primordium. The surgical experiments cited, because they involved principally the isolation of the primordium or primordium site from the central region of the shoot apex, could be interpreted as indicating that the most distal cells of the shoot meristem are the major source of the determining stimulus. Indeed, they have usually been so interpreted. This interpretation has been supported by Hicks and Steeves (1969) using shoot apices of *Osmunda* growing in nutrient culture. When all leaf primordia were removed from the apex, a new primordium quickly arose in the I_1 position, and within two weeks this primordium had been determined as a leaf that could be shown to be autonomous by explanting it separately. Thus, determination had occurred in the absence of any older leaf primordia. If the I_1 position was isolated from the center of the apex by a deep tangential incision into which a chip of mica was inserted, in 75 percent of the cases a shoot arose instead of a leaf. There is thus clear evidence of a major influence emanating from the distal cells of the apex.

Evidence pertaining to the nature of the stimulus, or at least to the mode of its transmission, has also been obtained in this series of experiments. If the isolating incision was made by a very fine knife with a drop of culture medium covering its surface to prevent desiccation and no mica chip was inserted, approximately 80 percent of the resulting outgrowths were leaves. Histological examination revealed that in most cases a grafting process had established a tissue continuity across the incision in the inner tissues of the apex, but not in the surface layer of large cells, and that this grafting process began within the first week after the incision. In the cases examined, there was a correlation between the failure of grafting to occur and the development of a shoot rather than a leaf. The influence that proceeds from the distal cells of the apex thus appears to require tissue continuity for its transmission, and its effective transmission through inner tissues of the apex where the surface layers have been severed supports the conclusion of Wardlaw and Cutter (1956) that a deep incision is much more effective in isolating a primordium that is a shallow one alone.

The fact remains, however, that in these experiments some 25 percent of the I_1 positions isolated from the center of the meristem by a nongrafting incision formed leaves even though it is known that this position is not determined as a leaf at the time of this incision. This suggests that other factors, or factors from other sources, may at least participate in the determination process, and there is other experimental evidence to support such a view. In a study of induced apogamy in the fern *Pteridium* Whittier (1962) observed that among the sporophytic structures that developed on haploid gametophytes without fertilization, there were a number of individual leaves. Careful examination of these structures and of their development failed to reveal the presence of a shoot apex at any stage of their ontogeny. An interesting similar phenomenon has been described by Sattler and Maier (1977) in *Begonia hispida* var. *cucullifera*. Here vegetative leaves bear small, leaflike outgrowths on their upper surfaces that, in their structure and development, closely resemble small leaves. As in the previous case, they are not initiated by a shoot apex but rather arise from tissues of the maturing leaf. Thus it is possible for a determinate and dorsiventral leaf to arise in a growing system without the participation of a shoot apex, and it is therefore difficult to visualize determination as a process that depends exclusively upon the meristematic cells of the apex.

This point of view has received considerable emphasis in experiments performed by Keuhnert (1967) in which third primordia (P_3) of *Osmunda*, which when explanted to nutrient culture normally give rise to leaves in only about 25 percent of the cases, produced leaves twice as frequently when cultured in contact with older and clearly determined primordia, notably P_{10} to P_{12}. It is presumed that some influence, probably chemical, passes from the older primordium to the younger where it exerts a determining influence, or at least a stabilizing influence upon determining processes already underway. In any event the exclusive role of the shoot meristem in determining leaf primordia is challenged by these observations.

The sum total of information pertaining to the origin and nature of the determining stimulus does not lend itself to a simple or clear-cut interpretation. Rather, it begins to appear that multiple factors may be involved and that in the intact shoot apex determination is the net re-

Figure 8.7. *Development of a leaf of* Solanum tuberosum *as a radially symmetrical organ. (a) Diagram of the shoot apex showing the incision that separated the site where the next leaf (I_1) would emerge from the terminal meristem and the two older leaf primordia P_1 and P_2. The radial shape of the I_1 leaf that will develop in the isolated site is indicated in the diagram. (b) External view of a shoot apex with a radially symmetrical I_1 primordium at the front left, and the larger dorsiventral leaves P_1 and P_2 behind. (c) Transverse section through the shoot tip shown in (b). The radial organ is at the bottom of the illustration, P_2 is on the left, and P_1 is on the right. (b) ×45, (c) ×140. (Sussex, 1955.)*

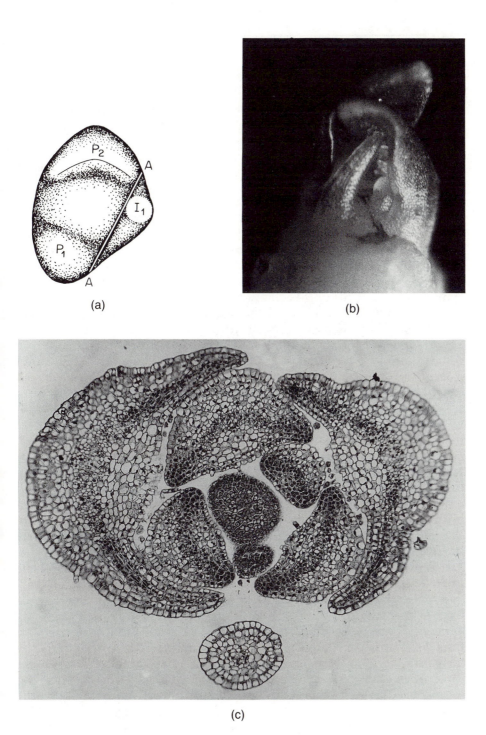

(a)

(b)

(c)

Figure 8.7. *(Caption on facing page).*

sult of a number of influences to which the primordium is subjected. Undoubtedly, the shoot meristem itself plays a prominent role, but in its absence other stimuli may be adequate to bring about determination. In a system such as this it is evident that experimental results must be interpreted with extreme caution.

DORSIVENTRALITY

Perhaps the most characteristic structural feature of most, although not all, leaves is their bilateral symmetry, or dorsiventrality. Developmentally the appearance of this feature is of great interest because the initially radial outgrowth of the leaf becomes flattened in a plane that is tangential to the circumference of the shoot meristem, suggesting that the one-sided relationship of the primordium to the meristem may be important in initiating this characteristic. Indeed, Wardlaw (1949) has suggested that an inhibitory influence from the meristem may retard growth on the adaxial side of the primordium as compared with the abaxial side, and that this growth inequality may be fundamental in the initiation of dorsiventrality. Experiments that he and later Cutter carried out on *Dryopteris*, some of which have been described, are consistent with this view. In potato Sussex (1955) was able to provide additional confirmation of this hypothesis when, after isolation of the I_1 position from the remainder of the meristem, he observed the development of organs that were determinate in growth but radially symmetrical, that is, centric leaves (Fig. 8.7). Centric leaves have also been obtained in *Epilobium* by the Snows (1959) and in *Sesamum* (sesame) following a similar operation. In *Sesamum*, Hanawa (1961) obtained both a dorsiventral leaf and a centric leaf from the same primordium by incising a P_1 tangentially across its apex. The dorsiventral leaf arose from the half primordium in continuity with the apex, and the centric leaf developed from the half on the abaxial side of the incision (Fig. 8.8).

However, there is evidence supporting another interpretation of the origin of bilateral symmetry in leaves. The experiments of Whittier (1962) in which perfectly dorsiventral leaves arose as apogamous outgrowths from fern prothalli, and those of Sattler and Maier (1977) in which leafy appendages were developed on the surfaces of foliage leaves, must be interpreted differently because no shoot meristem was present. Similarly, Steeves (1962) has reported that excised leaf primordia in culture sometimes undergo a period of development during which their symmetry is radial, following which they become dorsiventral, again in the absence of a shoot meristem (Fig. 8.9). In these cases the bilateral symmetry of the organ has been interpreted as a consequence of its determination as a leaf, that is, an inherent part of the leaf pattern of devel-

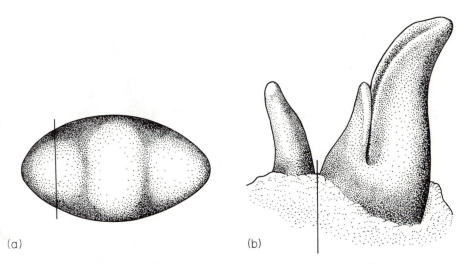

Figure 8.8. *Leaf dorsiventrality in* Sesamum. *(a) Diagram of seedling apex show-ing position of tangential incision of one of the* P₁ *leaf pair. (b) Development of a radial and a dorsiventral leaf from the incised primordium. (b) ×50. (Hanawa, 1961.)*

opment acquired in the process of determination, rather than as a special feature requiring its own induction. But in the experiments described on potato apices the development of determinate but centric leaves led to the conclusion that the primordia had been determined as organs of limited growth but had not undergone the induction of dorsiventrality at the time of the operation. In this second view the total determination of a leaf could be looked upon as consisting of more than one step, with various features of the leaf pattern being induced sequentially and per-haps independently.

From the preceding experiments it is clear that there are two inter-pretations of the timing of events in the determination of leaves. These alternative views, based as they are upon separate experiments on dif-ferent organisms, will only be reconciled by further investigations of a broader range of species.

BRANCHING

In considering the initiation of appendages that become shoots – that is, branches – it must be kept in mind that there appear to be two types of branching in the vascular plants. Terminal branching occurs when the shoot meristem becomes separated, equally or unequally, into two growing centers, each of which develops a shoot. It is found in the Psi-lopsida (*Psilotum* and *Tmesipteris*), Lycopsida (*Lycopodium* and *Selagi-*

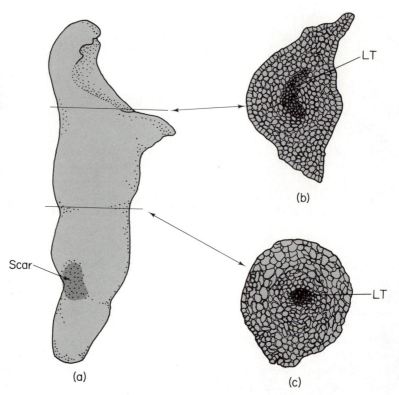

Figure 8.9. *A leaf primordium of* Osmunda cinnamomea *in which dorsiventrality developed after excision from the plant. (a) External view of the excised primordium after growth in sterile culture. The primordium was excised at the* P_3 *stage. The adaxial surface of the upper part of the leaf faces 90° away from the original adaxial face which is indicated by the scar on the base. A root is growing downward from the base near the scar. (b) Transverse section of the upper dorsiventral part. (c) Transverse section of the lower radial part of the leaf. The levels of the sections (b) and (c) are shown in (a). Key:* LT, *leaf trace. (a)* ×25, *(b,c)* ×45. *(Drawn from Steeves, 1962.)*

nella), many ferns and a few flowering plants, chiefly monocotyledons. If the division of the apex is equal and the subsequent growth of the two branches equivalent, an equal forking of the shoot results (Fig. 8.10a, b). On the other hand, if the subdivisions of the apex are unequal or subsequently develop unequally, smaller and larger shoots result from each branching, and in many cases the succession of larger branches may give the appearance of an axis with side branches (Fig. 8.10c). Typically in the seed plants branching is lateral; that is, lateral axes arise from buds situated in or near the axils of leaves (Fig. 8.10d). Lateral branch-

Figure 8.10. *Branching patterns in the shoots of vascular plants. (a) Equal, terminal branching in* Psilotum. *(b) Scanning electron micrograph of a branching shoot tip of* Psilotum. *Successive divisions of the apex are at right angles. (c) Unequal terminal branching in* Selaginella. *(d) Axillary branching in* Shepherdia *(buffalo berry). Key: SA, shoot apex; L, Leaf primordium. (a)* $\times \frac{2}{3}$, *(b)* $\times 37$, *(c)* $\times \frac{1}{3}$, *(d)* $\times \frac{1}{4}$. *(b) Photograph by T. C. Lacalli.*

ing also occurs in some ferns, the buds often being located in interfoliar positions.

Terminal branching

Terminal branching has been described as resulting from the separation of the shoot meristem into two equal or unequal portions, each of which becomes the apex of a branch axis. Although this is easy enough to visualize descriptively, it is quite another matter to understand how this kind of subdivision can occur if the meristem is a single, integrated growth center, as experimental studies reviewed earlier have indicated. Some of these experiments in fact have been revealing as to the mechanism of terminal branching. In many vascular plants, if the shoot meristem is surgically subdivided, the various portions reorganize and form complete meristems that initiate branch shoots. In several ferns it has been shown that puncturing of the apical cell, together with the attendant necrosis of a group of immediate derivatives, will produce the same result. Thus, if the integration of the shoot meristem is artificially broken down, a kind of terminal branching results, but there is nothing to suggest that any injury or related phenomenon is involved in the natural occurrence of terminal branching.

There is very little information concerning the structural changes that take place in terminal branching and the evidence suggests that there may be more than one type of process involved. In *Osmunda* where terminal branching is equal but infrequent, early stages of branching show two active growth centers on the flanks of the meristem, whereas in the center the original apical cell and its immediate derivatives can still be found, but subdivided into smaller cells in the same way that peripheral cells of the prismatic layer normally become subdivided. A similar process has been described in *Selaginella willdenovii* (Cusick, 1953) where, instead of a single apical cell, there is a band of enlarged initial cells across the summit of the apex. Branching occurs when several of these cells near the center of the band begin to differentiate, leaving two unequal growth centers that form branch shoots. This method of terminal branching would be equivalent to that induced by surgery if it could be shown that the differentiation of the most distal cells precedes the outgrowth of lateral flanks and presumably causes it, or at least permits it, to occur. However, although no precise information is at hand, it seems more likely that the establishment of the lateral growth centers occurs first and that their combined influence promotes the differentiation of the distal cells. Such a mechanism would be consistent with the observation in some lower plants, including both *Osmunda* and *S. willdenovii,* that there is a progressive increase in apical size as the plant develops that is limited by the occurrence of successive branchings. It might be postulated that the meristem remains as a single unit in its organization up to some maximum size, at which point two new organizational centers replace the original.

Although seed plants characteristically branch by means of axillary buds, there are a few species of angiosperms in which terminal branching has been well documented. Where the actual branching process has been investigated, as in *Flagellaria indica* (Tomlinson and Posluszny, 1977) it appears to be very similar to that described for *Osmunda* and *S. willldenovii*. Prior to branching, the apex enlarges; two new meristems are organized on the flanks of the enlarged apex and a furrow gradually appears, separating the two new apices, as the original apical center loses its characteristic organization.

On the other hand, in some lower vascular plants that branch terminally, the apical cell of the original meristem is described as continuing and one of its relatively recent derivatives becomes the initial of a smaller branch. In the fern *Pteridium* (Gottlieb and Steeves, 1961) this type of branch formation has been observed in the initiation of short lateral branches on the main rhizome, whereas the equal branching of the main axis occurs as has been described for *Osmunda*. The mechanism by which a second apical cell can become established and initiate a new apex while still in proximity to the original apical cell is difficult to visualize; furthermore, it is not apparent why such an outgrowth should not develop as a leaf. From the foregoing it is evident that the phenomenon of terminal branching has many interesting aspects, particularly concerning the nature of apical integration and how it is maintained or broken down in an orderly way, but until more than a few isolated cases have been described in detail, it is pointless to attempt any generalizations.

Lateral branching

For those plants that do not branch terminally, but rather initiate branches from buds that appear to develop at some distance from the shoot meristem, the histological phenomena have been much more fully investigated. Wardlaw (1943) has shown that in several ferns lateral shoots, occupying what have been called interfoliar positions, arise from groups of undifferentiated cells, *detached meristems*, that did not undergo development while leaf primordia were emerging around them, but also did not differentiate. These detached meristems have a developmental continuity with the apical meristem of the shoot of which they are in fact portions that are somehow inhibited in the region of the apex and may even remain inhibited in the mature shoot. In some cases they never develop unless the main apex is removed, in which case they give rise to small lateral shoot apices, or they may develop in this way at some distance from the main apex without its removal.

Lateral branches in seed plants typically arise in axillary positions, and in the majority of those investigated the origin has been found to be remarkably similar to that described by Wardlaw in ferns. Garrison (1955) has described axillary bud origin in a number of dicotyledons of diverse taxonomic affinities. As successive leaf primordia are formed at the shoot apex, it may be seen that the cells in their axils do not undergo enlargement and vacuolation like the cells around them, but remain as detached meristems (Fig. 8.11). In the majority of species investigated this meristem is delimited adaxially by a narrow, curved zone of cambiform cells often referred to as the shell zone (Shah and Patel, 1972). As detached meristems are left behind by continued growth of the shoot apex, they enlarge and become organized into shoot apices with leaf primordia. Commonly at this time the dominance of the main apex is expressed in the inhibition of further development of the lateral apices and they remain as axillary buds, often for long periods and sometimes permanently, unless the main apex is removed.

In certain respects the origin of lateral branches from detached meristems that have a developmental continuity with the shoot meristem of the original axis resembles unequal terminal branching in that a portion of the meristem initiates a small branch while the major part of the meristem continues its development. The main difference is one of timing, in that the detached meristem is restrained from further development until growth of the main axis has left it outside the region of the shoot apex. There is also the further difference that detached meristems show a more or less constant relationship to leaf primordia, a relationship that is lacking in terminal branching. It has been suggested that the detached meristem pattern may be derived phylogenetically from that of unequal terminal branching.

On the other hand there are some reports (e.g., Majumdar, 1942) of a different method of initiation of lateral buds in seed plants. In these it is stated that axillary buds do not arise directly from the shoot apex as detached meristems; rather, they are initiated at some distance from the apex from cells in the leaf axil that are already partially or completely differentiated. These reports deserve to be more fully investigated and an attempt should be made to determine whether this pattern has a widespread distribution. However, in view of the fact that adventitious buds are well known to arise from a variety of mature tissues in stem, leaf, and root, as well as from wound callus, it is not startling to contemplate such an origin from differentiated cells for axillary buds. Only their regular occurrence and constant placement in relation to leaves would pose problems of interpretation.

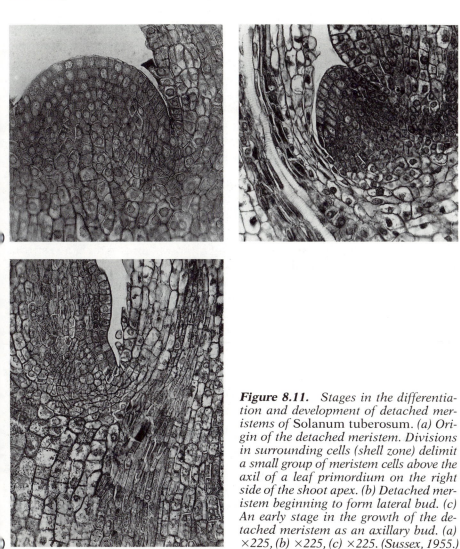

Figure 8.11. *Stages in the differentiation and development of detached meristems of* Solanum tuberosum. *(a) Origin of the detached meristem. Divisions in surrounding cells (shell zone) delimit a small group of meristem cells above the axil of a leaf primordium on the right side of the shoot apex. (b) Detached meristem beginning to form lateral bud. (c) An early stage in the growth of the detached meristem as an axillary bud. (a) ×225, (b) ×225, (c) ×225. (Sussex, 1955.)*

Several intriguing problems concerning the origin and development of buds arise because of the interesting relationships between buds and leaves. In the first place, there is the question of why there should be two types of appendages; that is, if the conditions in the shoot apex are such as to determine some appendages as leaves, why do they not determine all appendages in this way? In addition, the nearly constant spatial relationship to leaves suggests that there may be functional interactions between the two. These problems have been explored to some extent by experimental methods. In experiments discussed previously, Wardlaw and Cutter have shown that incipient leaf positions – and even emergent leaf primordia in certain ferns – are capable of initiating buds if they are suitably isolated from their surroundings, particularly from the influence of the shoot meristem itself. In other experiments in which the phyllotactic sequence has been altered by surgery, Wardlaw (1949) has shown that leaves can be developed in locations that may be identified as the normal sites of detached meristems, and hence buds. Thus, any part of the shoot meristem seems to be able to initiate either a bud or a leaf.

In the majority of cases investigated buds do not grow out from the intact meristem like leaves; rather, they are delayed and emerge in the subapical region. Thus, they are not exposed to the factors that determine other outgrowths that develop immediately as leaves. It is at least possible that the different morphology of leaves and buds is a result of this difference in timing of development and that the key to understanding this difference lies in the mechanism that inhibits the development and the differentiation of certain areas of the meristem, the detached meristems, until they are out of the apical region. The close association of buds and leaves might suggest that leaf primordia may be involved in this delay. But because in *Dryopteris* Wardlaw has shown that buds will grow out in the region of the meristem if the shoot apical cell is punctured while the adjacent leaves remain intact, the postulated role of leaf primordia must be more involved with the delay in the differentiation of the detached meristem than with the inhibition of its growth.

Other experiments that bear upon the relationship between bud and leaf in the angiosperms may now be considered. Various experimental procedures have been shown to alter the positions at which leaf primordia arise, that is, to change the phyllotactic pattern. It has been noted repeatedly in such cases that the axillary bud is associated with the leaf primordium in its unaccustomed position, indicating that the normal association is not one of chance. Because in many cases the leaf primordium distinctly precedes the bud in origin, the leaf must be considered the primary component in this association. In *Epilobium* and several species of *Salvia* the Snows (1942) have shown that if a young leaf primordium (P_1) is suppressed by incising it so that it fails to develop, its associated lateral bud, although not damaged by the operation, fails to appear. If, however, the leaf primordium is merely cut off at about the level of the meristem, the remaining base is often adequate to promote the formation of the bud. Similarly, if the leaf is isolated from the center of the meristem by a tangential incision immediately adjacent to the primordium so that the bud position is left intact, the bud likewise fails to develop. The Snows have interpreted these experiments as demonstrating the important role of the leaf in determining its own axillary bud. Although the experiments do indicate that the leaf is involved, they could equally well be interpreted as revealing the role of the leaf in preventing the differentiation of the potential detached meristem.

Evidence in support of this latter interpretation has been provided by a study of lateral bud development in *Arctostaphylos uva-ursi*, in which buds are formed in the axils of foliage leaves and of some transitional leaves but not of bud scales (Remphrey and Steeves, 1984). Examination of early leaf primordial stages, however, indicates that rudimentary bud meristems are present in the axils of incipient bud scales. As development proceeds, these rudimentary meristems become differentiated as tissues of the shoot axis. It thus appears that the more restricted growth of a bud scale primordium is correlated with a lack of the capacity to retard differentiation in its axil, whereas the expression of this capacity by a foliage leaf primordium permits a bud to develop in its axil.

Although an important role of the subtending leaf in permitting axillary bud development seems to be established, there is essentially no information on the mechanism by which this influence is exerted. Wardlaw has suggested the distribution of tensile stress resulting from the development of leaf primordia as a possible mechanism for their action in this respect. This interesting proposal has received some support from a recent study by Lintilhac and Vesecky (1980) that demonstrated in a photoelastic model simulating a shoot apex with leaf primordia that there are distinctive stress patterns in the leaf axils. Specifically there was a correspondence between the orientation of the major stress lines in the axil and the cell-division pattern of a typical shell zone. If such physical forces do have an influence upon the division pattern that results in the formation of a shell zone, they could play a role in bud development. It has frequently been proposed that the shell zone functions to isolate the detached meristem from the influence of the shoot apex, thus permitting it to develop in its characteristic manner.

There is, however, evidence that the regulation of bud initiation is, or at least can be, more complex than is indicated from the experiments thus far reported. This evidence is derived from a study of the distribution of bud primordia in the apices of certain angiosperms, a few examples of which will serve to illustrate the point (Cutter, 1966). In some plants buds are initiated in the axils of certain leaves only, and often there is a regular pattern of their distribution. Thus, whatever the factors that determine the initiation of a bud, they cannot be such as to be present in every leaf axil. There are other instances, as in *Nymphaea*, in which buds arise in place of leaf primordia in the phyllotactic sequence so that leaf-determining and bud-determining stimuli must be operative in close proximity in the same apex. In *Hydrocharis* a conspicuous bud primordium originates in the apical meristem itself in the axil of

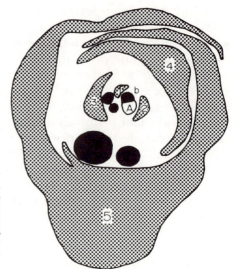

Figure 8.12. *Diagrammatic cross section of the shoot tip of* Hydrocharis. *Buds, shown in black, occur in the axils of alternate leaves and arise directly on the apical meristem. Each bud meristem quickly divides into two. ×50. (Adapted from E. G. Cutter,* Am. J. Botany *51:318, 1964.)*

every alternate leaf primordium (Fig. 8.12). These examples, and others like them, need not indicate complications in bud determination that are applicable to all plants, that is, to the basic mechanism of induction, but they certainly must be explained before any sweeping generalizations are offered.

GENERAL COMMENT

The phenomenon of leaf determination is one that invites a rigorous analytical investigation, particularly since the leaf primordium is an easily isolated developmental unit. The work completed to date has done little more than frame the questions that need to be answered. The evidence presently available does not suggest that a single, highly specific stimulus is responsible for bringing about determination, yet in different experiments both the shoot meristem and older leaf primordia have been shown to exert an influence, probably by chemical means. Meanwhile, it would be unwise to overlook the possible importance of the overall environment of the young primordium in inducing the leaf pattern. What, for example, is the consequence of the fact that the primordium is a rapidly growing center surrounded by areas that are inhibited from growing out in the same way? Any surgical interruption of such a system would disrupt this environment and could lead to erroneous conclusions about specific stimuli and their source. It may thus be worthwhile to consider whether the field theory, which offers a framework for the interpretation of both the integration of the apical meristem and the regular placement of leaf primordia, could also provide a basis for an understanding of the determination of lateral organs. Further progress in this important area of morphogenesis may well depend, therefore, upon precise information about the nature of fields – physical, chemical, or both. Highly refined techniques will be required to obtain this information, but the results would certainly justify the effort expended in perfecting them.

At the same time it is equally important to know precisely what happens to the leaf primordium when it is determined. Clearly, changes occur that do not appear to be structural and that quickly become independent of the inducing stimuli. Indeed, it is a pattern of development that is induced rather than a particular structural feature. Again, the fact that primordia can be isolated easily and even allowed to develop in isolation should facilitate this investigation, however exacting it may be technically. An important question here is the extent to which the leaf primordium is determined, that is, how much of its potential morphological complexity is established by the determination. Chapter 9 will analyze this problem in some detail.

REFERENCES

Caponetti, J. D. 1972. Morphogenetic studies on excised leaves of *Osmunda cinnamomea*: Developmental capabilities of excised leaf primordial apices in sterile culture. *Bot. Gaz.* 133:331–5.

Cusick, F. 1953. Experimental and analytical studies of pteriodophytes. XXII. Morphogenesis in *Selaginella willdenovii* Baker. 1. Preliminary morphological analysis. *Ann. Botany (London) (NS)* 17:369–83.

Cutter, E. G. 1956. Experimental and analytical studies of pteriodophytes. XXXIII. The experimental induction of buds from leaf primordia in *Dryopteris aristata* Druce. *Ann. Botany (London) (NS)* 20:143–65.

———. 1958. Studies of morphogenesis in the Nymphaeaceae. III. Surgical experiments on leaf and bud formation. *Phytomorphology* 8:74–95.

———. 1966. Patterns of organogenesis in the shoot. In *Trends in Plant Morphogenesis*, ed. E. G. Cutter, 220–34. London: Longmans.

Garrison, R. 1955. Studies in the development of axillary buds. *Am. J. Botany* 42:257–66.

Gottlieb, J. E., and T. A. Steeves. 1961. Development of the bracken fern, *Pteridium aquilinum* (L.) Kuhn. III. Ontogenetic changes in the shoot apex and in the pattern of differentiation. *Phytomorphology* 11:230–42.

Haight, T. H., and C. C. Kuehnert. 1969. Developmental potentialities of leaf primordia of *Osmunda cinnamomea*. V. Toward greater understanding of the final morphogenetic expression of isolated set I cinnamom fern leaf primordia. *Can. J. Botany* 47:481–8.

———. 1971. Developmental potentialities of leaf primordia of *Osmunda cinnamomea*. VI. The expression of P_1. *Can. J. Botany* 49:1941–5.

Hanawa, J. 1961. Experimental studies of leaf dorsiventrality in *Sesamum indicum*. L. *Bot. Mag. Tokyo* 74:303–9.

Hicks, G. S., and T. A. Steeves. 1969. In vitro morphogenesis in *Osmunda cin-*

namomea. The role of the shoot apex in early leaf development. *Can. J. Botany* 47:575–80.

Kuehnert, C. C. 1967. Developmental potentialities of leaf primordia of *Osmunda cinnamomea*. The influence of determined leaf primordia on undetermined leaf primordia. *Can. J. Botany* 45:2109–13.

Lintilhac, P. M., and T. B. Vesecky. 1980. Mechanical stress and cell wall orientation in plants. I. Photoelastic derivation of principal stresses with a discussion of the concept of axillarity and the significance of the "arcuate shell zone." *Am. J. Botany* 67:1477–83.

Majumdar, G. P. 1942. The organization of the shoot in *Heracleum* in the light of development. *Ann. Botany (London) (NS)* 6:49–81.

Pellegrini, O. 1961. Modificazione delle prospettive morfogenetiche in primordi fogliari chirurgicamente isoluti del meristema apicale del germoglio. *Delpinoa NS* 3:1–12.

Remphrey, W. R., and T. A. Steeves. 1984. Shoot ontogeny in *Arctostaphylos uva-ursi* (bearberry): Origin and early development of lateral vegetative and floral buds. *Can. J. Botany* 62:1933–9.

Sachs, T. 1969. Regeneration experiments on the determination of the form of leaves. *Israel J. Botany* 18:21–30.

Sattler, R., and U. Maier. 1977. Development of the epiphyllous appendages of *Begonia hispida* var. *cucullifera:* Implications for comparative morphology. *Can. J. Botany* 55:411–25.

Shah, J. J., and J. D. Patel. 1972. The shell zone: Its differentiation and probable function in some dicotyledons. *Am. J. Botany* 59:683–90.

Snow, M., and R. Snow. 1942. The determination of axillary buds. *New Phytol.* 41:13–22.

———. 1959. The dorsiventrality of leaf primordia. *New Phytol.* 58:188–207.

Steeves, T. A. 1961. A study of the developmental potentialities of excised leaf primordia in sterile culture. *Phytomorphology* 11:346–59.

———. 1962. Morphogenesis in isolated fern leaves. In *Regeneration*, 20th Growth Symposium, ed. D. Rudnick, 117–51. New York: Ronald.

———. 1966. On the determination of leaf primordia in ferns. In *Trends in Plant Morphogenesis*, ed. E. G. Cutter, 200–19. London: Longmans.

Steeves, T. A., and I. M. Sussex. 1957. Studies on the development of excised leaves in sterile culture. *Am. J. Botany* 44:665–73.

Sussex, I. M. 1955. Morphogenesis in *Solanum tuberosum* L.: Experimental investigation of leaf dorsiventrality and orientation in the juvenile shoot. *Phytomorphology* 5:286–300.

Tomlinson, P. B., and U. Posluszny. 1977. Features of dichotomizing apices in *Flagellaria indica* (Monocotyledones). *Am. J. Botany* 64:1057–65.

Wardlaw, C. W. 1943. Experimental and analytical studies of pteridophytes. II. Experimental observations on the development of buds in *Onoclea sensibilis* and in species of *Dryopteris*. *Ann. Botany (London) (NS)* 7:357–77.

———. 1949. Experiments on organogenesis in ferns. *Growth* 9 (Suppl.) 93–131.

Wardlaw, C. W., and E. G. Cutter. 1956. Experimental and analytical studies of pteridophytes. XXXI. The effect of shallow incisions on organogenesis in *Dryopteris aristata* Druce. *Ann. Botany* (London) (N.S.) 20:38–56.

Whittier, D. P. 1962. The origin and development of apogamous structures in the gametophyte of *Pteridium* in sterile culture. *Phytomorphology* 12:10–20.

9

Organogenesis in the shoot: later stages of leaf development

The stages of leaf development discussed in the two previous chapters – the primordial stages, which culminate in a simple outgrowth at the margin of the shoot meristem, somewhat flattened on its adaxial face – give little indication of the diverse morphology of the mature leaves of various groups of vascular plants. The multipinnate frond of a fern, the needle leaves of many conifers, and the diverse simple and compound leaves of the angiosperms are remarkably similar in the period immediately following their inception. Thus, the diversity of leaf form can be interpreted best through an understanding of the later, as opposed to primordial, stages of development. On the other hand the evidence here cited indicates that in the primordial stages the leaf undergoes a determination that confers upon it a considerable degree of autonomy in its later development. Does this imply that all of the diverse morphology of leaves must be thought of as originating in a process of determination at a relatively undifferentiated stage? The answer to this question is not an easy one, and the issues involved may best be exposed by examining some of the events of later leaf development and some of the experiments that have sought to interpret them.

DEVELOPMENT OF FERN LEAVES

There is a substantial body of information about later stages of leaf development in ferns, much of it collected from species that have also been used for experimental analysis, so that descriptive and experimental data may be correlated. One such example is the leaf of *Osmunda cinnamomea*, the early development of which has already been considered. In this fern the full developmental sequence of the leaf requires five growing seasons for its completion (Steeves, 1963) (Fig. 9.1). This is probably an extreme case, but undoubtedly it is not unique. Apical growth is long continued, even though it is limited, and is accomplished by a terminal meristem with a central apical initial. The growth of the apex is not uniform, however, and several phases may be recognized. An abaxial–adaxial disparity in growth, which reflects the initiation of

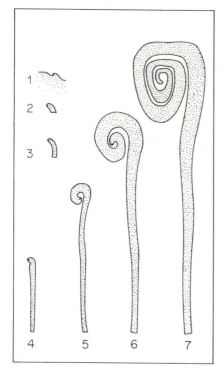

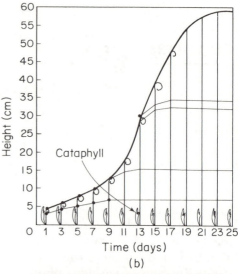

(a)

(b)

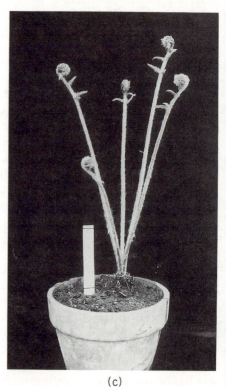

(c)

Figure 9.1. *Development of the frond of* Osmunda cinnamomea. *(a) Frond development from inception to the end of the fourth growing season: (1) soon after inception, (2) end of the first season, (3) end of the second season, (4) end of the third season, (5–7) developmental stages during the fourth season. The coiled crosier develops entirely during the fourth season (×1.5). (b) Graphic record of the development of a vegetative frond of* Osmunda *during the fifth growing season. The cataphyll heights indicate the length of the original apical bud. The frond was marked at regular intervals, but only a few of the marks placed on the frond are plotted in this graph. Observe that the mark placed at the top of the crosier does not change position during the period of crosier elevation, and that the crosier begins to uncoil only between days 9 and 11. (c) An* Osmunda *plant showing fronds in various stages of crosier uncoiling. Lines on pot label 10 cm apart. (Steeves, 1963)*

dorsiventrality, causes an early curvature in the primordium that is retained at or just behind the leaf apex, giving it the form of a hook. This
results from an inequality in cell number, at least in the ground tissue
of the primordium, but this inequality is continuously equalized at the
base of the hook so that a straight axis is produced. In the fourth growing season of a leaf, apical growth appears to be accelerated to such an
extent that the equalizing process does not keep pace and the result is
the rather rapid formation of a coiled axis, or crosier (Fig. 9.1a). As the
crosier is formed there is enlargement of the outer coils producing a
loosening that allows space for the still-forming inner coils. This is accomplished by extensive cell division at considerable distance from the
leaf apex with no net cell elongation. Surprisingly enough, this cell division maintains the imbalance in cell number set up just behind the
apex so that the crosier does not uncoil at this time. As the leaf apex
begins to form the crosier, it also starts the initiation of pinnae in acropetal sequence along each side of the axis. At the end of the fourth growing season, the apical meristem of the leaf ceases to function as such,
the apical cell differentiates, and ultimately a terminal pinna is formed.
Thus, after a prolonged period of activity, the determinate character of
the leaf apex is expressed.

The expansion of the leaf out of the apical bud and its final maturation occur in the fifth growing season immediately after a period of dormancy (Fig. 9.1b, c). In this phase, while cell division continues in the
more apical coiled regions of the leaf, a wave of cell elongation begins
to extend acropetally in the ground tissue of the leaf axis leading to
enormous increases in length and to final maturation. As this process
extends into the crosier, the coiled axis begins to straighten and the
crosier uncoils. Just before cell division ceases at any level in the crosier
in advance of the wave of elongation, the long-perpetuated imbalance
in cell number abaxially and adaxially is removed by the occurrence of
extra divisions on the adaxial side. Thus, when elongation is completed,
the rachis is straight with cell number and cell size equal on the two
sides.

This brief account of the later phases of leaf development in *Osmunda*
emphasizes the complexity as well as the long duration that may be
found in the postprimordial stages. The precise timing of the sequential
developmental events that are necessary for the successful completion
of the leaf points to a rather elaborate controlling system and gives focus to the question of whether all of this pattern can be considered as a
consequence of determination at the primordial stage. Until other species of ferns have been investigated in similar detail, it cannot be known
how nearly typical the *Osmunda* pattern is, but what is known of other
ferns suggests that the differences are largely ones of the duration and

distinctness of phases. Certainly in some ferns the complete development of a leaf from initiation to maturity is accomplished in one growing season or less and, even in *Osmunda*, juvenile leaves of the young sporophyte develop very rapidly. The study of later stages of leaf development in angiosperms, on the other hand, is complicated, not only in the interpretation of controlling mechanisms, but even at the descriptive level by the enormous diversity of this large plant group.

DEVELOPMENT OF ANGIOSPERM LEAVES

The investigator of angiosperm leaf development is confronted by a bewildering array of sizes and shapes of foliage leaves and, in addition, must deal with a variety of other leaf forms, such as bracts, cataphylls, spines, and insectivorous leaves, often occurring on the same plant along with more typical foliar organs. Although there have been many descriptive studies of angiosperm leaf development, it has become apparent during the past decade, with the application of more precise analytical methods, that much of this earlier work may have misinterpreted the actual growth pattern. Reliable information is therefore available for only a few species, and generalizations are difficult to make. Before considering what is known of the factors that regulate leaf form in angiosperms, it is desirable to review a few pertinent facts about postinitiation development.

By the time a leaf primordium has become a peg-like outgrowth, it is ordinarily dorsiventral but it does not have a lamina. This is initiated by marginal meristematic activity located along the two lateral edges of the primordium at the junctions of adaxial and abaxial faces (Fig. 9.2). Conventionally these meristematic strips have been called marginal meristems and have been described as consisting of marginal initials that initiate the epidermis by anticlinal divisions and submarginal initials that, by divisions in several planes, give rise to the inner tissues of the lamina. These initials were considered to be of limited duration but were believed to establish a relatively precise pattern of layers in the unexpanded lamina while they functioned. The major size increase of the lamina then occurred in a final phase of expansion and maturation during which both cell division and cell enlargement contributed to a great increase in laminar area.

There are several lines of evidence that indicate that this interpretation, which stresses the role of initial cells in the leaf lamina, is in error. Careful analyses of the frequency and orientation of mitoses in the leaf margin of *Xanthium* (cocklebur) (Maksymowych, 1973) gave no evidence for the existence of marginal or submarginal initials. Maksymowych therefore concluded that the cell division patterns he observed at

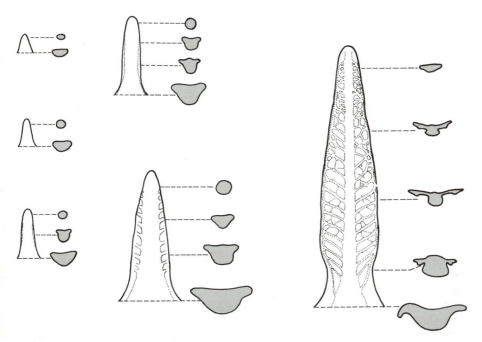

Figure 9.2. *Early stages in the development of a vegetative leaf of* Nicotiana tabacum. *At each stage the leaf is shown in face view on the left and in transverse section at several levels. The primordium first develops as a small peglike outgrowth widest at the base. Development of the blade begins later, and because blade development is more intensive above the base, the outline of the leaf becomes ovate in face view. (G. S. Avery, Am. J. Botany 20:565, 1933.)*

or near the margins of the lamina do not correspond to the regular cell lineages reported by other workers. In developing leaves of tobacco, Dubuc-Lebreux and Sattler (1980) determined mitotic frequency and mitotic index across the lamina from midrib to margin at four developmental stages. On the basis of their data, which indicated that mitotic frequency was highest in the region intermediate between midrib and margin, they concluded that the marginal zone played no special role in generating the lamina. They favored, therefore, the concept that the lamina develops by a generalized meristematic activity across the whole organ.

In spite of the doubts raised by these investigations, they do not entirely undermine the concept of marginal and submarginal initials. Such initials had been thought to function only for a brief period and, in any event, need not have an elevated division frequency in order to fulfill the role of initials. However, convincing confirmation of the absence of

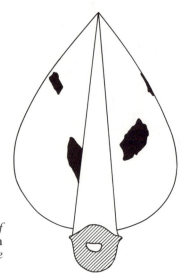

Figure 9.3. *Diagram showing the distribution of clones in the palisade layer of* Nicotiana tabacum *leaves irradiated just prior to the initiation of the lamina. (Poethig and Sussex, 1985.)*

such initials has now been provided by the method of clonal analysis in plants in which chlorophyll mutations can be induced by x-irradiation. In contrast to clonal analysis in the shoot apex (see Chapter 5) in which relatively permanent chimeras were used, clonal analysis in the leaf is carried out by irradiating plants so as to mutate cells in leaves at specific stages of development and subsequently assessing the position and extent of their clonally derived progeny in the mature organ. In tobacco Poethig and Sussex (1985) showed that if cells were mutated just prior to initiation of lamina growth, no clones in the mature leaf that resulted extended from the margin to the midrib as would be expected if initials functioned in the classical manner. Rather, three kinds of sectors were found: those that extended from the midrib part way into the lamina, those that were limited to the leaf margin, and those that were intercalary, isolated from both midrib and margin (Fig. 9.3). This distribution of clones is consistent with a pattern of generalized meristematic activity throughout the developing lamina but not with one in which initial cells at the margin give rise to the entire lamina.

Because the shape of a primordium before laminar initiation bears little resemblance to the ultimate form of the leaf, it is particularly important to analyze the manner in which the lamina grows and how it gives the leaf its final shape. This growth involves a combination of cell division and cell enlargement and is complicated by the fact that these processes do not occur uniformly throughout the organ at any one time. One of the most prominent features of the development of the lamina is

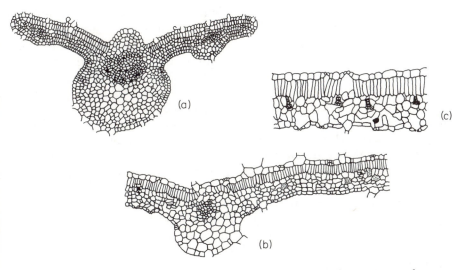

Figure 9.4. *Development of the blade in vegetative leaves of* Nicotiana tabacum. *(a) The blade is unexpanded and the precise pattern of layering is visible, especially on the left side. The palisade mesophyll cells form a vertically elongated layer under the upper epidermis. (b) Part of the blade of a leaf that is 150 mm long. Cells are expanding and intercellular spaces are present. The palisade cells are also expanding but they elongate at right angles to the plane of the blade. (c) Part of the blade of a leaf that is 210 mm long and has reached about two-thirds of its final size. Palisade cells are tightly packed together, but in the lower spongy mesophyll large intercellular spaces have formed where cells are pulled apart by expansion of the blade. (G. S. Avery, Am. J. Botany 20:565, 1933.)*

the early establishment of a layered pattern of cell arrangement that results from, and is maintained by, a predominance of cell divisions in a plane at right angles to the surfaces of the leaf. The layers are maintained with remarkable constancy while the lamina expands to many times its original area. They are disrupted in localized regions by the differentiation of the vascular tissue of the veins. Moreover, it is reported that in some species the number of layers is increased by cell divisions parallel to the leaf surface (Maksymowych, 1973). The layered organization is important because it is related to the internal organization of the mature leaf and also because the layers show differences in the timing of developmental events that are suggestive of the operation of precise controls. This pattern of cell division has led to the designation of the lamina at this stage as a *plate meristem*. Mitotic activity in the plate meristem continues until the leaf is surprisingly advanced in its development, in some cases having attained one-half to three-fourths of its final length. Although meristematic activity is high, cell size remains small, but as the division frequency declines, a phase of net cell enlargement, which completes laminar development, begins. In a number of species it has been shown that this change occurs first at the tip of the leaf and proceeds in a basipetal direction until the lamina is entirely mature.

The layered pattern of laminar growth assumes particular importance as the change to net cell enlargement begins to occur because the intensity and duration of both cell division and cell enlargement differ in different layers (Fig. 9.4). These differences are responsible for much of the functionally significant internal structure of the lamina. In general, division ceases first in the two epidermal layers and in the abaxial layers that will become the spongy mesophyll but continues for some time in the adaxial layer that will become the palisade tissue. Cell enlargement stops first in the spongy mesophyll layers, and these cells are pulled apart and to some extent distorted by the growth of other laminar tissues. The palisade cells continue to keep pace with the general expansion because they are still dividing but they elongate somewhat in a vertical direction. These cells ultimately stop dividing and undergo a limited enlargement that is completed shortly before the end of expansion of adjacent epidermal cells. As a result the intercellular spaces, which are necessary for gas exchange in photosynthesis, are formed by cell separation. The regulatory mechanisms that maintain control over the development of parallel layers in an expanding lamina must be exceedingly precise, and thus far there is little to indicate their nature. Where a petiole is formed, it ordinarily appears late in the development of the leaf and arises by intercalary growth in the axis of the leaf below the lamina.

It has been noted that the transition to final expansion and maturation occurs first at the tip of the leaf. This is an aspect of the general basipetal direction of maturation in leaves that contrasts with the overall acropetal maturation in the stem and is a reflection of the determinate nature of leaf development. Clonal analysis in tobacco (Poethig and Sussex, 1985) has provided a striking illustration of basipetal maturation in leaves irradiated at different stages of development. This has shown that, as the leaf matures, clone formation disappears first at the tip and then progressively in a basipetal direction. Clones initiated at any stage of development are larger in the basal portion of the leaf than in the apical region, showing that a cell in the basal region at any stage will contribute proportionally more to the growth of the lamina than will a cell at the tip (Fig. 9.5).

This technique of analysis can be extended to the early stages of leaf development by examining clonal patterns in leaves that were represented only by presumptive leaf sites (I_1, I_2, and so on) at the time of

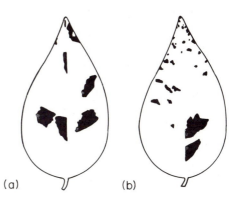

Figure 9.5. *Clones in the mesophyll of* Nicotiana tabacum *leaves irradiated at different stages: (a) 0.1 to 0.2 mm at irradiation, (b) 0.8 to 1.0 mm when irradiated. (Modified from Poethig and Sussex, 1985.)*

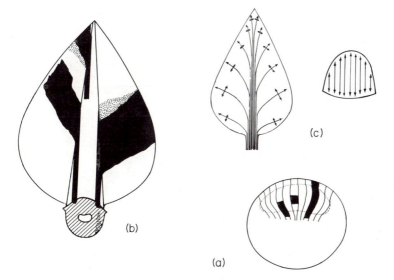

Figure 9.6. *Induction of clones in the mesophyll of* Nicotiana tabacum *by irradiation just prior to leaf initiation. In (a) the possible distribution of clones (black) at the time of leaf initiation is shown. (b) Distribution of clones in the palisade (black) and spongy mesophyll (stippled) layers of the mature leaf. (c) Diagrammatic representation of the derivation of cell lineages in the primordium (right) and expanding leaf (left). ([a,b] Poethig and Sussex, 1985. [c] Poethig, 1984.)*

irradiation. Some observations of this sort on tobacco (Poethig and Sussex, 1985) will be considered here rather than in Chapter 7 because they are more readily understood when the whole process of leaf development can be taken into account. The most revealing clones were those that occurred in leaves initiated fewer than three days after irradiation. When these clones are examined in the mature leaf it is observed that each occupies only a fraction of the leaf, extends along the midrib for some distance and then extends laterally into the lamina on one side (Fig. 9.6). This indicates that the primordium arises from parallel cell lines across its width rather than from a localized apical meristem for in the latter case clones would be expected to extend from tip to base of the leaf. An individual clone may be confined within the leaf, may extend into the tissues of the stem above or below the leaf, or may extend both above and below, depending upon the extent to which, at the time of leaf initiation, the clonal population was restricted to the leaf primordium or overlapped the axis above or below. On this basis it was concluded that the primordium arises from at least three horizontal tiers of cells on the flank of the meristem. The width of the clones indicates that each tier is 12 to 14 cells in horizontal extent, although perhaps fewer in the inner apical layers. Thus it was concluded that approximately 100 to 150 cells in four layers of the meristem participate in the initiation of a leaf in tobacco. These cells have been designated the "founder cells" of the leaf.

The development of compound leaves is more complex but, as far as is known, does not differ in its fundamental aspects from that of simple leaves. When a dorsiventral, peglike outgrowth has formed, instead of initiating laminar growth, it gives rise in turn to localized outgrowths in a palmate or pinnate arrangement. After some longitudinal extension, these secondary outgrowths initiate laminar development and become the leaflets of the compound leaf. In pinnately compound leaves the tip of the original outgrowth forms the terminal leaflet. In some cases the leaflets are initiated in an acropetal sequence but in many the sequence is basipetal.

The determinate nature of leaf development is well established and is characteristic not only of simple leaves but of compound leaves as well, even those in which the acropetal succession of leaflet initiation superficially resembles shoot growth. However, it appears that in some tropical plants with large (up to 6 m long), pinnately compound leaves the initiation of leaflets by the activity of a meristematic region at the leaf tip may extend over a period of years. In two species of *Guarea* of the mahogany family (Meliaceae) it is reported that the leaf apex remains meristematic and continues to initiate new leaflets periodically for more than two years. In fact it is not know how long this process may continue, and it is possible that such leaves may be indeterminate in their growth. Alternatively they may be comparable to those fern leaves in which the apex produces leaflets for an extended period but ultimately differentiates. It will be interesting to see whether further investigation of these and related species will, in fact, reveal a new pattern of angiosperm leaf development.

In monocotyledons the tendency of the young leaf primordium to encircle the shoot apex (see Chapter 7) by lateral extension is correlated in later development with the formation of a distinctive sheathing leaf base. This phenomenon has been extensively investigated in several members of the grass family (Poaceae) in which a separation into sheath and blade is an important feature of the mature leaf. The demarcation of these two regions occurs early in development and is indicated by the development of the ligule, a thin, membranous outgrowth of the adaxial protoderm. Subsequently the two regions are relatively independent in their growth. In the blade intercalary growth becomes progressively restricted to the zone just above the ligule, where it continues for some time, giving rise to the characteristic long, narrow lamina. The sheath generally lags behind the blade in its development but its growth is also intercalary and is progressively restricted to its base (Fig. 9.7). As successive leaves grow they emerge through the cylinder formed by the sheaths of older leaves. Clonal analysis of leaf development in maize has confirmed this general pattern of development and added some useful clarifications (Poethig, 1984). It has been deduced that the leaf arises from approximately 250 founder cells in two layers of the shoot apex. As in tobacco there is no suggestion of specialized apical or marginal meristems.

An extreme expression of sheath formation in the monocotyledons is found in *Musa* (banana). During vegetative development the stem undergoes essentially no internodal elongation. As each successive leaf is initiated, it encircles the shoot apex and undergoes extensive growth to form an elongated sheath topped by a petiolate lamina (Fig. 9.8). The cylinder of encircling leaf sheaths forms a "trunk," or pseudostem, which may be 6 m or more in height. Only during reproductive development does the stem, as an inflorescence axis, grow up through the center of the pseudostem and emerge from its top.

Although some monocotyledons have been shown to develop compound leaves in a manner comparable to that of dicotyledons, at least one group with large divided leaves, the palms, has a strikingly different pattern of leaf development. The leaf primordium consists of an ensheathing base surmounted by a hoodlike precursor of the blade. The two flanks of the blade region are thrown into folds or pleats by differential growth. This pleating, however, does not extend to the leaf mar-

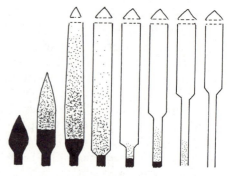

Figure 9.7. *Leaf expansion in* Zea mays *showing intercalary growth in both blade and sheath. Cell division is occurring in regions shown in solid black. Stippled zones are regions of cell expansion. (Adapted from B. C. Sharman,* Ann. Botany *6:245, 1942.)*

Figure 9.8. *Partially dissected pseudostem of* Musa *(banana) with a single leaf that has been removed being held on the right. The basal white portion is the sheath.*

Figure 9.9. *Scanning electron micrograph of a 7-mm-long leaf primordium of the palm* Chrysalidocarpus lutescens *showing the origin of leaflets by splitting. The arrow indicates the confluence of the unpleated margins with the hoodlike leaf apex. Scale bar = 500 μm. (D. R. Kaplan, N. G. Dengler, and R. E. Dengler,* Can. J. Botany *60:2939, 1982.)*

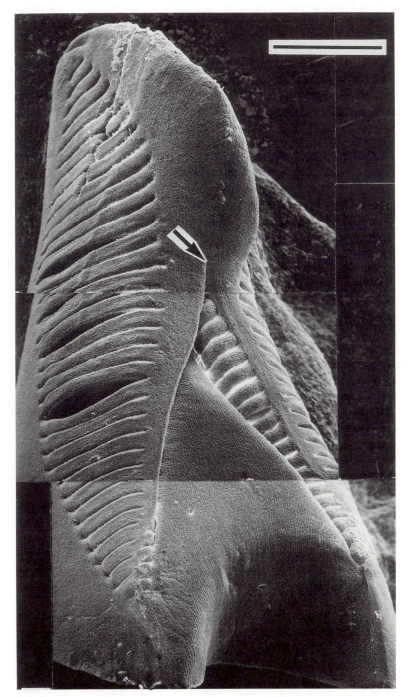

Figure 9.9. *(Caption on facing page).*

gins, which remain flat. After considerable deepening of the folds, tissue separation occurs along the tops of the ridges on one face or the other of the leaf (Fig. 9.9). The separated segments now develop into folded leaflets and the unpleated marginal strips are shed as the leaf expands (Kaplan, 1984). This phenomenon, when compared with compound leaf development in dicotyledons, provides a classic example of the way in which different developmental pathways can lead to similar mature structures.

ANALYSIS OF LEAF DEVELOPMENT

These descriptions of postprimordial stages of leaf development should serve to confirm the assertion made at the beginning of this chapter that much of the diverse morphology of leaves has its origin in these later stages. We may now turn to the question of whether this developmental diversity reflects determinative events at earlier primordial stages or emerges epigenetically as the leaf develops. The problem, then, is to understand how the later phases of development are regulated.

Experimental studies on *Osmunda*

In *Osmunda* some information on controlling mechanisms has been obtained by the use of experimental techniques including sterile culture in which leaves in various stages of development have been removed and allowed to continue in isolation. In this way the capacity of the leaf to regulate its own development from the stage at which it is excised can be determined and its dependence upon extrinsic factors can also be evaluated. Caponetti and Steeves (1963, 1970) have investigated primordia excised at the end of their third growing season but prior to the beginning of crosier formation. The isolated development of these fronds is remarkably similar to that of natural fronds and it duplicates the details of crosier formation and loosening, pinna initiation, cessation of apical growth, uncoiling, and final expansion and maturation (Fig. 9.10). The resulting mature leaf, however, is only about one-tenth the normal height, with one-third the normal number of pinnae, and development is completed in a fraction of the time taken by the normal frond. The reduced height results from reduction in cell number along the axis of the frond, because cell length measured in the ground tissue equals that of natural fronds.

Although the overall pattern of morphogenesis in cultured and natural fronds is remarkably similar, cellular processes show marked quantitative differences. Cell division is greatly diminished in the cultured fronds, whereas cell enlargement and maturation are essentially nor-

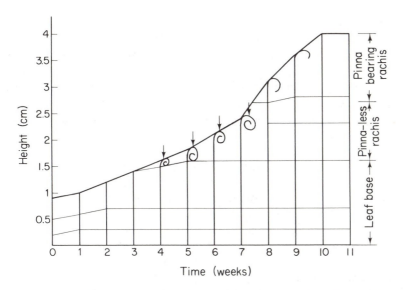

Figure 9.10. *Graphic record of the development of a frond of* Osmunda cinna-momea *excised from the plant at the end of the third growing season and grown in sterile culture to maturity. The crosier developed during the first three weeks in culture. It elevated in the next three weeks (note the arrow indicating the position of a mark placed on the top of the crosier), and uncoiled in the last three weeks. (Capo-netti and Steeves, 1963.)*

mal. This has several consequences for leaf development in vitro. Most of the reduction in duration of development just mentioned occurs in the crosier formation and loosening phase, where extensive cell division is involved, and the uncoiling and expansion phase is of almost normal duration. Moreover, in the cultured fronds cell enlargement is not delayed until the final expansion phase of the leaf as in the normal frond but actually participates in leaf growth during crosier formation. What seems to be involved here is a precocious maturation of the leaf when it is isolated from the parent plant. The basic morphogenetic control must remain intact, but the evidence is clear that some aspects of the normal regulation are lacking and that these have a great deal to do with the size and complexity of the leaf.

Some information is available concerning the nature of the influences that regulate these later stages of development in fern leaves. Some of these are clearly contained within the leaf itself, some give hints of regulation of a rather specific nature by the whole plant, and some suggest rather nonspecific influences of the whole plant system. It has been pointed out that the expansion and maturation phase of leaf development in *Osmunda* does not seem to be adversely influenced by isolation of the leaf. In other studies (Steeves and Briggs, 1960) it has been shown that cell elongation in the rachis of the normal frond, a major aspect of the expansion phase, is promoted by auxin produced in the pinnae during laminar development and transported in a polar fashion out of the pinnae and basipetally along the axis. Removal of the pinnae causes elongation to cease, but it can be restored by application of IAA. Because the uncoiling of the crosier is accomplished only when the wave of elongation extends into it, this process too is stopped by pinna removal and restored by IAA application. Because the mechanism by which this phase is regulated is shown to be within the leaf, it is not surprising that it is unimpaired by isolation. However, why it begins precociously in excised fronds is an unanswered question.

Because excised leaves complete their development in culture on a medium of simple composition, it has been customary to carry out experiments primarily on such a nutrient supply. Although leaves are photosynthetic organs, they are dependent upon the parent plant for their supply of nutrients, even carbohydrates, during the early stages of development. Experiments with excised leaves have shown a marked effect of carbohydrate supply in the medium upon leaf size and form. Caponetti (1972) succeeded in doubling the length of cultured third-season fronds of *Osmunda* by raising the concentration of sucrose in the medium from 2 to 6 percent, although it was necessary to transfer the leaves to 2-percent sucrose after the crosier had formed in order to permit uncoiling and expansion to occur. Histological analysis of the larger leaves produced under these conditions showed that the increased size was the result of an enhancement of cell division and that cell length was unaffected. Thus, the major defect of isolated leaves was partially overcome by an enhancement of nutrition. Considering that the absorption and translocation of nutrients in an isolated leaf must be greatly restricted in comparison to a normal leaf with intact vascular connections to a parent plant, it seems likely that much of the reduced cell-division activity could be explained in this way.

Although it is difficult to make rigid distinctions between nutrients and growth factors, especially when mechanisms of action are not understood, it is important to consider the question of specific hormonal regulation of leaf development in relation to the phenomenon of autonomy. It has been pointed out already that leaves of *Osmunda* and a number of other ferns have been shown to produce auxin in their pinnae, which promotes and is essential for the elongation of the rachis in the expansion phase. This mechanism, however, is leaf-contained and functions effectively in isolated leaves. The evidence for hormonal influences from outside the leaf suggests that they play at most a minor role in later leaf development. In cultured excised leaves the addition of various hormones or other growth-regulating substances to the medium has been shown to be without effect in a few cases and more commonly has been inhibitory. An exception to this general statement is the marked promotion of leaf apical growth by kinetin in the presence of adequate sugar (Sussex, 1964). This result seems to indicate that specific growth substances derived from the parent plant may play a role in the regulation of later stages of fern leaf development, but it cannot diminish the overriding importance of carbohydrate nutrition.

Alternative leaf forms – an introduction

The occurrence of more than one type of leaf on the same plant, a condition known as *heterophylly*, is of considerable developmental interest. It represents a situation in which the development of an organ can apparently follow different pathways, and thus it offers an opportunity to explore further the nature of organ determination. Two patterns of heterophylly have generally been recognized. In some cases there is a transition of leaf types, which may be gradual or more abrupt, in the ontogeny of the plant, as for example in cases in which juvenile forms give way to more adult types. This sequence is an expression of *heteroblastic* development, a term used to describe an ontogenetic progression in which early structures are markedly different from later ones. In other cases the production of more than one distinctive leaf type is a regular and often periodic feature of the adult plant, for example, the alternating formation of foliage leaves and the bud scales or cataphylls that surround a winter bud.

The instances of heterophylly are so numerous and diverse that it is difficult in a brief treatment to encompass the essential features of the phenomenon. Therefore, before examining some examples of heterophylly, it may be helpful to identify what seem to be the major questions of morphogenetic interest that are raised by this phenomenon. These may be summarized as follows:

1. At what stage in the development of a primordium does it diverge along a distinctive pathway?
2. What are the conditions or factors that are correlated with this divergence and may have a causal role?
3. Do these conditions or factors operate directly upon the primordium, even though transmitted through the plant, or do they act indirectly by modifying the shoot apex?

In the following three sections some examples of heterophylly will be examined in the light of these questions.

Heteroblastic leaf development in ferns

It is characteristic of ferns to produce a succession of leaves in the early stages of development of the sporophyte that show a progressive increase, not only in size, but in complexity as well. This phenomenon has long interested botanists because it is a clear example of the formation of leaves of different forms in a genetically uniform system and thus offers an opportunity for experimentation. Crotty (1955) has provided an excellent account of this phenomenon in the fern *Acrostichum danaeofolium* by comparing leaf development in a series of progressively older and larger plants. He concluded that the small, simple leaves of the youngest plants differ developmentally from the large, pinnate fronds of adult plants in their precocious maturation. The ultimate size and complexity of a leaf appears to depend upon the extent of meristematic growth that is accomplished before the onset of maturation growth, the final expansion and maturation of the leaf. Long ago Goebel (1908) suggested that carbohydrate supply has an important influence upon leaf form in ferns and that the explanation for the sequence of increasing size and complexity lies in the changing nutritional status of the plant. Wetmore (1953) provided experimental evidence that supports this interpretation. In *Adiantum* (maidenhair fern) young sporophytes grown on a medium containing sucrose at a concentration of 0.5 percent or higher rarely if ever formed juvenile, two-lobed leaves, and the heteroblastic transition was quickly disposed of. More strikingly, shoot apices of adult plants produced two-lobed leaves if grown on sugar concentrations of 0.1 percent or lower. Wetmore interpreted his observations as indicating a direct carbohydrate influence upon the leaf and felt that the morphologically simple leaf of the young sporophyte is the result of an energy deficit that restricts cell division.

Working with the aquatic fern *Marsilea* growing in sterile culture, Allsopp (1963) has obtained a similar effect of sugar upon the heteroblastic leaf sequence (Fig. 9.11). On the other hand, Allsopp has found a number of other substances that also influence the sequence and believes that the controls must be somewhat more complex, or at least, less direct. A variety of nutrients and hormones were found to influence the heteroblastic development of leaves in *Marsilea*. The substitution of ammonium ions or of urea for nitrate or the inclusion of gibberellic acid in the medium accelerated the rate of heteroblastic change, as did the addition of casein hydrolysate or certain amino acids. Indoleacetonitrile, which promoted petiole and internodal elongation, reduced the rate of change, and kinetin induced pronounced abnormalities in the leaves that obscured the sequence. In attempting to explain these results, Allsopp has favored an indirect influence of the various substances upon

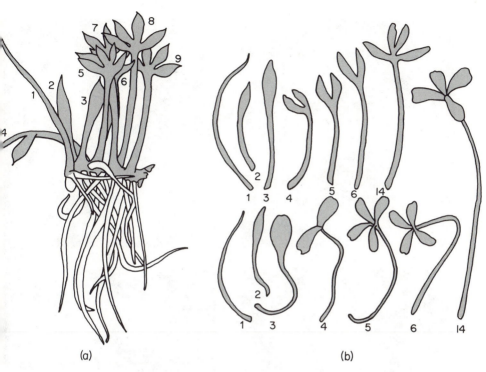

Figure 9.11. *Leaf shape variation in* Marsilea drummondii. *(a) A plant grown in sterile culture in a liquid medium containing 3 percent glucose. The leaves are numbered successively. The first-formed leaves (1–3) are all simple in outline but differ in shape; leaf 4 is bilobed; and leaves 5–9 are four-lobed. (b) Leaves from plants grown in media containing different glucose concentrations. The leaves from each plant are numbered sequentially. The leaves in the upper row are from a plant in 3 percent glucose; the transition from simple to bilobed shape was made at leaf 4, and from bilobed to four-lobed at leaf 14. The lower row of leaves are from a plant in 5 percent glucose. The progression through the various shapes is accelerated. Leaf 4 is the first bilobed leaf and leaf 5 is the first four-lobed leaf. Observe the difference in size and orientation of the pinnae in leaf 14 of the two treatments. (a) ×6, (b) ×4.5. (Allsopp, 1963.)*

leaf development operating through the whole plant. Recognizing that there is considerable enlargement of the shoot apex during development, he has suggested that it is apical size that is influenced by the nutritional and hormonal status of the plant, perhaps through direct effects upon protein synthesis, and that this in turn influences developmental process in the leaf primordium in some unspecified manner. In support of this interpretation, Allsopp has shown that several inhibitors of metabolism have a retarding effect upon heteroblastic development

(a)

(b)

(c)

(d)

Figure 9.12. *Development of excised leaf primordia of* Osmunda cinnamomea *in culture media containing different concentrations of sucrose. Leaves were excised from the plant when they were at the end of the second growing season and 1.6 mm high. (a) Small bilobed leaf that developed on a medium containing 0.006 percent sucrose. (b) Three-lobed leaf that developed on a medium containing 0.025 percent sucrose. (c,d) Larger pinnate leaves that developed on media containing 0.5 and 2.0 percent sucrose, respectively. ×2. (Sussex and Clutter, 1960.)*

or may even cause a reversion to the production of more juvenile leaf forms.

The question remains as to whether leaf form is regulated by direct action of nutrients upon the leaf as suggested by the excised leaf experiments described above or is influenced indirectly through the parent plant. Further leaf-culture experiments with *Osmunda* in which leaves were excised in their second season of development (rather than the third) support the direct influence interpretation. Interestingly enough, the influence of nutrition was even more profound in the case of these younger primordia. Sussex and Clutter (1960), by varying the sucrose concentration of the medium, were able to produce mature leaves showing a wide range, not only of size, but of morphological complexity as well (Fig. 9.12). On very low concentrations, small two- or three-lobed leaves were formed that were essentially comparable to those of a juvenile plant.

Experimental production of sporophylls

Further evidence of the ability of developing leaves to react directly to inductive stimuli is provided by experiments on the formation of sporangia by isolated leaves. Young sporophytes of the fern *Todea*, if grown in culture on very high concentrations of sucrose, not only form adult-type leaves but also, in many cases, initiate sporangia. These produce spores that are viable and may ultimately be germinated. Interestingly enough, if excised leaves are exposed to the same sugar concentrations in culture, they also initiate sporangia, but the development of sporogenous tissue is arrested at the spore mother cell stage just prior to meiosis and no spores are formed (Sussex and Steeves, 1958). This led to a tentative conclusion that the induction of meiosis requires a specific stimulus from the parent plant, a conclusion supported by some preliminary studies on *Osmunda* leaves. However, more recently Harvey and Caponetti (1972) have shown that even the apparent meiotic block in isolated leaves of *Osmunda* can be overcome. Sporangia that contained viable haploid spores were produced when the leaves were cultured with the appropriate conditions of light, temperature, sucrose concentration, and nitrogen source.

Heterophylly in angiosperms

Turning from ferns to the angiosperms, one is confronted by a rich expression of the phenomenon of heterophylly. Many cases of heterophylly offer an opportunity to compare divergent developmental pathways in presumably homologous organs with identical genetic constitution. There have been many descriptive studies of the development of contrasting leaf forms in the same plant, mostly concerned with the question of when in the course of development their pathways diverge. There is, however, very little experimental evidence that would make it possible to establish how the pathways are controlled – the extent to which they are expressions of predetermined patterns as opposed to epigenetic modifications of a common pattern. This issue is well illustrated by a consideration of the development of bud scales.

One of the early workers to adopt a developmental approach to the interpretation of bud scales was the German botanist Goebel (1905). He concluded that the primordial stages of bud scales and foliage leaves are equivalent and that it is only as a result of influences operating upon it at a later stage that the bud scale acquires its distinctive form. Referring back to the ferns, this interpretation is entirely applicable to the development of the distinctive bud scales of _Osmunda cinnamomea_ (Steeves and Wetmore, 1953). In this species there is no structural distinction among leaf primordia for approximately three years after their inception, until the season just prior to final maturation. Furthermore, if primordia which, because of their position in the leaf sequence, can be identified as prospective bud scales, are isolated in nutrient culture, they develop just as do the primordia of prospective foliage leaves.

Bud scales of the angiosperms cannot be so easily fitted into this scheme. In a series of studies on leaf differentiation in woody species, Foster (1932) showed that bud scale and foliage leaf primordia diverge in their developmental pathways at a very early stage and these pathways are so different that bud scales cannot be regarded merely as modified foliage leaves (Fig. 9.13). One of the earliest differences to be noted in many cases was a lower height-to-basal-width ratio in the bud scale primordium. Foster suggested that the developmental pathways are different because the two primordial types are initiated by shoot apices in different physiological states, a situation that carries the implication that they are determined in their developmental pathways at least at a very early stage. Of course, it is not known whether the determination of the developmental pathway coincides with the morphological expression, and this can only be established by experiment. The critical experiment would be to explant primordia of different developmental ages and allow them to express their inherent potentialities in isolation. Unfortunately, this experiment has not been done.

Recent studies of other heterophyllous systems have revealed a similar early divergence of developmental pathways between contrasting types of lateral organs. For example, Franck (1976) has compared the development of the juvenile and adult leaves of the insectivorous plant _Darlingtonia californica_. The juvenile leaves are relatively simple tubular structures that are open at the top, in contrast to the adult leaves,

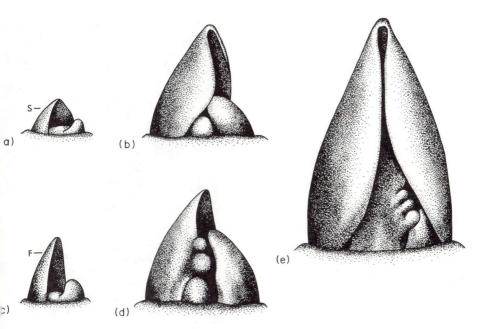

Figure 9.13. *A comparison of bud scale and foliage leaf early development in* Carya buckleyi. *The development of a bud scale (S) may be followed in the sequence (a), (b), (e), and that of a compound foliage leaf (F) in the sequence (c), (d), (e). ×60. (Modified from Foster, 1932.)*

which develop as complex, hooded insect traps. The growth patterns of the two leaf types were found to be different from the earliest stages that could be measured. Moreover, at the time of initiation of the simpler, juvenile leaves the shoot apex was smaller than the adult apex and lacked the distinctive cytohistological zonation of the latter. Thus it was concluded that the condition of the apex plays a major role in determining the type of leaf produced.

Particular interest is attached to another instance of heterophylly because in this case it is possible to rule out changes in the apex as a determining factor. Mueller and Dengler (1984) have described leaf development in the anisophyllous shoot of *Pellionia daveauana* in which there are two ranks of large ventral leaves and two of very small dorsal leaves. The two leaf types differ substantially in growth patterns and in histology, yet the leaves arise in pairs, one large and one small leaf arising simultaneously on opposite sides of the apex (Fig. 9.14). Thus, although determination at an early stage cannot be ruled out, it seems unlikely that changes in the status of the apex could be responsible for the divergent leaf types in *Pellionia*.

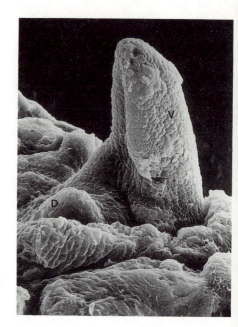

Figure 9.14. *Scanning electron micrograph of a shoot apex of* Pellionia daveauana *showing a large ventral leaf primordium (V) and a small dorsal primordium (D) arising simultaneously. ×275. (Mueller and Dengler, 1984.)*

The three cases of heterophylly in angiosperms reviewed here have all indicated an early divergence in the development of contrasting leaf types and hinted at an equally early determination of their developmental fates. There are, however, other cases of angiosperm heterophylly in which there is good evidence for a relatively late divergence in the development of different leaf forms. One such plant is *Proserpinaca palustris*, studied by Schmidt and Millington (1968). In this amphibious plant, leaves on submerged shoots are finely dissected while those on aerial shoots are lanceolate and entire if the plants are grown in long days (Fig. 9.15). Regardless of the conditions of growth, the primordia are structurally identical from inception until they have achieved a height of 500 to 600 μm, at which time differences begin to appear. If a shoot that is forming submerged leaves is placed in an aerial environment, immature leaves in the bud that would otherwise have developed as submerged leaves form a transitional series that increasingly resemble aerial leaves, after which typical aerial leaves are formed. Transfer from an aerial to a submerged environment produces a transition series in the reverse order. Changes in other environmental factors, such as photoperiod, light intensity, and temperature, also result in the formation of intermediate leaves. This experiment is significant for two reasons. First, it shows that the structural features of distinct leaf types may arise at a late stage in development. Second, leaves which have par-

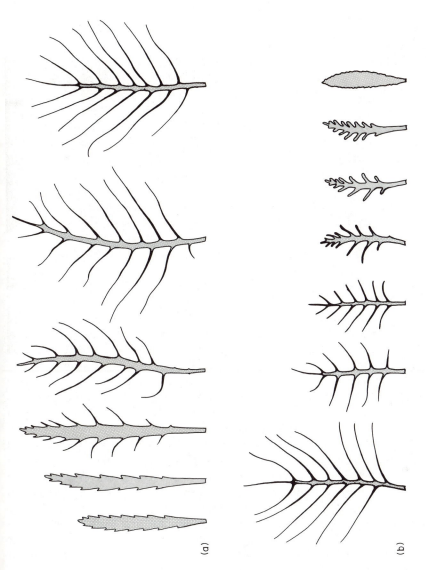

Figure 9.15. *Transitional leaf forms produced in Proserpinaca palustris as a result of an abrupt transfer (a) from an aerial to a submerged environment or (b) from a submerged to an aerial environment, under long-day conditions. ×1.5. (Schmidt and Millington, 1968.)*

tially developed along one pathway can be changed to a different one at relatively late stages of development. A further significant observation made in this study was that the size and structural organization of the apex were constant regardless of the leaf type being produced.

The change from juvenile to adult leaf form is perhaps the most widely recognized expression of heterophylly in the angiosperms. *Centaurea solstitialis* (yellow star thistle) is a rosette plant in which the juvenile leaves, which are small and simple, are replaced in a gradual transition by deeply lobed adult leaves. In an experimental study of this phenomenon Feldman and Cutter (1970a) reared seedlings in sterile culture on media containing various concentrations of sucrose and gibberellic acid. In the absence of sucrose, plants continued to form only simple leaves. On 1-percent sucrose the first lobed leaf developed after seven or eight simple leaves had formed, and on 4-percent sucrose after five simple leaves. Gibberellic acid, on the other hand, completely suppressed the transition to lobed leaves, whereas the gibberellic acid antagonist CCC accelerated the change. Histological examination did not reveal any consistent relationship between size and organization of the shoot apex and the ability to produce adult leaves. These results resemble in many respects those obtained with young fern plants in which carbohydrate supply appeared to play an important role in increasing leaf complexity in the heteroblastic sequence. The possible involvement of hormones is also suggested in both cases. In a further study of *Centaurea* Feldman and Cutter (1970b) were able to show that leaves isolated at the primordial stage from plants which were producing adult leaves are able to initiate lobes and thus achieve the adult form. However, on the limited range of culture media tested, only primordia older than P_3 were able to do this; younger primordia formed only simple juvenile leaves. Although this experiment was not designed to establish at what stage one or the other of the alternative patterns becomes fixed or the extent to which developing leaves can respond directly to regulatory factors, it does indicate that the leaf culture technique could be used to this end.

GENERAL COMMENT

The past few years have witnessed a significant deepening of our understanding of the way in which angiosperm leaves develop as a result of the application of more refined techniques. A number of long-held concepts have been severely challenged. Certainly it would be well to remember that, among the vast number of diverse species that constitute the flowering plants, only a very few have thus far been examined by methods like clonal analysis. The consistency and clarity of the new interpretations, however, suggest that they may be widely applicable, although not necessarily universally or uniformly. There is ample room for variation.

The new information about angiosperm leaf development, while it provides a clearer, and presumably more accurate picture of the events, has done little to improve our understanding of how these events are controlled. Indeed, the regulation of cellular processes within the leaf as they are now understood is, if anything, more difficult to visualize than was the case when localized meristematic regions seemed to be of major importance. There is also a further difficulty this new information has revealed. Although it has long been recognized that there are important developmental differences between the leaves of flowering plants and those of ferns, it is now becoming doubtful that there are any similarities at all in the later stages. It is not yet apparent what effect these findings will have upon the understanding of the regulation of leaf development in the two groups. It may also be suggested that further careful analysis of fern leaf development might be profitable. The lesson to be learned from angiosperm leaves is that things are not always what they appear to be.

In the previous chapter the term *determination* was used to describe the process whereby a primordial outgrowth acquires a developmental program different from that of the meristem from which it arose. Subsequent to this event the primordium appears to be capable of directing its own development to the formation of a mature leaf. In this chapter the discussion has centered upon processes whereby alternative types of leaves are produced by the same plant, in some cases at different times, in others almost simultaneously. The question now to be considered is what relationship exists between this latter process and the basic determination of the leaf.

Clearly, in some ferns the initial determination of a primordium as a leaf and the specification of the ultimate form it attains are separable in time. Although this does not necessarily mean that the processes are different in nature, it seems probable that the stimuli that induce them must be. In the angiosperms there are also cases in which the same kind of temporal separation occurs between initial determination and ultimate specification. Often, however, the specialized features of a particular leaf type make their appearance so early that it is difficult to separate them from the determination process. This overlapping of events had led many to the conclusion that different types of foliar organs arise from fundamentally different primordia. In cases in which two very distinctive leaf types are formed on the same plant so that the decision between them is of an either–or type, it is reasonable to consider such a possibility. However, when contrasting leaf types merely represent the extremes of a graded series of transitional forms, as in heteroblastic leaf development or in many cases of bud scale development, this interpretation is less plausible. In the present state of knowledge, the conservative view would be that there is a fundamental leaf-determination process that, however, may have imposed upon it at any stage, further specifications as to the developmental program. The question of determination and specification will arise again in the next chapter in relation to the development of floral organs.

REFERENCES

Allsopp, A. 1963. Morphogenesis in *Marsilea. J. Linn. Soc. London (Bot.)* 58:417–27.

Caponetti, J. D. 1972. Morphogenetic studies on excised leaves of *Osmunda cinnamomea:* Morphological and histological effects of sucrose in sterile nutrient culture. *Bot. Gaz.* 133:421–35.

Caponetti, J. D., and T. A. Steeves. 1963. Morphogenetic studies on excised leaves of *Osmunda cinnamomea* L. Morphological studies of leaf development in sterile nutrient culture. *Can. J. Botany* 41:545–56.

———. 1970. Morphogenetic studies on excised leaves of *Osmunda cinnamomea* L. Histological studies of leaf development in sterile nutrient culture. *Can. J. Botany* 48:1005–16.

Crotty, W. J. 1955. Trends in the pattern of primordial development with age in the fern *Acrostichum daneaefolium. Am. J. Botany* 42:627–36.

Dubuc-Lebreux, M. A., and R. Sattler. 1980. Développment des organes foliacés chez *Nicotiana tabacum* et le problème des méristèmes marginaux. *Phytomorphology* 30:17–32.

Feldman, L. J., and E. G. Cutter. 1970a. Regulation of leaf form in *Centaurea solstitialis* L. I. Leaf development on whole plants in sterile culture. *Bot. Gaz.* 131:31–9.

———. 1970b. Regulation of leaf form in *Centaurea solstitialis* L. II. The developmental potentialities of excised leaf primordia in sterile culture. *Bot. Gaz.* 131:39–49.

Foster, A. S. 1932. Investigations on the morphology and comparative history of development of foliar organs. III. Cataphyll and foliage-leaf ontogeny in the black hickory (*Carya buckleyi* var. *arkansana*). *Am. J. Botany* 19:75–99.

Franck, D. H. 1976. Comparative morphology and early leaf histogenesis of adult and juvenile leaves of *Darlingtonia californica* and their bearing on the concept of heterophylly. *Bot. Gaz.* 137:20–34.

Goebel, K. 1905. *Organography of plants. Part II. Special Organography.* Oxford: Oxford University Press.

———. 1908. *Einleitung in die experimentelle Morphologie der Pflanzen.* Leipzig and Berlin: Teubner.

Harvey, W. H., and J. D. Caponetti. 1972. In vitro studies on the induction of sporogenous tissue on leaves of cinnamon fern. I. Environmental factors. *Can. J. Botany* 50:2673–82.

Kaplan, D. R. 1984. Alternative modes of organogenesis in higher plants. In *Contemporary Problems in Plant Anatomy*, eds. R. A. White and W. C. Dickison, 261–300. New York: Academic Press.

Maksymowych, R. 1973. *Analysis of Leaf Development.* Cambridge: Cambridge University Press.

Mueller, P. A., and N. G. Dengler. 1984. Leaf development in the anisophyllous shoots of *Pellionia daveauana* (Urticaeae). *Can. J. Botany* 62:1158–70.

Poethig, S. 1984. Cellular parameters in leaf morphogenesis in maize and tobacco. In *Contemporary Problems in Plant Anatomy*, eds. R. A. White and W. C. Dickison, 235–59. New York: Academic Press.

Poethig, S., and I. M. Sussex. 1985. The cellular parameters of leaf development in tobacco: A clonal analysis. *Planta* 165:170–84.

Schmidt, B. L., and W. F. Millington. 1968. Regulation of leaf shape in *Proserpinaca palustris*. *Bull. Torrey Botan. Club* 95:264–8.

Steeves, T. A. 1963. Morphogenetic studies of fern leaves. *J. Linn. Soc. London (Bot.)* 58:401–15.

Steeves, T. A., and W. R. Briggs. 1960. Morphogenetic studies on *Osmunda cinnamomea* L. The auxin relationships of expanding fronds. *J. Exp. Botany* 11:45–67.

Steeves, T. A., and R. H. Wetmore. 1953. Morphogenetic studies on *Osmunda cinnamomea* L.: Some aspects of the general morphology. *Phytomorphology* 3:339–54.

Sussex, I. M. 1964. The permanence of meristems: Developmental organizers or reactors to exogenous stimuli? *Brookhaven Symp. Biol.* 16:1–12.

Sussex, I. M., and M. E. Clutter. 1960. A study of the effect of externally supplied sucrose on the morphology of excised fern leaves *in vitro*. *Phytomorphology* 10:87–99.

Sussex, I. M., and T. A. Steeves. 1958. Experiments on the control of fertility of fern leaves in sterile culture. *Bot. Gaz.* 119:203–8.

Wetmore, R. H. 1953. Carbohydrate supply and leaf development in sporeling ferns. *Science* 118:578.

10

Determinate shoots: thorns and flowers

Previous chapters have considered how the basic plan of the vascular plant shoot is initiated and elaborated. It will be recalled that shoot development occurs in two relatively distinct phases. An initial phase involves terminal meristem activity in which the tissues and organs are laid down. There follows a phase of expansion growth in the subapical part of the shoot during which the previously formed structures enlarge and mature. Chapter 11 will examine how variations in the extent of the expansion phase could produce shoots of widely differing morphology. However, there are other developmental variations in the basic body plan of the shoot in which the phase of terminal meristem growth is principally involved, and these are the subject of this chapter.

It should be expected that if terminal meristem activity is modified there might be cases of extreme modification in the kind of organs produced and in the extent and pattern of their subsequent growth and development, and indeed this is so, as any student of plant taxonomy or morphology knows. These modifications have been the subject of extensive researches in which the question has been the degree of homology between the modified organs and more usual organs of the shoot. However, in this chapter attention will be confined to some examples that have proved to be especially amenable to developmental analysis and about which relatively recent information is available. What these examples have in common is that they represent cases in which the shoot meristem ceases its organogenetic and histogenetic activities. In other words, it becomes determinate.

DETERMINATE MERISTEMS – THORNS

One of the most striking departures from typical shoot morphology is the thorn, supposedly protective in function, which is a shoot – terminal or axillary, branched or unbranched – whose apex has ceased to grow and has become more or less hardened and sharply pointed. In terms of developmental modification the critical feature of thorns is that they are determinate organs. In some cases the shoot thorn is little mod-

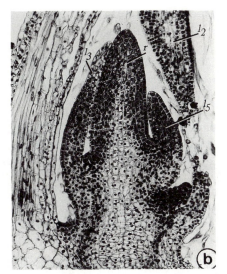

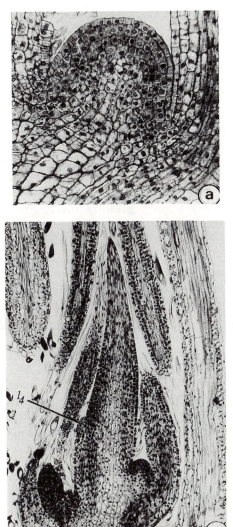

Figure 10.1. *Thorn development in* Ulex europaeus. *(a) Axillary vegetative shoot meristem. (b) Intermediate stage in thorn development in which the apex has elongated as a result of extensive rib meristem activity. (c) Later stage in development showing further narrowing of the shoot and the beginning of differentiation of the cells at the tip. Key:* l, *leaf;* r, *rib meristem. (a)* ×200, *(b)* ×100, *(c)* ×80. *(Bieniek and Millington, 1967.)*

ified from an ordinary vegetative branch. For example, in *Hymenan-thera alpina*, a shrub of dry mountainsides, the thorn is scarcely more than an arrested shoot with a blunt tip. Frequent reversion to vegetative growth occurs, but acropteal extension of cork cambial activity ultimately seals off the tip and prevents further development (Arnold, 1959). In *Gleditsia*, on the other hand, Blaser (1956) has shown an interesting transition in the organization of the shoot meristem as it changes from the potentially indeterminate vegetative condition to the determinate state characteristic of thorns. The branched thorns arise ordinarily as axillary buds and initially possess a dome-shaped apex like that of any vegetative shoot. As the transition occurs, there appears to be an acceleration of mitotic activity throughout the meristem and the apex becomes narrower and more elongated. The leaves produced are scale leaves, and two or more of them subtend lateral shoots that also become thorns. As the transformation of the apex begins, leaf production ceases. The acceleration of apical mitotic activity is soon followed by total cessation of divisions, and elongation and differentiation of cells convert the former apex into a narrow, pointed thorn. Much of this differentiation produces sclerenchyma, causing the thorn to be extremely hard and resistant. An interesting feature of the maturation of the thorn is that tissues remain meristematic in the subapical region to a late stage of development and lead to considerable elongation of the thorn after the tip has become mature.

In *Ulex europaeus* (gorse), studied by Bieniek and Millington (1967), almost all lateral shoots are transformed into thorns (Fig. 10.1). The axillary meristem produces only a few scale leaves before undergoing extremely rapid elongation, much as in *Gleditsia*. The terminal cells of the attenuating tip elongate and differentiate relatively early and maturation then proceeds basipetally. Sclerification is extensive, especially in the terminal portion of the shoot.

There is an interesting study of thorn development that emphasizes its similarity to flower development. In *Carissa grandiflora* Cohen and Arzee (1980) have shown that thorns and flowers are parts of the same lateral branch system and, in fact, appear to be alternative expressions of determinate growth.

REPRODUCTIVE MORPHOLOGY

Somewhat more complex than thorns are the determinate shoots that are involved in reproduction in various groups of vascular plants. Reproduction in the vascular plants involves the production of spores that subsequently develop as the sexual, haploid phase of the life cycle, the gametophyte. The production of spores is accomplished in diverse manners in various groups of vascular plants with varying degrees of modification of the shoot. Such modifications are perhaps best known in the angiosperms, where they constitute the flowering process and often occur in response to specific environmental conditions, such as the photoperiod. Very much less is known about the developmental basis of comparable phenomena in the gymnosperms and the vascular cryptogams, although on phylogenetic grounds one might expect to find the underlying physiological mechanisms less complicated by accessory phenomena.

The varying extent to which the shoot system participates in reproduction in different groups of vascular plants is revealed by consideration of some examples. In the ferns the shoot apex is not directly involved in reproduction. Only the leaves are involved, and in many cases a spore-bearing leaf is not significantly different from a vegetative frond. In the horsetails and most clubmosses – members of Sphenopsida and Lycopsida, respectively – terminal cones are produced that bring vegetative growth to a close in the shoots on which they occur. In the gymnosperms cone production likewise involves the conversion of an entire shoot to a determinate reproductive structure. In the flowering plants a variety of relationships exists between the individual flower and the shoot system. A single flower may arise through modification of an individual vegetative shoot apex, as in tulip. More often a vegetative shoot is transformed into a flowering shoot or inflorescence that contains several or many individual flowers. In such cases the flowers often arise directly from the inflorescence meristem without a preceding vegetative phase. Reproductive development of the angiosperms, like that of other vascular plants, is generally characterized by determinate growth of the meristems involved, but there are well-known cases, both normal and abnormal, in which there is a return to indeterminate vegetative growth.

DEVELOPMENT OF THE FLORAL APEX

Where a single flower is produced it is possible to follow the transformation of a vegetative shoot apex into a reproductive axis in a relatively uncomplicated situation. Several investigators, beginning with Satina and Blakeslee (1941), have described the changes that take place when a vegetative shoot apex of *Datura stramonium* is transformed into a flower (Fig. 10.2). The vegetative apex has the form of a low dome. Internally two tunica layers may be distinguished from the underlying corpus and there is an indistinct zonation. However, in terms of mitotic activity there is a zonation with a central zone in which cell cycle duration is approximately twice that in the surrounding peripheral zone (see Table 5.1). With the onset of flowering, the apex becomes enlarged and more highly mounded. This is associated with an increase in mitotic frequency throughout the apex but particularly in the central region, where previously it had been lowest. While this enlargement is occurring, five sepal primordia are initiated sequentially around the apex. The apex then flattens and broadens and five petals arise simultaneously in positions alternating with those of the sepals. At about the same time that

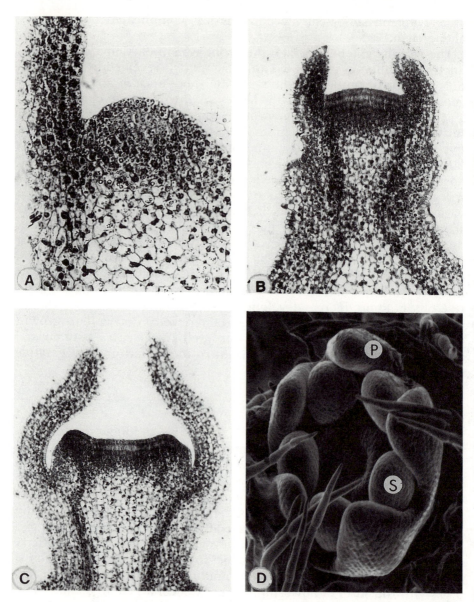

Figure 10.2. *Flower development in* Datura. *(A) Vegetative shoot apex. (B) Floral apex at the stage of sepal initiation. (C) Sepals well developed, petals beginning to form. (D) Scanning electron microscope view of floral apex from which sepal primordia have been removed. Petal and stamen primordia are well developed and two carpel primordia are beginning to form in the center. Key:* **P**, *petal,* **S**, *stamen. (A) ×335, (B,C) ×195, (D) ×100.*

the petals appear, five stamens are initiated, also simultaneously, alternating with the petals. Finally, two crescentic carpel primordia are initiated and the remaining distal portion of the apex mounds up to form the internal tissues of the ovary. As a result of this rapid initiation of organs, without a corresponding growth of the apex, the floral meristem is, in effect, used up and the axis has become determinate.

The sepal and petal primordia are dorsiventral in symmetry and in general resemble leaf primordia in the early stages of development. However, in later stages both the sepals and the petals become joined to form calyx and corolla tubes, respectively. The stamen and carpel primordia, on the other hand, show very little resemblence to leaf primordia. The two carpel primordia are joined to surround the internal tissues of the ovary, which bear the ovules. Their tops grow upward to form the style and stigma.

It is evident from the preceding description, that the meristem that gives rise to a flower represents a continuation with modifications of a vegetative shoot meristem. This raises interesting questions regarding the nature of the lateral organs it produces, questions related to issues raised in the previous chapter. Is there a basic floral organ primordium of which the sepal, petal, stamen, and carpel represent different expressions, or is each of these a distinct organ type? Further, since the floral meristem is a transformed vegetative shoot meristem, are these organs a further elaboration of the diversity of leaf types described earlier? These questions should be kept in mind as the process of floral development is analyzed.

There are many cases in which the shift to reproductive function by a vegetative shoot results in the formation, not of a single flower, but of a floral shoot, or inflorescence, that includes several to many individual flowers. The transformation of a vegetative shoot apex to an inflorescence meristem may be illustrated by examining the development of the head, or capitulum, of *Helianthus annuus,* a compact inflorescence that superficially resembles a flower but in fact contains several hundred individual florets. This has recently been described by Marc and Palmer (1982). Here, as in the development of an individual flower, the apex enlarges substantially, becoming more dome-shaped. Mitotic activity is accelerated, particularly in the previously quiescent central zone, and the distinct cytological zonation of the vegetative apex, described in Chapter 5, is lost. Organ primordia are initiated in rapid succession from the periphery to the center of the apex, but they are not the incipient stages of floral organs. Rather, they develop into reduced leaves, or bracts, in the axils of which the florets are initiated without a preceding vegetative stage (Fig. 10.3).

The examination of the formation of reproductive apices in several

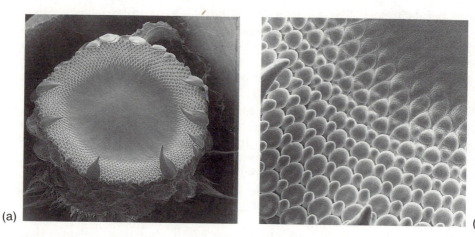

(a)   (b

Figure 10.3. *(a) Inflorescence apex of* Helianthus annuus *showing the process of floret initiation proceeding toward the center. In (b), at higher magnification, it may be seen that as each mound emerges it differentiates into a bract with a floret primordium in its axil. (a) ×10, (b) ×50. (Photographs by D. Jegla.)*

species by transformation of vegetative meristems has revealed the rather surprising fact that, as far as the nature of the transformation is concerned, it makes little difference whether the resulting structure is to be a single flower or an inflorescence. Wetmore, Gifford, and Green (1959) have emphasized this point in a comparative study in which the essential features of the transformation were the same in the compact head of *Xanthium pensylvanicum*, the branching inflorescence of *Chenopodium album*, and the single flower of *Papaver somniferum*. Although the changes associated with the onset of flowering are ordinarily regarded as representing reproductive development, it seems reasonable to ask whether, in fact, the fundamental change is not associated with the onset of determinate growth. In this connection, the similarity between the changes of developing flowers and inflorescences and those that occur in thorns, determinate shoots having nothing to do with reproduction, is of considerable interest. In all these cases there appears to be an initial acceleration of mitotic activity that heralds the end of meristematic development in the apex. Observations like this do not give support to the méristème d'attente hypothesis of apical organization discussed in Chapter 5, in which the central region is considered to be a meristem specifically destined to produce the reproductive parts of the flower.

The development of the flower of *Datura* has illustrated the major events in the transformation of a vegetative shoot apex to a reproductive one. In this case the axis undergoes very little elongation; there is a single cycle of each type of floral organ, and organ initiation progresses acropetally. There are, however, many examples of floral meristems that differ to a greater or lesser extent from the vegetative meristem than does the flower of *Datura*. A few of these will be considered to provide a basis for a better understanding of the scope of this transformation.

The development of *Michelia fuscata*, which was described by Tucker (1960), is of interest because of the prolonged activity of the floral meristem (4 to 5 months), the great ultimate length of the floral axis, and the large number of appendages (80 to 90) produced in this primitive flower. In the extended growth of the floral meristem, although the overall pattern of change in the apex is similar to that described previously, there are periodic fluctuations in size comparable to those occurring in a vegetative apex, so that a greater similarity between vegetative and reproductive development is revealed. Plastochronic changes in size are characteristic of vegetative apices, where they are associated with leaf initiation but are not ordinarily found in floral apices. However, in *Michelia* there are periodic intervals between the initiation of groups of primordia during which conspicuous increases in size of the apex occur. Ultimately, however, after carpel initiation is complete, only a small plate of meristem is left that differentiates as parenchyma.

An acropetal sequence of leaf initiation is a primary characteristic of the meristems of indeterminate vegetative shoots. The determinate floral meristem, although it may adhere to this pattern, in many instances does not. For example, Soetiarto and Ball (1969a), in describing floral development in *Portulaca grandiflora*, have reported a general acropetal succession of organ formation until the carpels have been initiated. Then, however, two additional cycles of stamens are initiated basipetally to the original cycle. This seemingly anomalous pattern of development is representative of nonacropetal sequences, which are in fact relatively common in floral development. Sattler's atlas of floral organogenesis (1973) includes many illustrations of basipetal initiation of one or more of the types of floral organs. Phenomena like this emphasize the extent to which the floral apex may depart from the patterns of regulation characteristic of the vegetative apex.

In considering the departure from the patterns of vegetative development shown by the floral apex, it must be recognized that there may be more than one type of departure on the same plant, that is, more than one pattern of floral development. An example of this is the phenomenon in which one type of flower undergoes typical anthesis while the other is cleistogamous, that is, remains closed and is obligately self-pollinated. Recognized as a case of alternative developmental pathways within the same genetic framework, cleistogamy has become the object

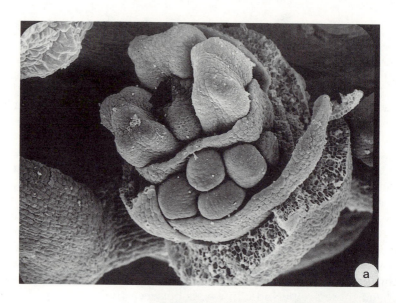

Figure 10.4. *Development of unisexual flowers in Zea mays. (a) Spikelet with two staminate florets. In the upper one, three anthers are well developed and the central gynoecium (arrow) is degenerating. The lower floret is less advanced. (b) Spikelet with two pistillate florets, the upper more developed than the lower, which will ultimately abort. Arrows indicate stamens in both florets. (a) ×195, (b) ×225. (P. C. Cheng, R. I. Greyson, and D. B. Walden,* Am. J. Botany *70:450, 1983.)*

of renewed morphogenetic interest. In *Collomia grandiflora* Minter and Lord (1983) found that both open and cleistogamous flowers develop in the same inflorescence. The flowers are indistinguishable until all organs have been initiated, and they do not begin to diverge until the precursors of the pollen mother cells can be detected in the anthers. At that time the growth rate of the anthers of cleistogamous flowers decreases, while that in open flowers remains high. Slightly later, the growth rate of open flower petals increases, a phenomenon associated with anthesis, while that of cleistogamous flower petals remains essentially unchanged. In other cases that have been investigated the divergence of cleistogamous and open flowers occurs at an earlier stage of development but the general pattern of quantitative differences is similar.

Another common type of flower dimorphism is that which results in the development of unisexual flowers. In these there appears to be considerable variation in the stage at which the divergence of developmental pathways occurs. In some, such as *Zea mays*, both stamen and pistil primordia are initiated, but at a later stage one set or the other is arrested (Fig. 10.4). In other cases, for example *Cannabis sativa* (hemp), where the two types of flowers are on separate individuals, the flowers appear to be committed from the outset and only one type of reproductive organ is initiated.

ANALYSIS OF FLORAL TRANSFORMATION

The examples just considered indicate that in the process of flowering a vegetative shoot apex undergoes a major alteration in its developmental program. The question arises as to whether this alteration is the consequence of a process of determination comparable to that which fixes the pattern of leaf development. It is, therefore, appropriate to analyze floral transformation with this question in mind.

Histological descriptions of floral meristems suggest in many cases that a major reorganization has occurred, but it is not clear to what extent histogenesis and organogenesis have been altered. Fortunately there is some evidence derived from clonal analysis that bears on this question. In several plants that have been investigated, the derivation of tissues in the floral axis and the lateral organs indicated that the apical layers that functioned in vegetative development continued in the transformed apex (Stewart, 1978). It has been possible, also, to trace the contributions of the various apical layers to the development of the floral organs. In the development of sepal primordia it has been found that the apical layers contribute in approximately the proportions characteristic of foliage leaves. The remaining floral organs, however, do not follow the leaf pattern in this respect. Nevertheless, these observations indicate that there must be a basic similarity of function between vegetative and floral meristems.

In order to analyze floral transformation, it is important to identify the earliest events in this process as well as those that follow from them. Some of the most careful analyses of the changes associated with the onset of reproductive growth are those that have been carried out on plants in which there is a precise photoperiodic control of flowering. In such cases it is possible to follow the time course of floral initiation and development with considerable accuracy. This is a useful experimental device as long as the experimenter remembers that most angiosperms do not require such an environmental stimulus to flower.

There is a voluminous and ever-growing literature documenting cytological and biochemical events that follow photoperiodic floral induction in a wide variety of sensitive species. The current status of this research has recently been evaluated by Bernier, Kinet, and Sachs (1981) and by King (1983). It is apparent that workers in this field have not found it necessary to make a distinction between the responses that occur in long-day and short-day plants or between those that occur in the initiation of single flowers and inflorescences.

Perhaps the most generally recognized feature of the vegetative to floral transformation is, as has been pointed out, an increase of mitotic frequency. This is accomplished by a general reduction in the duration of the cell cycle, which is particularly significant in the central zone. In view of the acceleration of mitotic activity, it is not surprising that labeling experiments reveal increased synthesis of DNA, again particularly noticeable in the central zone (Fig. 10.5). It is interesting to note, however, that in *Sinapis alba*, which can be induced to flower by a single long-day photocycle, the rise in mitotic activity precedes the increase in DNA synthesis by some hours (Bernier, 1969). This indicates the presence in the meristem of cells that are in the stage of the cell cycle immediately following DNA synthesis and are thus ready to divide without further DNA synthesis. The effect of this early division is to bring many cells of the meristem into the same phase of the cell cycle. This possibly facilitates a synchronized response to further inductive stimuli. In any event, it is distinct in function from the overall increase in mitotic frequency associated with growth and organogenesis of the flowering apex. There are also a few reports of excess DNA synthesis in floral meristems, and it has been suggested that this may represent amplification of repetitive sequences and DNA coding for ribosomal RNA. Associated with increased DNA synthesis is a corresponding increase in synthesis of RNA, revealed by autoradiography, which results in an increase in the content of RNA, particularly in the cytoplasm, as shown by specific staining techniques. Furthermore, in contrast to its localized

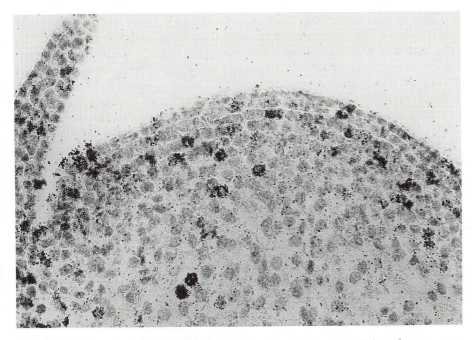

Figure 10.5. *An early stage of inflorescence development in* Helianthus annuus *(sunflower). The shoot tip has been labeled with tritiated thymidine and autoradiographed. Compare with Fig. 5.1, the corresponding treatment of a vegetative apex, and note the uniform distribution of labeled nuclei in the central and the peripheral regions of the inflorescence meristem. ×390. (T. A. Steeves, M. A. Hicks, J. M. Naylor, and P. Rennie,* Can. J. Botany 47:1367, 1969.)

distribution in the vegetative apex, RNA is rather uniformly distributed throughout the reproductive apex. Shortly after the increase in RNA there is a rise in the total protein content of the cells of the meristem, which may be a further biochemical step following from the increase in RNA. In several species it has been found that, apart from the general increase, there is a very early and transitory increase in both RNA and protein synthesis at about the time that the floral stimulus is believed to reach the meristem. This early synthesis is believed to be essential for floral evocation because, if it is blocked by inhibitors, floral evocation does not occur. This discovery, if it proves to be general, may lead to important insights into the determination process.

During the floral transformation, striking changes in cellular characteristics at the ultrastructural level have been reported. Very early there is a significant increase in the number of mitochondria and plastids throughout the meristem. At the same time there is a marked decrease in the size and number of vacuoles. Following these early extranuclear events, there is an increase in the proportion of euchromatin to heterochromatin in the nuclei and a considerable enlargement and reorganization of the nucleoli. Both of these features have been associated with increased transcriptional activity.

The observations reported here indicate that when a vegetative apex is transformed to the reproductive state, there are rather substantial structural and physiological changes in the meristem. Many of the changes are suggestive of kinds of modifications that could be brought about by the sequential expression of genes. Future work, possibly using floral mutants, may be able to identify genes that act at specific times in the flowering process. In the meantime, it is useful to consider these changes as a model for what may occur in the determination of organs in the vegetative shoot. In this connection attention may be turned to some other features of the floral meristem that can best be revealed by experimental study.

EXPERIMENTAL ANALYSIS OF FLORAL MERISTEMS

The reproductive apex, with its predictable sequence of appendages, contrasts sharply with the repetitive pattern of growth exhibited by the vegetative apex. Several workers have applied the same sort of experimental methods used in the study of the vegetative apex to elucidation of the reproductive apex. Not surprisingly the results have often differed. No attempt will be made here to explore the vast literature dealing with the production, translocation, and action of the flowering stimulus. Rather, attention will be focused on the experiments that deal with the reproductive apex as a developing system and seek to understand its organization and the interactions among its parts.

It is evident that in the initiation of reproductive development, a vegetative shoot apex responds dramatically to a stimulus. Although the response in this case involves an entire shoot rather than a single organ, it may be reasonable to ask whether the induction of a flowering apex is not comparable to the determination of a leaf primordium. In both cases meristematic cells capable of producing a vegetative shoot are altered in their development to the extent that they produce an entirely different structure. In both cases the altered meristem proceeds through a series of well-defined stages to complete maturation. Experiments discussed in Chapters 8 and 9 have shown that early in its development a leaf primordium is determined and subsequent to this is relatively autonomous in completing its developmental destiny. If plants of *Chenopodium album* are subjected to two inductive short-day cycles, they will flower even if they are returned to long days for the remainder of their development. This would suggest that some sort of determination occurs, but it is necessary to distinguish clearly between the induction of a plant to flower and the actual determination of one or more of its meristems. The necessity for such a distinction is particularly clear in plants that continue to grow vegetatively even while they are flowering. Although the plant as a whole is induced to flower, only some of the apices are determined as floral meristems; the remainder continue development as vegetative meristems.

In *Chenopodium*, Wetmore, Gifford, and Green (1959) have specifically tested the determination of the individual apex by excising it from the plant six days after the two inductive cycles were given and growing it in isolation in sterile culture. Under these conditions excised apices continued to develop and produced small, but seemingly normal, inflorescences. Thus there seems to be a stage of development in the reproductive apex, as in the leaf, beyond which no further external stimulus or specific control is required.

Comparable results have been reported for several other species, including tobacco (Hicks and Sussex, 1970), in which individual flower meristems were able to complete their development in vitro when cultured on a suitable medium. The common feature of all of these experiments is that the meristem seems to be determined prior to the initiation of any floral organs.

Some observations on intact plants also indicate the existence of a determination process. In *Chrysanthemum*, Schwabe (1959) has described experiments in which inflorescence development in this short-day plant is arrested by transfer to long days, by reducing the light intensity, or by applying auxin. The significant point regarding determination was that if a particular apex had passed a critical stage, indicated in this case by the development of the carpels in the marginal flowers of the inflorescence, it continued its development regardless of the fate of other apices of the same plant. The apices that had not reached the critical stage at the time of removal from inductive conditions did not revert to the vegetative condition but expanded into enlarged, receptacle-like, determinate shoot tips and were arrested in their development. If the arrested apices remained viable, subsequent return to inducing conditions brought about renewed activity and the completion of inflorescence development. Thus it is possible that determination is a two-step process, with determination of the receptacle preceding the stage in which flower initiation is induced. This system would be an extremely favorable one for the application of excision and culture techniques, such as have been employed for *Chenopodium*, for in this way the developmental capacity of the meristem at successive stages could be assessed accurately.

In contrast to these observations that suggest that determination manifests a high degree of permanence in a meristem, mention should be made of other cases in which there appears to be natural or experimentally induced reversion to the vegetative state in meristems that had entered upon reproductive development. For example, King and Evans (1969) reported that in plants of *Pharbitis nil* that had been induced to flower but were subsequently exposed to abnormally high temperatures, floral buds that had produced sepal, petal, and stamen primordia reverted to the vegetative condition. Meristem reversion has been rather frequently reported for plants bearing apices in early transition stages. A case of reversion of a well-developed floral meristem was reported in *Impatiens balsamina* by Krishnamoorthy and Nanda (1968). When plants were subjected to four or more inductive short days to induce flowering and then returned to long days, all of the meristems could revert to the vegetative condition even after the formation of several or all the whorls of floral organs was completed (Fig. 10.6). If there is a developmental stage in the meristem of this plant that could be called determination, it appears that it is very labile. In view of this result considerable interest is attached to the surgical experiments of Wardlaw (1963) on inflorescence meristems of *Petasites hybridus*. Working with a series of apices in the transition from the vegetative to the flowering condition, Wardlaw variously punctured them at the center, bisected them, or isolated the meristem from adjacent subapical tissues by vertical incisions. When these operations were carried out on relatively advanced stages, inflorescence development continued from portions of the original apical surface, indicating a determination in the meristem. When the experiments were performed on early transition stages, however, there was a more or less complete return to the vegetative condition. Thus, determination of the reproductive meristem becomes progressively more fixed as development proceeds. An interesting aspect of these experiments was that direct application of solutions of gibberellic acid to the regenerating apex significantly favored the retention and further elaboration of reproductive development.

There is also evidence in some cases that flowering involves an alteration in the plant that is not restricted to the transformed meristems. In certain nonphotoperiodic varieties of tobacco, stem segments (Aghion-Prat, 1965) or even strips of tissue consisting of the epidermis and several underlying cell layers (Tran Thanh Van, 1973) excised from the stems of plants that were in flower showed a remarkable retention of a tendency to form flowers when they regenerated in culture (Fig. 10.7). Explants were removed along the length of the shoot from apex to base. Explants from the base produced only vegetative buds and there was a progressive increase in the proportion of flower buds acropetally, reach-

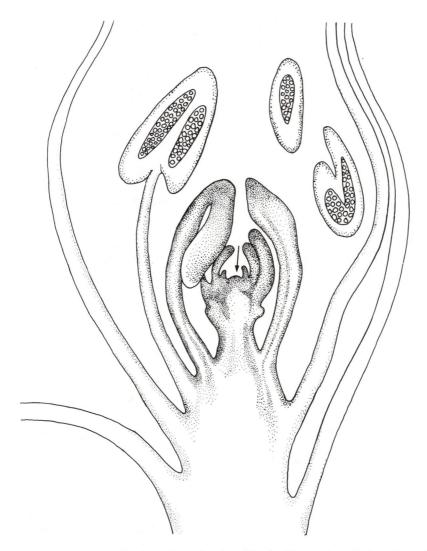

Figure 10.6. *Longitudinal section of a floral bud of* Impatiens balsamina *that has reverted to vegetative development. The vegetative shoot apex (arrow) may be seen inside the ovary at the tip of the placenta. Stamens containing mature pollen grains had been formed. (Adapted from Krishnamoorthy and Nanda, 1968.)*

ing 100 percent in explants taken near the top of the plant. In experiments of this sort, the stem segments gave rise to callus from which the vegetative and floral buds arose, whereas the tissue slices gave rise directly to buds. In a rather startling extension of these experiments, Kon-

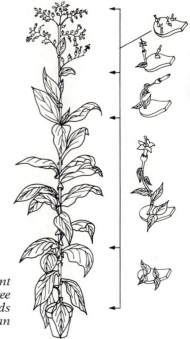

Figure 10.7. *Diagram showing levels on a plant of* Nicotiana tabacum *from which explants of three to six cell layers are excised and the kinds of buds formed by these explants in sterile culture. (Tran Thanh Van, 1973.)*

stantinova et al. (1969) demonstrated that callus derived from flower-producing stem segments continued to give rise to flower buds and that even in the third subculture 40 percent of the regenerated buds were floral. These experiments show that the flowering process must involve persistent changes at the cellular level that are not necessarily confined to the floral meristem.

Up to this point in the consideration of experimental investigations, attention has been focused upon the early events in the transformation of a vegetative to a floral meristem. It is also important to give some attention to the further development of the transformed meristem and particularly the initiation and development of the different kinds of floral organs. The reproductive apex appears to pass through a series of relatively distinct morphological stages during which lateral organs of different types are produced. Attempts to explore the basis of this succession have included surgical experiments, particularly the bisection of floral apices at various stages in their development. Work of this sort was begun in 1956 by Cusick on flowers of *Primula bulleyana*, with particular attention paid to the nature of the organs produced by the regenerating half-apices. More recently similar studies have been carried out on three additional species, with results that confirm Cusick's findings with only minor differences of detail.

Experiments of this type on tobacco are of particular interest because the operated apices were grown in sterile culture and thus were isolated from outside influences (Fig. 10.8) (Hicks and Sussex, 1971). Following the bisection, organ formation continued normally on the unoperated side of each half-apex while the operated side underwent a certain amount of regeneration. Particular interest is attached to the organs that formed on the regenerating side. If the operation was performed at the presepal stage, that is before any floral organs had been initiated, in some cases each half-apex regenerated a complete flower with all appendages present; in others no sepals were formed on the regenerating side of the apex. If the bisection was performed after the sepal primordia had been initiated, petals, stamens, and carpels, but no sepals, were formed in the regenerating portion of the apex. Operation at the petal stage resulted in the absence of all organs except carpels in the regenerating region. At the stamen stage new carpels were sometimes formed and sometimes not, and at the carpel stage the result was two half-flowers with no regeneration. These experiments emphasize the progressive restriction of developmental potentiality of the floral meristem as it advances through the sequence of stages, a restriction associated with the determinate nature of the flower and contrasting markedly with the repetitive development of the indeterminate vegetative shoot. In view of the importance of this concept of floral development, it should be noted that these results are consistent, except for differences in detail, with findings for flowers of *Portulaca grandiflora* (Soetiarto and Ball, 1969b) and of *Aquilegia* (Jensen, 1971).

As the floral apex proceeds through the successive stages of its development it gives rise to a succession of different organ types. It seems reasonable to expect that these organs, as they are formed, in turn might participate in the regulation of the developmental sequence. There is some evidence suggesting that this may be the case. McHughen (1980), again with floral meristems of tobacco grown in sterile culture, was able to show that the excision or suppression of a particular set of floral organs interfered with the initiation of successive sets. For example, the removal of the presumptive sepal-forming areas from an early-stage floral meristem completely suppressed the formation of petals, prevention of petal formation suppressed the initiation of stamens, and carpel initiation depended upon the presence of stamen primordia. Experiments of this sort are in agreement with a theory of floral morphogenesis proposed by Heslop-Harrison (1964) that invovles specific interactions between the organs of successive whorls (Fig. 10.9). He supposes that substances released by the first-formed whorl activate genes that lead to

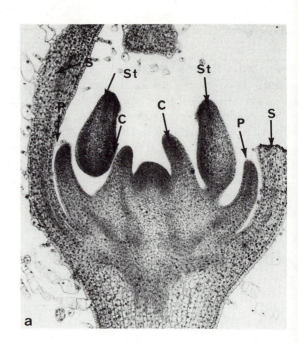

Figure 10.8. (Caption on facing page).

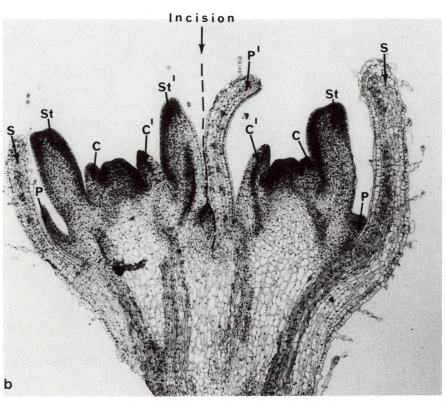

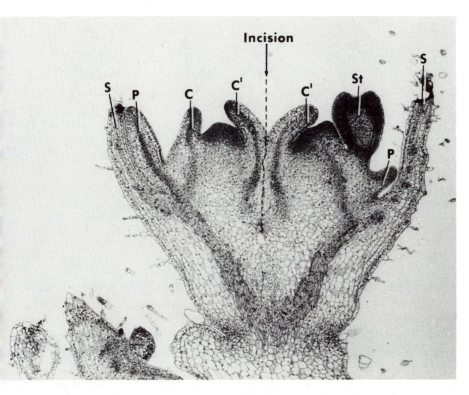

Figure 10.8. *Surgical bisection of the floral meristem of* Nicotiana tabacum. *(a) Longitudinal section through an intact apex after all the floral organs have been initiated. (b–c) Development of floral meristems that had been vertically bisected, then excised from the plant and grown in sterile culture. (b) Meristem bisected before sepal primordia were initiated. On the outer side of each flower organ formation has proceeded normally. On the inner face of each flower, where organ regeneration has occurred, petals, stamens, and carpels have been initiated in their correct positional relationships, but no sepals have been formed. (c) Meristem bisected after sepals and petals had been initiated. The ability to regenerate stamens and petals on the inner face of each flower has been lost, but carpels are still regenerated. Key: C, carpel; P, petal; S, sepal; St, stamen. The preceding are the normal organs that formed on the outer side of the flower. C′, regenerated carpel; P′, regenerated petal; St′, regenerated stamen. ×65. (Hicks and Sussex, 1971.).*

the development of the next type of organ, and so on. Though there is no information on the nature of substances that might be involved in such interactions, this approach provides a useful background against which to view experiments in which floral development has been modified.

It also appears that there are developmental interactions among the

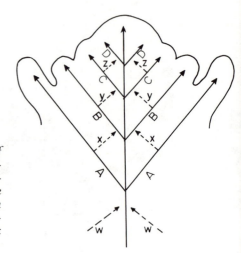

Figure 10.9. *A model for the control of the sequential events in floral development. A stimulus W initiates the transition to flowering and activates gene complex A, which controls the initiation of sepals and the production of an inducer X, which activates gene complex B, and so on. (Heslup-Harrison, 1964.)*

floral organs subsequent to their initiation. McHughen (1982) found that the growth of the pistil (composed of two carpels) in tobacco was enhanced by the presence of stamens but retarded by the sepals in experiments in which flower buds were cultured with various components removed. Antagonistic interactions among various appendages have also been reported by Tepfer and his associates (1963) in cultured floral buds of *Aquilegia formosa*. The sepals were found to exert an inhibitory influence on the development of all other floral parts and, in fact, had to be removed to obtain satisfactory growth of the remainder of the flower. In this connection it may be noted that the different floral organs in *Aquilegia* responded distinctively to hormones incorporated in the medium. It was found that gibberellic acid stimulated the growth of all floral organs except the stamens, and that the carpels were especially sensitive to IAA, too low or too high a concentration resulting in failure of growth. These distinctive hormonal responses are also of interest in comparison to the reports of progressive restriction of organ-forming capacity in the developing floral meristem, suggesting that the altered capacities may result from a changing hormonal status in the meristem. The possible significance of the hormonal status of the floral meristem in the orderly sequence of development is also suggested by the often drastic modifications of floral morphology induced by treatment with growth-active substances (Vieth, 1965).

It is also possible that interactions between organs may be involved in the development of unisexual flowers. In *Cucumis sativus* (cucumber) there are three types of flowers produced, male, female, and herma-

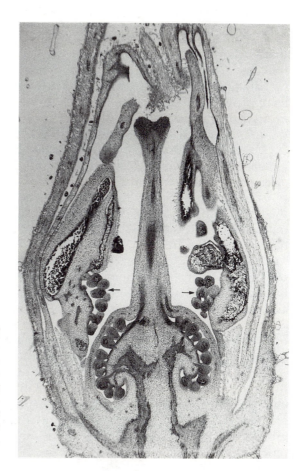

Figure 10.10. *Longitudinal section of a flower of the stamenless-2 mutant of* Lycopersicon esculentum *showing abnormal stamens bearing external ovules (arrows). Normal ovules may be seen in the ovary. ×35. (V. K. Sawhney and R. I. Greyson,* Am. J. Botany *60:514, 1973.)*

phroditic, depending upon the genetic constitution of the plant and the growth conditions. They are indistinguishable in their early development, and if they are excised from the plant at this time and grown in sterile culture, it is possible to convert potentially male flowers into female flowers by including IAA in the culture medium (Galun, Jung, and Lang, 1963). Gibberellic acid was found to nullify the auxin effect. In no case did flowers in culture become hermaphroditic, suggesting that there is an antagonism between the development of stamens and of carpels, and no treatment was found that would convert potentially female flowers into male flowers.

As with leaf formation by the vegetative shoot apex, the question of organ determination in the floral meristem has prompted some experimental investigation. It is of interest to know at what stage in their development primordia of floral organs are determined and whether there might be an initial phase during which they reveal latent vegetative characteristics. Experiments in which petal, stamen, or carpel primordia of tobacco were excised and cultured soon after their emergence have provided a provisional answer to these questions. In all cases the excised primordia gave rise to small but relatively normal organs, although, of course, the characteristic postinitiation fusion did not occur. Determination at an early stage as specific organ rudiments thus appears to be a characteristic of the floral apex, but the limited evidence presently available is hardly adequate to permit generalization.

An interesting aspect of the experiments just described was that, in the case of the stamens, determination did not lead to developmental inflexibility (Hicks, 1979). In these there was a development of stigma-like outgrowths from the stamens indicating an expression of partial carpelloid characteristics. Further indication of the developmental flexibility of stamens is provided by an examination of certain male-sterile lines of tomato. In the stamenless-2 mutant of tomato normal stamens are replaced by abnormal structures that bear naked external ovules or are carpelloid in form (Fig. 10.10). In the early stages of development, up to a height of about 100 μm, the primordia of normal and mutant stamens are not distinguishable structurally. Evidence that the divergence of developmental program occurs at about this time is provided by experiments in which the mutant plants were treated with gibberellic acid. In response to this treatment, primordia of less than the critical length gave rise to entirely normal stamens. As the application of gibberellic acid was delayed, the resulting stamens were progressively more abnormal (Sawhney and Greyson, 1979).

GENERAL COMMENT

In the development of thorns and of reproductive apices, the characteristic zonation of the vegetative apex disappears. This is accomplished particularly by changes in the central zone, including an acceleration of mitotic activity. So striking are the changes in the central zone that one group of investigators regarded this region as a meristem that specifically initiates the reproductive portions of the flower. Careful time-course studies, however, have shown that the whole apex is involved in the transformation to the reproductive state, and the similarity of the changes in apices being converted to thorns and to reproductive meristems argues against the former interpretation. An important consequence of this conclusion is that it demonstrates unequivocally that, unlike the situation in animals in which the gametes arise from a germ line segregated early in development, the reproductive cells of seed plants have their origin in a population of cells that have been involved in the construction of the vegetative body. Certain features of the vegetative meristem persist in the flowering meristem. In particular it appears that the meristem layers of the vegetative shoot apex continue to function after the transformation, and the initiation of lateral appendages in predictable positions continues.

It has already been pointed out that reorganization of the meristem in flowering seems related more to the onset of determinate growth than to reproductive activity specifically, and this then points to the probable significance of the central zone as a mechanism that serves to maintain continued meristematic activity of the vegetative apex. Thus, the shoot apex is provided with a reservoir of centrally located cells from which its differentiating regions are continually supplied and that remain meristematic by virtue of the fact that they do not divide frequently. Conditions that lead to increased mitotic activity of these cells are associated with determination of the meristem, which then may proceed through an elaborate succession of distinctive stages but ultimately becomes fully differentiated and nonmeristematic.

These considerations call attention to the most important difference between vegetative and flowering meristems. The indeterminate vegetative meristem is repetitive in its activity and seemingly does not change over protracted periods of development. By contrast the floral meristem appears to pass rapidly through a programmed sequence of changes culminating in final complete differentiation. In this respect it resembles the lateral determinate appendages of the vegetative shoot and the term *determination* has been applied to the process that commits the flowering apex, as well as to that which commits the leaf primordium. In both cases a program is activated that carries the determinate structure through to a differentiated final form. On the other hand, the development of the floral meristem has some similiarity to that of the embryo in that a patterned sequence of events may be observed, with the exception, of course, that the embryo is not determinate because of the formation of the apical meristems (see Chapter 2). Since there is good evidence that sequential development results from sequential gene activation in the embryo, it is probable that a similar mechanism operates in the floral apex.

At this point another question interjects itself into the discussion. What is the nature of the process that establishes the developmental patterns of the floral appendages if the meristem that produces them is itself determined? Are these organs determined from the outset only to be floral organs because they arise from a floral meristem, or might they, if excised early enough, produce something resembling a vegetative leaf? If determined to be floral, are they initially committed to be one or the other of the organ types? The very limited information derived from experiments suggests the floral organ primordia are determined as floral very early in development and, in general, as specific organ types. However, with respect to this latter point, some flexibility of expression among organ types seems to be possible.

Some students of floral development have found it useful to account for this flexibility by introducing the concept of "canalisation." According to this interpretation, an outgrowth of the meristem is determined as a floral organ at or soon after its inception. It then proceeds along one of several possible developmental pathways to become, for example, a stamen or a carpel. This latter process, in which the specific fate of the organ is fixed, is called canalisation (Sawhney and Greyson, 1979). This concept has not been applied generally in plants except in relation to floral development.

REFERENCES

Aghion-Prat, D. 1965. Néoformation de fleurs *in vitro* chez *Nicotiana tabacum* L. *Physiol. Végét.* 3:229–303.

Arnold, B. C. 1959. The structure of spines of *Hymenanthera alpina. Phytomorphology* 9:367–71.

Bernier, G. 1969. *Sinapis alba* L. In *The Induction of Flowering. Some Case Histories*, ed. L. T. Evans, 305–27. Melbourne: Macmillan.

Bernier, G., J.-M. Kinet, and R. M. Sachs. 1981. *The Physiology of Flowering. Transition to Reproductive Growth*, vol. 2. Boca Raton, Fla.: CRC Press.

Bieniek, M. E., and W. F. Millington. 1967. Differentiation of lateral shoots as thorns in *Ulex europaeus. Am. J. Botany* 54:61–70.

Blaser, H. W. 1956. Morphology of the determinate thorn-shoots of *Gleditsia. Am J. Botany* 43:22–8.

Cohen, L., and T. Arzee. 1980. Two-fold pathways of apical determination in the thorn system of *Carissa grandiflora. Bot. Gaz.* 141:258–63.

Cusick, F. 1956. Studies of floral morphogenesis. I. Median bisections of flower primordia in *Primula bulleyana* Forrest. *Trans. Roy. Soc. Edinburgh* 63:153–66.

Galun, E., Y. Jung, and A. Lang. 1963. Morphogenesis of floral buds of cucumber cultured *in vitro. Devel. Biology* 6:370–87.

Helsop-Harrison, J. 1964. Sex expression in flowering plants. *Brookhaven Symp. Biol.* 16:109–25.

Hicks, G. S. 1979. Feminized outgrowths of the stamen primordia of tobacco *in vitro. Plant Sci. Lett.* 17:81–9.

Hicks, G. S., and I. M. Sussex. 1970. Development *in vitro* of excised flower primordia of *Nicotiana tabacum. Can. J. Botany* 48:133–9.

——— 1971. Organ regeneration in sterile culture after median bisection of the flower primordia of *Nicotiana tabacum. Bot. Gaz.* 132:350–63.

Jensen, L. C. W. 1971. Experimental bisection of *Aquilegia* floral buds cultured *in vitro*. I. The effect on growth, primordia initiation and apical regeneration. *Can. J. Botany* 49:487–93.

King, R. W. 1983. The shoot apex in transition: Flowers and other organs. In *The Growth and Functioning of Leaves*, ed. J. E. Dale and F. L. Milthorpe, 109–44. Cambridge: Cambridge University Press.

King, R. W., and L. T. Evans. 1969. Timing of evocation and development of flowers in *Pharbitis nil*. *Austral. J. Biol. Sci.* 22:559–72.

Konstantinova, T. N., N. P. Aksenova, T. V. Bavrina, and M. K. Chailakhyan. 1969. On the ability of tobacco stem calluses to form vegetative and generative buds in culture *in vitro*. *Doklady Botan. Sci.* 187:82–5.

Krishnamoorthy, H. N., and K. K. Nanda. 1968. Floral bud reversion in *Impatiens balsamina* under non-inductive photoperiods. *Planta* 80:43–51.

Marc, J., and J. H. Palmer. 1982. Changes in mitotic activity and cell size in the apical meristem of *Helianthus annuus* L. during the transition to flowering. *Am. J. Botany* 69:768–75.

McHughen, A. 1980. The regulation of floral organ initiation. *Bot. Gaz.* 141:389–95.

——— 1982. Some aspects of growth characteristics of tobacco pistils *in vitro*. *J. Exp. Botany* 33:162–9.

Minter, T. C., and E. M. Lord. 1983. A comparison of cleistogamous and chasmogamous floral development in *Collomia grandiflora* Dougl. ex Lindl. (Polemoniaceae). *Am. J. Botany* 70:1499–508.

Satina, S., and A. F. Blakeslee. 1941. Periclinal chimeras in *Datura stramonium* in relation to development of leaf and flower. *Am. J. Botany* 28:862–71.

Sattler, R. 1973. *Organogenesis of flowers. A Photographic Text-Atlas*. Toronto: University of Toronto Press.

Sawhney, V. K., and R. I. Greyson. 1979. Interpretations of determination and canalisation of stamen development in a tomato mutant. *Can. J. Botany* 57:2471–7.

Schwabe, W. W. 1959. Some effects of environment and hormone treatment on reproductive morphogenesis in the chrysanthemum. *J. Linn. Soc. London (Bot.)* 56:254–61.

Soetiarto, S. R., and E. Ball. 1969a. Ontogenetical and experimental studies of the floral apex of *Portulaca grandiflora*. 1. Histology of transformation of the shoot apex into the floral apex. *Can. J. Botany* 47:133–40.

——— 1969b. Ontogenetical and experimental studies of the floral apex of *Portulaca grandiflora*. 2. Bisection of the meristem in successive stages. *Can. J. Botany* 47:1067–76.

Stewart, R. N. 1978. Ontogeny of the primary body in chimeral forms of higher plants. In *The Clonal Basis of Development*, ed. S. Subtelny and I. M. Sussex, 131–60. New York: Academic Press.

Tepfer, S. S., R. I. Greyson, W. R. Craig, and J. L. Hindman. 1963. *In vitro* culture of floral buds of *Aquilegia*. *Am. J. Botany* 50:1035–45.

Tran Thanh Van, K. 1973. Direct flower neoformation from superficial tissue of small explants of *Nicotiana tabacum* L. *Planta* 115:87–92.

Tucker, S. C. 1960. Ontogeny of the floral apex of *Michelia fuscata*. *Am. J. Botany* 47:266–77.

Vieth, J. 1965. Étude morphologique et anatomique de morphoses induites par voie chimique sur quelques Dipsacacées. Thèse. Dijon.

Wardlaw, C. W. 1963. Experimental investigations of floral morphogenesis in *Petasites hybridus*. *Nature* 198:560–1.

Wetmore, R. H., E. M. Gifford, Jr., and M. C. Green. 1959. Development of vegetative and floral buds. In *Photoperiodism and Related Phenomena in Plants and Animals*, ed. R. B. Withrow. *Am. Assn. Adv. Sci. Publ.* 55:255–73.

11

The development of the shoot system

In order to understand developmental processes in a complex system like a plant, it is necessary to analyze parts of the system individually. Thus, in the preceding chapters the shoot apex as the initiating center of the shoot has been examined in detail, as have the initiation and development of the lateral organs to which it gives rise. The full significance of the processes that occur in these parts, however, can only be appreciated in the context of the integrated system in which they occur. It is appropriate now, therefore, to consider the development of the whole shoot system. Reflection upon the enormous diversity of shoot forms might seem to make this an impossible task, but fortunately the emergence of concepts of shoot architecture has established a framework for the analysis of varied patterns of shoot ontogeny. A relatively small number of developmental processes, occurring in various combinations, provide interpretations of widely divergent shoot forms.

This chapter will consider first the sequence of events by which the individual shoot is elaborated to its final form. It will then examine different developmental potentialities that may be expressed by shoots, often within the same shoot system. The ways in which different shoot expressions fit together into integrated shoot systems will be examined in terms of architectural concepts. Finally, these concepts will be used to interpret some of the major plant growth forms, such as trees, shrubs, and herbs.

PATTERNS OF SHOOT EXPANSION

Although the shoot apex clearly does function as the ultimate source of the cells of the shoot, it is equally true that the apical meristem is by no means the direct source of all or even most of the cells; the multiplication of the immediate derivatives of the meristem in the subapical region amplifies and augments the cellular contribution of the meristem. Nowhere is this more evident than in the phenomena associated with shoot expansion, where numerous divisions, coupled with extensive cell enlargement, elaborate the minute structures initiated by the meristem

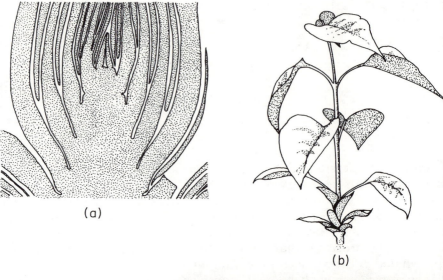

(a)

(b)

Figure 11.1. *(a) Longitudinal section of the terminal bud of* Syringa vulgaris *(lilac). The internodes are unexpanded. (b) The shoot that results from the expansion of a bud like that in (a). (c) Stages in the expansion of buds of* Acer *(maple). (a) ×20, (b) ×½. ([a,b] Adapted from R. Garrison, Am. J. Botany 36:205, 1949. [c] Courtesy of C. Wilson.)*

(c)

into the recognizable features of the mature shoot. Much of the growth of interest to plant physiologists is, in fact, accomplished outside the meristem during the expansion phase of development.

The stem in the vascular plants is a segmented structure, consisting of a sequence of leaf-bearing nodes and internodes. Whatever its ultimate form may be, at the shoot apex the internodes are foreshortened so that the immature leaves are closely crowded around the base of the apical meristem (Fig. 11.1a). In other words, in the shoot apex the leaf primordia and the tissues of the stem are initiated but the internodes are not expanded. The subsequent expansion of the internodes, which may be extensive or insignificant in different cases, represents a second and relatively distinct phase of development. The nature of this second phase appears most dramatically in the shoots of some perennial species in which the two phases are separated by a period of dormancy. In such a plant, if the apical bud is dissected during the period of dormancy, the foreshortened condition of the internodes and the crowding of young leaves around the meristem are readily observed. With the breaking of dormancy, the internodes begin to elongate, the leaves are separated along the extending axis as their subjacent nodes are pushed apart, and the shoot attains its mature form (Fig. 11.1b, c). In plants whose growth is not interrupted by periods of dormancy the distinction between the initiation and expansion phases is not so obvious, because internodes at the base of the apical bud expand successively while new leaf primordia, nodes, and internodes are continually added at the tip.

The segmental nature of the shoot axis profoundly influences the pattern of growth in the expansion phase, in that each internode appears to develop as a distinctive unit. Thus, while the overall growth of the shoot may be observed or plotted, it is also possible to carry out similar analyses of the component internodes. Wetmore and Garrison (1966) have given an account of this kind for the process of shoot elongation in the annual plant *Helianthus annuus*. At the time that it is entering the expansion phase, a young internode is approximately 1 mm in length. Initially elongation occurs throughout the internode, but as it lengthens, growth is progressively restricted to the upper regions (Fig. 11.2). This was demonstrated by placing evenly spaced India ink marks on one side of the expanding internode when it was long enough to permit this manipulation and observing the relative spacing of the marks at successive time intervals. At any one time, most of the growth in the shoot was concentrated in a single internode, and each successive internode began its most active elongation only when the preceding one was nearing the completion of growth.

Internally it is evident that both cell division and cell enlargement are involved in the elongation of *Helianthus* internodes (Table 11.1).

Wetmore and Garrison found that as an immature internode increased to sixty-five times its original length there was approximately a fivefold increase in cell number in longitudinal files in the pith and a thirteenfold increase in average cell length. Increase in cell number was detected throughout the entire period of growth in an internode, but cell division was progressively restricted to the upper regions of the internode. It is evident that there is no sharp separation between cell division and cell expansion phases in the *Helianthus* internode. In this study attention was devoted only to growth of the ground tissues. Growth in the vascular regions is undoubtedly of a very different type because, during the elongation of the internode, the procambial elements achieve a considerable length, in part because they divide less frequently in the transverse plane. This *symplastic growth*, in which different tissues keep pace in an elongating organ by different mechanisms, is one of the more interesting phenomena of plant development.

Wetmore and Garrison also investigated internodal elongation in *Syringa vulgaris* (lilac), a perennial plant with shoots that, unlike those of *Helianthus*, expand in a flush of growth following a dormant period. Here, although the growth pattern in individual internodes is essentially as in *Helianthus*, there is a marked overlapping of internodal activity, and several contiguous internodes develop contemporaneously, although they are not all in the same developmental stage. Another important difference found between the two species was in the relative contributions of cell division and cell enlargement in the ground tissues to the elongation process. In *Syringa* cell numbers in the longitudinal files increased sixty-five times the original number, whereas cell length increased only by a factor of three. The enormous difference between these proportions and those in *Helianthus* emphasizes the necessity of investigating the cellular basis of growth in particular cases in conjunction with attempts to elucidate the physiological controls.

Considering the great diversity of vascular plants, it is perhaps not surprising to find that there are still other patterns of internodal elongation. In plants like *Equisetum*, *Ephedra*, and many monocotyledons, including grasses, the growth sequence in elongating internodes is the reverse of that just discussed. Here the uppermost region of the internode matures first and growth is progressively restricted to the base of the internode. This pattern often results in a persistent growth zone located just above the node that bears the next-older leaf, and this zone may function as a persistent isolated meristem bounded both above and below by mature tissues. Such a region often is referred to as an *intercalary meristem*.

A shoot that develops in the manner just described is referred to as a *long shoot* (Fig. 11.3a). The use of this term in itself implies the existence

(a)

Figure 11.3. *Long shoot and short shoots of* Ginkgo biloba. *(a) Long shoot show-ing the leaves separated by expanded internodes. (b) Short shoots with crowded leaves and unexpanded internodes. (a) ×⅙, (b) ×¼. (Courtesy of R. H. Wetmore.)*

of an alternative mode of development and, in fact, *short shoots* are widespread among the vascular plants (Fig. 11.3b). In a short shoot, when the leaves are expanded and mature, the internodes are unelon-gated so that leaves remain crowded much as in an apical bud. The short-shoot habit may be characteristic of a plant throughout its life, as in cycads and many ferns. On the other hand, it may be restricted to certain stages in the life of the plant, giving way to the long-shoot habit at a later time, as in the bolting of rosette plants in the reproductive phase. This transition is notable in those plants where elongation is lim-ited to a few internodes, or in some cases apparently to one, followed by a return to the short-shoot habit in the reproductive structures. In *Musa* (banana) an inflorescence stalk many feet in length consists of only a few internodes, and in *Gerbera* expansion is limited to a single inter-node. Finally, many plants produce both long shoots and short shoots contemporaneously, a condition seen clearly in *Ginkgo* (Fig. 11.3) and in some conifers and woody dicotyledons. The main axes of the plant are long shoots bearing lateral short shoots, and interconversion of the two types can occur under certain conditions.

REGULATION OF SHOOT EXPANSION

Since the expansion of a shoot involves both the elongation of the stem and the enlargement of the leaves it bears, it might be expected that these two processes are correlated. In fact there is good evidence for the dependence of stem expansion upon the expanding leaves. In the studies of *Helianthus* described earlier it was found that removal of the pair of

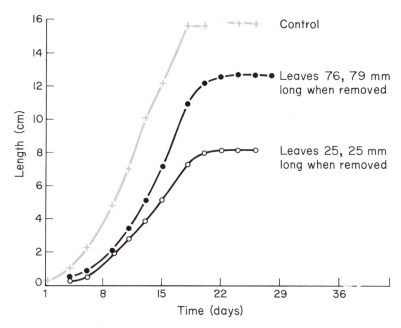

Figure 11.4. *The effect of removing the second pair of leaves on the growth of the second internode of* Helianthus annuus. *The graph shows the growth rate and the final length of the internode of intact plants (+–+), in plants where the second leaf pair removed on day 10 (●–●) and in plants where the second leaf pair was removed on day 1 (○–○). (Wetmore and Garrison, 1966.)*

leaves at a particular node had a marked depressing effect upon elongation of the subjacent internode so long as the leaves were excised before they were half grown (Fig. 11.4). The effect was sharply limited to a single internode and little influence was noted on development of the internode below or above. Thus, the leaf and its subjacent internode seem almost to consititue a growth unit in development of the shoot, a matter of interest in relation to the step pattern of elongation of the axis previously discussed. It is important, however, to note that removal of the pair of leaves did not completely suppress internodal elongation in the subjacent internode, suggesting that the process is not completely dependent upon the associated leaves for the stimuli required for development. Such a response to leaf removal would suggest the operation of a hormonal mechanism, and some evidence relating to this will be presented later in this chapter.

On the other hand there is not always a direct relationship between leaf and stem expansion, as is shown by nonelongating short shoots. Before considering this question it is necessary to examine the structure of the dormant bud from which a shoot expands in woody perennials with periodic growth. In some species – for example, *Aesculus* (horse chestnut) or *Fraxinus* (ash) – a dormant bud contains inside the cataphylls, or bud scales, all of the leaves that will expand during the next growth period in a partially developed state. During shoot expansion the apex becomes active and initiates a new set of leaf primordia, the outer of which develop as cataphylls surrounding the new terminal bud while the inner are the presumptive foliage leaves for the next growth period. These will pass through the resting period in a partially differentiated state. They are referred to as *preformed* leaves. By contrast, in other species the preformed leaves are supplemented by additional leaves, initiated while the shoot is expanding, which mature in the same growth period in which they were initiated. These are called *neoformed* leaves. Although the initiation of leaf primordia continues, the development of neoformed foliage leaves for the current season is ended by the differentiation of primordia as cataphylls that will surround the new terminal bud. After the complement of cataphylls has been formed, the preformed leaves that will expand in the next season are initiated. In some species, but certainly not in all, neoformed leaves can be distinguished morphologically from those that were preformed. This leads to a condition of heterophylly, considered in Chapter 9.

The importance of recognizing the existence of preformed and neoformed leaves in the study of shoot expansion is illustrated by Critchfield's (1970) investigation of *Ginkgo biloba* (maidenhair tree), a species with very distinct long and short shoots (Fig. 11.3). In the short shoots, in which there is essentially no internodal elongation, all of the leaves that expand in a particular growth period were preformed in the unexpanded bud. In the long shoots the expansion of preformed leaves similarly occurs without internodal elongation, resulting in a basal leaf cluster. However, new leaf primordia are quickly initiated and the expansion of these neoformed leaves is associated with extensive internodal elongation. A very similar relationship has been demonstrated in *Populus trichocarpa* (western balsam poplar) (Critchfield, 1960) and in a number of other woody plants.

These observations could suggest that there is a causal relationship between preformed leaves and the nonelongation of internodes. However, such a relationship cannot exist in the long shoots of trees like ash in which there are only preformed leaves. Furthermore, even in species in which both preformed and neoformed leaves are found there need not be such a close correlation between leaf type and shoot elongation as is found in the examples discussed above. In some cases – for example, *Ligustrum* (privet) – both preformed and neoformed leaves are associated with stem elongation, while in others – for example, *Larix* (Clausen and Kozlowski, 1970) – the internodes associated with early

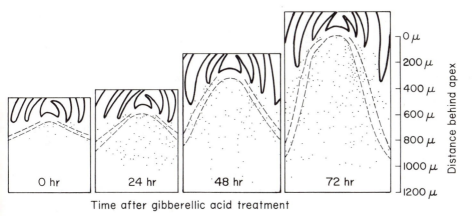

Time after gibberellic acid treatment

Figure 11.5. *The effect of gibberellic acid on subapical meristematic activity in the biennial rosette plant* Hyoscyamus niger. *Mitotic figures, each represented in the figure by a dot, were counted in median sections of the stem tip at each stage. (Sachs, Bretz, and Lang, 1959.)*

preformed leaves do not elongate but those subtending later preformed leaves do.

The physiological basis of shoot expansion has been investigated to some extent in herbaceous plants, particularly with a view to elucidating the role of hormones in leaf–stem interactions. The possible involvement of auxin from expanding leaves in promoting internodal elongation has often been suggested. In *Coleus*, where the production of auxin by developing leaves has been well documented, Jacobs and Bullwinkel (1953) have shown a similar reduction in internodal elongation following leaf removal as in *Helianthus*, but in this case they have demonstrated an almost complete recovery following the application of the hormone indoleacetic acid (IAA) in a concentration equivalent to that normally produced by attached leaves. Thus, the evidence argues for production of auxin by developing leaves having a promoting influence upon elongation in the subjacent internode of the stem.

Hormones other than auxin have also been implicated in stem elongation. In sunflower, where leaf removal decreases internodal elongation, Jones and Phillips (1966) showed that gibberellic acid application could restore elongation, that IAA application was without similar effect, and that the immature leaves of the apical bud were the probable source of the active hormone. The role of gibberellins has also been investigated in rosette plants like *Samolus* and *Hyoscyamus* (henbane) (Sachs, Bretz, and Lang, 1959), which are converted to long shoots with the onset of flowering (Fig. 11.5). These substances have a marked promoting effect on subapical meristematic activity – that is, upon the cell-division component of stem expansion – but they do not influence mitotic activity in the terminal meristem of the shoot. That this action of gibberellins is not restricted to rosette plants is demonstrated by some ingenious experiments on the caulescent plant *Chrysanthemum morifolium*, in which treatment with the substance AMO-1618, which has antigibberellin activity, effectively converted these plants into rosettes (Sachs et al., 1960). The induced rosette habit could then be reversed by application of gibberellic acid just as in natural rosette plants.

Internodal expansion in *Pisum sativum* (pea) is gibberellin dependent (Eckland and Moore, 1968), but there are indications that the response to the hormone is complex. These come from studies of mutants that fail to synthesize adequate amounts of gibberellin during shoot development and are dwarfed as a result of decreased expansion of the internodes. In one study of such mutants (Reid, Murfet, and Potts, 1983) it was found that in the internodal epidermis cell length was reduced by 71 percent and cell number by 43 percent, while in the outer cortex cell length was reduced by 37 percent and cell number by 73 percent. In the inner cortex and xylem the total reduction was in cell number. When dwarf mutants of pea are treated with gibberellins, they form elongated internodes. In one histological comparison of internodes of dwarf and gibberellin-treated dwarf plants it was shown that the increase in cell length was much greater than the increase in cell number in the epidermis while in the outer cortex cell number increased much more than did cell length (Arney and Mancinelli, 1966). Thus it appears that epidermis and outer cortex respond to gibberellin by both cell elongation and cell division but in different proportions. Other tissues, such as inner cortex and xylem, seem to respond only by cell division.

Other studies on rosette plants have indicated that two or more hormones acting in combination may be involved in stem extension. In *Gerbera jamesonii*, for example, Sachs (1968) has found that elongation of the scape, or inflorescence stalk, is dependent upon the presence of the terminal floral head. If this is excised, scape growth stops but may be partially restored by terminal applications of either indoleacetic acid or gibberellic acid. When these were applied together, growth was restored to nearly the values obtained in intact control plants, but histological examination of the ground parenchyma cells revealed that hormone-induced growth had resulted almost entirely from cell elongation and very few divisions had occurred. If, however, only the flowers were removed from the inflorescence, leaving the receptacle and surrounding bracts, combined indoleacetic acid and gibberellic acid treatment again restored scape growth to control levels; cell division was maintained at a normal level and a scape of normal morphology as regards cell number and size was produced.

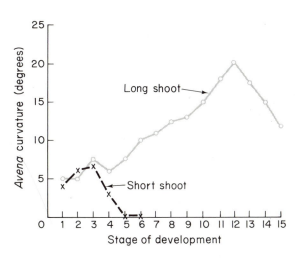

Figure 11.6. *Auxin production by long and short shoots of* Cercidiphyllum japonicum *at successive stages of development in the spring. Auxin yield is expressed in degrees of curvature in the standard* Avena *coleoptile bioassay. (Titman and Wetmore, 1955.)*

There have been relatively few studies of the physiology of shoot expansion in woody plants and these have not resulted in a clear understanding of the process. In several studies it has been demonstrated that removal of immature leaves from expanding long shoots brings about a significant reduction of internodal length. There is also a pattern of auxin production and distribution that has been shown in some cases to have an interesting correlation with the pattern of shoot expansion. In *Cercidiphyllum* (Titman and Wetmore, 1955) and in *Ginkgo* (Gunckel and Thimann, 1949), both plants that produce distinctive long and short shoots, the difference in growth between the two shoot types could be correlated with differences in auxin distribution in the stem. The pattern was the same in the two species (Fig. 11.6). Long and short shoots both yield diffusible auxin from their terminal buds as they turn green and begin to expand after a period of dormancy, but as the bud opens auxin production falls off to a low level. In the short shoots there is no further yield of auxin even though the attached leaves are expanding. However, as the long shoots begin to elongate, there is a second rise in auxin yield to a level considerably above the earlier level (Fig. 11.6). In both cases there is considerable evidence that the auxin in the elongating stem does not have its origin in the leaves but rather is produced in the elongating internodes. In *Ginkgo*, however, if the leaves are removed two days prior to the auxin assay, an operation that reduces growth, the quantity of auxin is also diminished. This leads to the conclusion that the leaves contribute something essential for auxin synthesis rather than the hormone itself. None of this serves to explain why the internodes of short shoots, or the basal internodes of long shoots, do not elongate, nor why there should be a correlation between preformed leaves and non-elongation in some cases. The probable involvement of other hormones may provide an interpretation in the future, but this has not been investigated. Alternatively, the answer may lie in the loss of the ability to elongate by certain internodes, in some cases those that have been preformed along with the leaves they subtend.

These several studies emphasize that the dependence of internodal elongation in the stem upon the attached leaves is a variable one and may differ in herbaceous and woody plants. They also suggest that the hormonal aspects of regulation are probably complex and cannot be interpreted in terms of a single controlling agent.

PERIODICITY OF SHOOT GROWTH

At the beginning of this chapter the distinctive roles of the shoot meristem and the region of subapical growth in giving rise to the final form of the shoot were emphasized. Obviously, however, the activities of these two formative regions must be closely coordinated if shoot growth is to be sustained over a prolonged period as in perennial plants. Although the structural relationship of the shoot apex to its subapical derivatives is of necessity unvarying, investigation has shown a high degree of diversity in the timing of their respective activities. This leads to a consideration of the phenomenon of growth periodicity, which is usually regarded as commonplace in markedly seasonal climates but poses intriguing questions in more uniform environments (Fig. 11.7).

The simplest situation that could be imagined is one in which leaf initiation proceeds continuously and there is a correspondingly uninterrupted expansion of leaves and their associated internodes (Fig. 11.7A). Continuous growth of this type does, in fact, occur in some tropical trees, particularly palms (Hallé, Oldeman, and Tomlinson 1978). In a detailed study of *Rhizophora mangle* (red mangrove) in Florida, Gill and Tomlinson (1971) found that leaf expansion is matched by leaf production throughout the year but that the rate of both processes varies at different times of the year, being slower in the cooler season (Fig. 11.7B). However, this pattern does not appear to be a common one even in tropical trees.

Far more common in all climates is the occurrence of periodicity in shoot expansion or both expansion and leaf initiation. In *Camellia sinensis* (tea) it is reported that leaf initiation occurs continuously but with some variation in rate (Fig. 11.7C). However, shoot expansion occurs in separate growth "flushes," up to four per year, separated by periods in which no external growth is evident. Thus, the number of immature leaves in the terminal bud rises and falls periodically throughout the

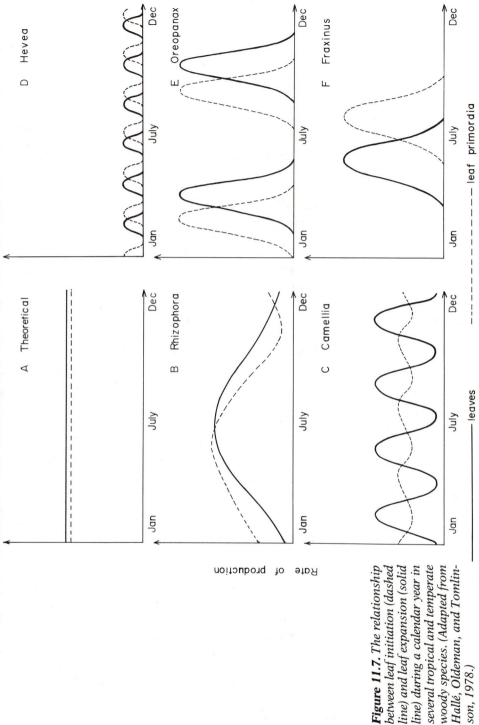

Figure 11.7. The relationship between leaf initiation (dashed line) and leaf expansion (solid line) during a calendar year in several tropical and temperate woody species. (Adapted from Hallé, Oldeman, and Tomlinson, 1978.)

A Theoretical

B Rhizophora

C Camellia

D Hevea

E Oreopanax

F Fraxinus

Rate of production

——— leaves

– – – – – leaf primordia

year. In other cases the initiation of leaf primorida, as well as shoot expansion, is periodic. In *Hevea* (rubber tree) there are up to six growth episodes each year (Fig. 11.7D). In each of these, shoot expansion begins while the meristem is inactive and preformed leaves are expanded. During this time new primordia are initiated, and this process continues after shoot expansion is completed, forming a new terminal bud. Then it too ceases. By contrast, in *Oreopanax* (Borchert, 1969) it has been found that leaf initiation begins before each episode of shoot expansion and leaves expand without interruption in their growth (Fig. 11.7E). At the end of the growth period there is essentially no reservoir of immature leaves. There are two such growth flushes each year.

Although these rhythmic changes in growth activity can sometimes be correlated with environmental conditions like temperature or moisture, consistent evidence of external control of changes in activity has only rarely been obtained (Borchert, 1978). This has led to the recognition that there are endogenous controls that determine the rhythm of shoot growth in tropical trees. Further support for this conclusion is provided by observations that the trees of a single population may not be synchronized in their growth activity, that portions of a large tree may likewise be asynchronous, and that many plants exhibit rhythmic growth even in constant, controlled environments. It is important to recognize the endogenous control of growth periodicity in tropical woody plants because they constitute so large a part of the earth's vegetation.

In temperate regions, where investigation has been much more extensive, the correlation between periodic growth and seasonal environmental changes is generally very close. There is typically one episode of shoot growth in which leaf initiation begins sometime after the onset of shoot expansion and continues for varying periods beyond its cessation (Fig. 11.7F). As has already been noted, all of the leaves and internodes that expand in a particular growth episode may be preformed in the preceding period, or some may be preformed and some neoformed. If only preformed leaves expand, the first primordia formed at the shoot apex begin to differentiate as cataphylls that will surround the next terminal bud. Subsequently the next set of preformed leaves is initiated. If neoformed leaves are involved, these are initiated at the outset and continue to be produced until cataphyll differentiation begins the formation of a new terminal bud that will also include the next set of preformed leaves. The importance of environmental factors like temperature and photoperiod in controlling, or at least signaling, the beginning and ending of growth phases seems well established (Wareing and Phillips, 1981). However, the different durations of the growth period in long and short shoots of the same plant, the varying duration of neoformed growth in shoots of differing vigor, and the not infrequent occurrence of second or even multiple growth flushes are all indicative of an endogenous component in the regulation of shoot growth.

THE COMPONENTS OF THE SHOOT SYSTEM

The foregoing descriptions have concentrated to a large extent upon the development of individual shoots. When attention is turned to the whole shoot system, it becomes apparent that the same system usually contains more than one expression of shoot morphology. Before considering how different shoot types may be integrated into a system, it may be useful to review briefly some of the major forms of shoot expression.

The shoot system begins with a primary axis that typically grows in an upright, or orthotropic, manner from the embryo. This initial shoot may, if the growth of its apical meristem is indeterminate, become the principal axis of the plant. On the other hand, its fate may be to be replaced, as a result either of the conversion of its meristem to a determinate state in the formation of a flower, or inflorescence, or of the spontaneous abortion of the shoot apex. Growth of the axis is then continued by the orthotropic growth of a shoot developed from one of the upper lateral meristems. The principal axis of such a plant is then a sympodium composed of a succession of units or modules. In other cases the same result is achieved when the original shoot apex, although remaining indeterminate, becomes subordinate to a dominant lateral shoot.

In some plants the original axis seldom or never branches (many tree ferns), but more commonly additional axes are formed, often in great profusion. In some lower vascular plants, and a few seed plants, branching occurs by subdivision of the original meristem, but most branches arise laterally from meristems located in leaf axils (see Chapter 8). Potentially a lateral meristem can occur in every leaf axil, and many plants nearly achieve this potentiality. Others, however, initiate buds only in certain positions, sometimes a very small number of those that are available. The distribution of lateral bud meristems is one of the factors that contribute significantly to the form of the shoot system. Further, there is variation in the pattern of subsequent development of lateral meristems into shoot branches. Few or many may grow out, these may be well spaced or clustered, and they may develop into branches soon after their initiation or be long delayed. These patterns of lateral branch abundance, position, and timing, which strongly influence plant form, are indicative of precise endogenous control mechanisms, among which apical dominance is considered to be of particular importance.

Further clear evidence of internal correlations in the shoot system is found in the diversity of lateral branch forms. The important differences between long and short shoots (discussed above) often reflect the dominance of a vigorous main shoot over subordinate laterals. Frequently, however, the subordinate status of lateral branches is expressed as reduced vigor that does not show the extreme form of a short shoot. Another manifestation of branch differentiation is seen in the common occurrence of horizontal (plagiotropic) lateral branches borne on an upright (orthotropic) main axis. In some cases these lateral branches differ little from erect shoots in organization, the main differences being a tendency for further branching to be horizontal and for leaves to be secondarily oriented parallel to the plane of branching. In extreme forms, however, the plagiotropic laterals have an inherent dorsiventral symmetry including a distinctive distichous phyllotaxy and anisophylly. In some cases these features have been shown to be so fixed that plagiotropic shoots, when removed and rooted as cuttings, are unable to produce a plant with an orthotropic axis but retain their dorsiventral symmetry. In many species the development of reproductive structures is restricted to subordinate lateral branches. Finally, lateral branches may be drastically modified to form highly specialized structures, such as thorns or tendrils.

THE ONTOGENY OF THE SHOOT SYSTEM

It is a fact of simple observation that every plant species, in spite of considerable variability, has a characteristic form that is genetically determined. This typical form is achieved progressively as the plant passes through stages from seedling to adult status. Particularly in the case of large and long-lived plants, simple observation often does not reveal the genetically determined program of this progression. In part this is because plants tend to be plastic in their response to environmental conditions and hence show a variability that masks the inherent developmental program. A recognition of the importance of the total system within which developmental processes occur has led to attempts to identify the developmental program of individual plant species in terms of components, such as have been described earlier in this chapter. Rather surprisingly, in view of the enormous diversity of plant forms, a relatively small number of ontogenetic programs or models appear to be applicable to a wide range of plants. Perhaps the most efficient way to introduce this approach is to consider a specific example.

Terminalia catappa, a tree native to the coastal region of southeast Asia but widely planted and naturalized in the tropics and subtropics elsewhere, has a distinctive overall form in which tiers of horizontal lateral branches are borne on an orthotropic main axis. This pattern, sometimes called the "pagoda" form, is found in many tropical trees (Fig. 11.8a) and is so clearly shown in *T. catappa* that this species has

a)

Figure 11.8. *The branching pattern of* Terminalia. *(a) A 2-meter tall sapling of* T. lucida *illustrating the "pagoda" form. (b) The developmental pattern of a lateral branch of* T. catappa *showing the symbodial organization. Each unit is extended by the elongation of a single internode* (i) *and inflorescences* (infl) *are borne in axillary positions on the upturned nonelongated portion of each unit. (b)* ×⅓. *([a] Courtesy of J. Fisher. [b] Adapted from Fisher, 1978.)*

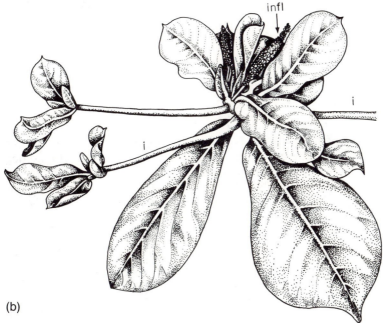

(b)

been the object of intensive study (Fisher, 1978). The main apex of the seedling is indeterminate and grows indefinitely by means of a series of growth flushes producing an orthotropic axis or trunk. Near the end of each growth flush internodal elongation in the leader is reduced in extent, and in the axils of four to five leaves in this region lateral branches develop. Since the buds in other leaf axils do not develop, the branch tiers are distinctly separated. The lateral branches are oblique or nearly horizontal in orientation and the first internode undergoes extensive elongation (Fig. 11.8b). Succeeding internodes, however, are much shorter, and the shoot apex assumes a nearly vertical orientation. The branch is, in effect, converted into a short shoot that continues to produce leaves in periodic flushes but does not increase the horizontal extent of the branch. Inflorescences are borne in axillary positions on these short terminal segments. Continued horizontal extension of a branch is accomplished by the outgrowth of one or usually two branches from leaf axils in the curved zone where the orientation changes from horizontal to vertical. Each of these repeats the developmental sequence of the original branch unit, and ultimately they give rise to two additional units. Thus, the branch is extended horizontally in a sympodial fashion and at the same time increases in lateral extent.

The essential elements of this growth plan may be summarized as follows (Fig. 11.9c):

1. an indeterminate principal meristem that grows rhythmically producing an orthotropic axis
2. the periodic formation of lateral branches in tiered clusters
3. plagiotropic and sympodial branches producing lateral reproductive structures

A growth program of this kind, which expresses the inherent genetic potentiality of the plant, has been called its *architectural model* (Hallé, Oldeman, and Tomlinson, 1978) (Fig. 11.9). Models like this, when identified, have been found to be applicable to many species that may or may not be taxonomically related. Hallé et al. (1978), using such growth criteria, have recognized 23 architectural models, largely based on tropical trees but also including nontropical trees and shrubs and even herbs. Considering the immense diversity of plant forms, it is surprising to find that these models encompass the majority of species that have been examined thus far. The model represented by *Terminalia* has been identified in 19 different families of dicotyledons. Another model in which the terminal meristem produces an unbranched indeterminate axis of the palm or tree fern type is found among the ferns, gymnosperms, dicotyledons, monocotyledons, and even some extinct fossil forms (Fig. 11.9a).

Although many different species can be shown, upon analysis, to have the same architectual model, they obviously are not identical in detailed form. Indeed, each species has its own distinctive form. When species are examined that have the same architectural model as *T. ca-*

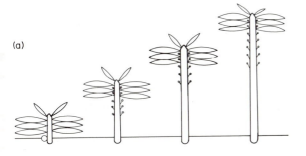

(a)

(b)

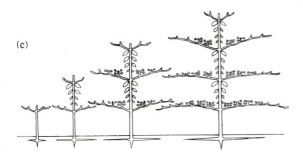

(c)

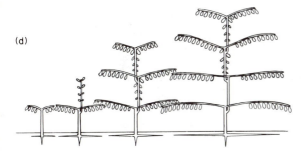

(d)

Figure 11.9. *Diagrammatic representation of four architectural models illustrating developmental stages leading to reproductive maturity. The essential features of each are as follows. (a) Indeterminate principal meristem, which grows continuously. No lateral branches. Reproductive structures lateral. Example:* Cocos nucifera *(coconut palm). (b) Indeterminate principal meristem, which grows rhythmically. Periodic formation of lateral branches, which repeat pattern of main axis. Reproductive structures lateral on lateral branches. Example:* Quercus rubra *(red oak). (c) Pattern illustrated by* Terminalia catappa. *Description in text. (d) Determinate principal meristem producing a sympodial trunk. Periodic initiation of plagiotropic lateral branches, which are either monopodial or sympodial. Reproductive structures lateral on trunk or branches. Example:* Theobroma cacao *(cacao). (Adapted from Hallé, Oldeman, and Tomlinson, 1978.)*

tappa but show differences in the extent to which the tiered or pagoda form is reconizable, it is found that they differ from *T. catappa* in rather minor quantitative features (Fisher and Hibbs, 1982). These include the tendency of the lateral branches to depart from a horizontal orientation, the occurrence of internodal elongation in the orthotropic tips of sympodial units and the tendency for more than two lateral shoots to be involved in each sympodial extension of a lateral branch. The repetition of these differences as the tree increases in size ultimately can lead to major differences in overall form. These minor characteristics, like the components of the basic model, are genetically determined and, with those of the model, are referred to as *deterministic*.

A further consideration of particular importance in large, long-lived species is the fact that the form may change significantly as a plant ages. There are many instances in which a young tree has a distinctly pyramidal (excurrent) shape that gives way to a more spreading (decurrent) form with age. In *Ginkgo* it has been shown (Gunckel, Thiman, and Wetmore, 1949) that, whereas in young trees the terminal shoots are nearly all long shoots, in mature trees (thirty-five to forty years old) a large proportion are short shoots and many laterals become long shoots. This tends to reduce vertical growth and enhance lateral spreading. Although such changes appear to be an inherent part of the growth plan, it is evident that they may also be subject to environmental influence, as is shown by the earlier change in trees of some species growing on certain sites (Zimmerman and Brown, 1971).

The question of environmental influence introduces a major problem into the recognition of architectural models. In spite of genetic determination, it is obvious that most plants display a high degree of plasticity in their development, often in response to the environment in which they are growing. This fact is particularly important in influencing the form of large, long-lived plants that may accumulate responses to environmental stress and accidents over a period of many years. For example, in old trees of *T. catappa*, particularly in exposed locations, it is common to find that the vertical parts of lateral branch units are converted into long shoots and function like the leader of a young tree, producing a succession of tiered, plagiotropic laterals. New leaders can also arise from previously inactive lateral buds on the main axis, particularly if the original leader is damaged. When many new leaders are present, they tend to diverge outward, possibly because of a shading effect, and depart from a typical vertical orientation. The lateral branches formed by the new leaders tend to fill in the gaps between the tiers of the original axis, and the pagoda form is thus obscured (Fisher and Hibbs, 1982). These new axes in many respects repeat the growth pattern or

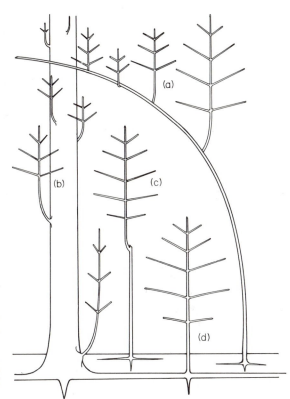

Figure 11.10. *Diagrammatic representation of several examples of reiteration. (a) On an arched trunk. (b) As stem suckers on an old tree. (c) As a mechanism of regeneration. (d) As root suckers. (Adapted from Hallé, Oldeman, and Tomlinson, 1978.)*

model of a seedling and constitute an example of the phenomenon known as *reiteration* (Fig. 11.10). Although the plasticity of a plant clearly must be genetically determined, developments of the reiterative type are often referred to as *opportunistic* to distinguish them from the more specifically determined features of the basic model.

As the concept of architectural models has been extended to plants of increasingly diverse habit, it has become apparent that there may be some for which no structural model can be discerned. In these plants developmental plasticity is the dominant characteristic from very early stages of ontogeny and it is difficult to recognize with confidence any deterministic growth characteristics. An illustration of this difficulty is provided by *Arctostaphylos uva-ursi* (bearberry) a prostrate or creeping shrub forming dense patches from which it colonizes adjacent areas (Remphrey, Steeves, and Neal, 1983). The pattern of branching and shoot extension in this plant was found to be so variable that no typical growth plan could be recognized. Instead, the variable features, such as fate of the terminal apex, number and position of branches, types of branches, and branching angle, were identified and the probablility of occurrence of each possible expression of each characteristic was calculated from actual measurements. It was thus possible to arrive at a statistical expression of the growth pattern. The validity of this analysis was tested by simulations based on the calculated probabilities. These simulations closely resembled growth patterns observed in the marginal colonizing zone of a bearberry patch. The set of rules used in the simulations was presented in this study as an alternative to a structural model (Remphrey, Neal, and Steeves, 1983). This study suggests that the concept of architectural models is sufficiently flexible to encompass growth patterns of extreme plasticity.

TREES, SHRUBS, AND HERBS

The architectural models summarized by Hallé, Oldeman, and Tomlinson (1978), although conceived primarily in relation to trees, are held to be equally applicable to plants of smaller stature. Indeed, many shrubs and a few herbs have been described in terms of models. On the other hand, shrubs and herbs are not simply small trees and it is worthwhile to consider some specific features of these forms. In general a tree is recognized by its ability to form a trunk, that is, a single main axis that is dominant over all others. This may be produced by a single terminal meristem or by a succession of meristems that replace the original. A shrub, on the other hand, lacks this ability and consists of a cluster of equivalent axes resulting from continual sprouting from the basal region. Although many factors may be involved in this difference, one developmental phenomenon seems to be of particular importance. In trees there is a tendency for buds closest to the apex to grow out whenever conditions permit sprouting. In shrubs this tendency is reversed and it is the basal buds that usually sprout. However, in some species the preeminence of basal buds is a seasonal phenomenon and only if sprouting occurs at the correct time does the shrublike habit result. Such plants may begin to show a treelike habit if the normal seasonal periodicity is artificially altered (Champagnat, 1978).

Herbs are essentially plants that do not develop a long-lasting, woody shoot system above ground, although they may have underground structures that persist indefinitely. There are two important features that seem to distinguish herbs from woody plants, although undoubtedly other factors are also involved. The first of these is a limitation of cambial activity, which in some cases is so extreme that no secondary tissues are formed at all. The second is the precocious development of reproductive structures that terminate shoot growth and have the effect of limiting the size and the duration of the shoot. This is the more important from the architectural point of view. The extreme expression of both of these tendencies is found in annuals and ephemerals, which complete their entire life span in one, often brief, growing season.

It is to be expected that, of the mechanisms that limit shoot growth in herbaceous plants, those related to the onset of reproductive development would be among the most important. Photoperiodic induction of flowering has been extensively documented in herbaceous plants. However, there are indications of other mechanisms that bring about determinate growth through flowering. For example in one cultivar of tobacco which is not sensitive to photoperiod, it has been found that flowering is initiated when the shoot apex has produced a set number of nodes from the nearest root. Thus, if rooting is induced along the stem, or if the tip of the plant is removed and rooted, the apex remains vegetative until the required number of nodes has been formed (McDaniel, 1980). It remains to be seen whether such internal mechanisms limiting shoot growth are widespread in occurrence.

There is evidence for the existence of even more restrictive limitations of shoot growth in some cases in which it appears that cells destined to initiate reproductive structures are identifiable in the embryo before seed germination. Coe and Neuffer (1978) have used the method of clonal analysis (see Chapter 5) to demonstrate that the shoot apex of the *Zea mays* embryo is compartmentalized into horizontal tiers, each of which is destined to form a particular segment of the shoot. The most apical of these compartments is the one that will give rise to the tassel, thus terminating shoot growth. In such a plant the shoot must be considered to be nearly as determinate as an animal embryo.

GENERAL COMMENT

This chapter, if nothing else, has revealed the incompleteness of our understanding of shoot development when considered in the context of the diversity of plant form. The operation of apical dominance, whether mediated through hormones or some other mechanism, has been extremely useful in interpreting the branching patterns of herbaceous plants but is severely challenged by many phenomena in woody plants, such as periodic release of lateral buds, the specific release of either distal or basal buds in different species, the frequently observed outgrowth of laterals on the most vigorously growing current shoots, and the fact that laterals are often modified in their development rather than inhibited. These phenomena offer a real opportunity to explore the role of hormones, and particularly the interactions of hormones with one another and with other factors like nutrient or water supply. It may be, however, as suggested by Brown, McAlpine, and Kormanik (1967) that apical dominance should be viewed as part of a broader phenomenon that might be designated *apical control*. Certainly the consideration of a wide range of plant types requires a broadening of current concepts of the regulation of plant growth.

Fortunately the renewed interest in shoot architecture, while calling attention to the often ignored diversity of plant form, has also revealed that this seemingly infinite variation can be encompassed in a relatively small number of developmental models. In essence these models identify the specific developmental processes that account for the form of a plant and thus expose them to analysis. Even in extremely plastic species, which are difficult to reduce to a structural model, it is possible to pinpoint the developmental variables and describe them quantitatively as a basis for analysis. Thus, the concept of architectural models constitutes a significant breakthrough in the field of plant development. It is interesting, and perhaps prophetic, that this important advance began with a study of tropical plants and has revealed much that had been overlooked in more familiar temperate species.

REFERENCES

Arney, S. E., and P. Mancinelli. 1966. The basic action of gibberellic acid in elongation of "Meteor" pea stems. *New Phytol.* 65:161–75.

Borchert, R. 1969. Unusual shoot growth pattern in a tropical tree, *Oreopanax* (Araliaceae). *Am. J. Botany* 56:1033–41.

——— 1978. Feedback control and age-related changes of shoot growth in seasonal and nonseasonal climates. In *Tropical Trees as Living Systems,* ed. P. B. Tomlinson and M. H. Zimmermann, 497–515. New York: Cambridge University Press.

Brown, C. L., R. G. McAlpine, and P. P. Kormanik. 1967. Apical dominance and form in woody plants: A reappraisal. *Am. J. Botany* 54:153–62.

Champagnat, P. 1978. Formation of the trunk in woody plants. In *Tropical Trees as Living Systems,* ed. P. B. Tomlinson and M. H. Zimmermann, 401–22. New York: Cambridge University Press.

Clausen, J. J., and T. T. Kozlowski. 1970. Observations on growth of long shoots in *Larix laricina. Can. J. Botany* 48:1045–8.

Coe, E. H., and M. G. Neuffer. 1978. Embryo cells and their destinies in the corn plant. In The *Clonal Basis of Development,* ed. S. Subtelny and I. M. Sussex, 113–29. New York: Academic Press.

Critchfield, W. B. 1960. Leaf dimorphism in *Populus trichocarpa. Am. J. Botany* 47:699–711.

——— 1970. Shoot growth and heterophylly in *Ginkgo biloba. Bot. Gaz.* 131:150–62.

Ecklund, P. R., and T. C. Moore. 1968. Quantitative changes in gibberellin and RNA correlated with senescence of the shoot apex in the "Alaska" pea. *Am. J. Botany* 55:494–505.

Fisher, J. B. 1978. A quantitative study of *Terminalia* branching. In *Tropical Trees as Living Systems*, ed. P. B. Tomlinson and M. H. Zimmermann, 285–320. New York: Cambridge University Press.

Fisher, J. B., and D. E. Hibbs. 1982. Plasticity of tree architecture: Specific and ecological variations found in Aubreville's model. *Am. J. Botany* 69:690–702.

Gill, A. M., and P. B. Tomlinson. 1971. Studies on the growth of red mangrove (*Rhizophora mangle* L.) 3. Phenology of the shoot. *Biotropica* 3:109–24.

Gunckel, J. E., and K. V. Thimann. 1949. Studies of development in long shoots and short shoots of *Ginkgo biloba* L. III. Auxin production in shoot growth. *Am. J. Botany.* 36:145–51.

Gunckel, J. E., K. V. Thimann, and R. H. Wetmore. 1949. Studies of development in long shoots and short shoots of *Ginkgo biloba* L. IV. Growth habit, shoot expression and the mechanism of its control. *Am. J. Botany* 36:309–18.

Hallé, F., R. A. A. Oldeman, and P. B. Tomlinson. 1978. *Tropical Trees and Forests. An Architectural Analysis.* New York: Springer-Verlag.

Jacobs, W. P., and B. Bullwinkel. 1953. Compensatory growth in *Coleus* shoots. *Am. J. Botany* 40:385–92.

Jones, R. L., and I. D. J. Phillips. 1966. Organs of gibberellin synthesis in light-grown sunflower plants. *Plant Physiol.* 41:1381–6.

McDaniel, C. N. 1980. Influence of leaves and roots on meristem development in *Nicotiana tabacum* L. cv. Wisconsin 38. *Planta* 148:462–7.

Reid, J. B., I. C. Murfet, and W. C. Potts. 1983. Internode length in *Pisum*. II. Additional information on the relationship and action of loci *Le, La, Cry, Na* and *Lm. J. Exp. Botany* 34:349–64.

Remphrey, W. R., T. A. Steeves, and B. R. Neal. 1983. The morphology and growth of *Arctostaphylos uva-ursi* (bearberry): An architectural analysis. *Can. J. Botany* 61:2430–50.

Remphrey, W. R., B. R. Neal, and T. A. Steeves. 1983. The morphology and growth of *Arctostaphylos uva-ursi* (bearberry): An architectural model simulating colonizing growth. *Can. J. Botany* 61:2451–7.

Sachs, R. M. 1968. Control of intercalary growth in the scape of *Gerbera* by auxin and gibberellic acid. *Am. J. Botany* 55:62–8.

Sachs, R. M., C. F. Bretz, and A. Lang. 1959. Shoot histogenesis: The early effects of gibberellin upon stem elongation in two rosette plants. *Am. J. Botany* 46:376–84.

Sachs, R. M., A. Lang, C. F. Bretz, and J. Roach. 1960. Shoot histogenesis: Subapical meristematic activity in a caulescent plant and the action of gibberellic acid and AMO-1618. *Am. J. Botany* 47:260–6.

Titman, P. W., and R. H. Wetmore. 1955. The growth of long and short shoots in *Cercidiphyllum. Am. J. Botany* 42:364–72.

Wareing, P. F., and I. D. J. Phillips. 1981. *Growth and Differentiation in Plants*, 3d ed. Oxford: Pergamon Press.

Wetmore, R. H., and R. Garrison. 1966. The morphological ontogeny of the leafy shoot. In *Trends in Plant Morphogenesis*, ed. E. G. Cutter, 187–99. London: Longmans.

Zimmermann, M. H., and C. L. Brown. 1971. *Trees: Structure and Function.* New York: Springer-Verlag.

CHAPTER 12

The root

The previous eight chapters have dealt with postembryonic development of the shoot. Consideration must now be given to the subsequent development of the other meristem initiated in the embryo, the root apical meristem. The organ system that develops from this meristem during the ontogeny of the plant is as extensive as the shoot system and in many cases exceeds the aerial system in size. Moreover, root systems show considerable morphological diversity and are by no means stereotyped in form or in development. Unfortunately, however, this is not generally appreciated because root systems are inaccessible to direct and sequential observation of the type that can easily be made on the shoot. Without elaborate excavation, root systems cannot be studied except in special cases, and even when exposed, they can hardly be observed ontogenetically in anything approaching normal circumstances. It is, therefore, regrettable, but not surprising, that much of our knowledge of root development is based upon laboratory-cultured seedlings of annual crop plants.

The remarkable extent of certain individual root systems has been revealed by excavation and measurement. Ecologists have long recognized that in a plant community there is usually a stratification of root systems comparable to the multiple stories of shoots. Such a layering tends to minimize competition among species for water and nutrients in the soil, and it has been shown to have an important bearing upon survival under adverse conditions. What is perhaps more significant in the present context is the morphological complexity of root systems. Root systems often consist of both deeply penetrating roots and shallower horizontally extending roots. The horizontal roots may rather abruptly turn downward and assume a vertical direction of growth (Fig. 12.1). Differentiation into vigorously growing long roots and lateral short roots, which are limited and often evanescent, also appears to be a common phenomenon. Such morphological complexity suggests the existence of correlative mechanisms perhaps comparable to those found in the shoot, but little is known of these mechanisms in roots. For example, there is evidence indicating that the long and short root pattern reflects

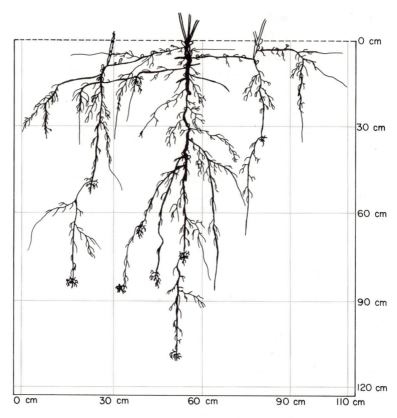

Figure 12.1.   *The root system of* Euphorbia esula *(leafy spurge). The root system is morphologically complex. Some roots penetrate vertically downward; others grow horizontally and turn down at differing distances from their point of origin. Horizontally growing roots may give rise to buds that grow upward as shoots. Some lateral roots branch profusely and remain stunted in growth. These are distinguished as short roots. (M.V.S. Raju, T. A. Steeves, and R. T. Coupland,* Can. J. Botany *41:579, 1963.)*

the operation of a kind of apical dominance, but the hormonal or other basis of the phenomenon remains obscure.

To a greater extent than is often recognized root structure may be modified from the typical pattern in relation to specialized functions. Although less common than shoot thorns, there are well-documented cases in which aerial roots, and more rarely subterranean roots, are transformed into thorns by the development of the root apex into a hardened point (Fig. 12.2) (Gill and Tomlinson, 1975). In those angiosperms that are parasitic upon other species, the haustorium, which

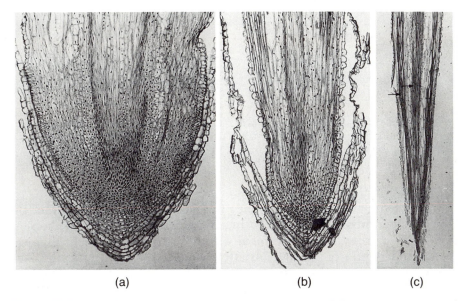

(a)   (b)   (c)

Figure 12.2. *Development of aerial root thorns in* Cryosophila guagara, *a Central American palm, as observed in median longitudinal sections of the root tip. (a) Unmodified root apex. (b) A stage in the elongation and attenuation of the root apex. The arrow indicates the progressive extension of epidermal differentiation into the region of the meristem. (c) Mature thorn tip. Arrows point to cortical and stelar sclerenchyma. Epidermis and outer cortex are being sloughed and the root cap has been shed. (a, b) ×60. (c) ×35. (I.C.S. McArthur and T. A. Steeves,* Can. J. Botany *47:1377, 1969.)*

penetrates the host plant, is clearly a modified root in some cases and is interpreted as a root in other instances in which the highly altered structure makes this relationship less obvious. The roots of a great many, in fact probably most, land plants form intimate associations with soil microorganisms, the physiological significance of which is becoming increasingly apparent. These associations often result in highly modified patterns of root development.

The individual root differs from the shoot in certain developmental features, and these must be recognized in any analysis of root growth. The apical meristem of the root is covered distally by a cap of mature tissue and thus is subterminal rather than terminal, as in the case of the shoot apical meristem. Consequently the meristem is surrounded by its differentiating derivatives. Moreover, the root meristem does not produce any lateral appendages comparable to leaf primordia, and the segmentation both in mature structure and in differentiation of the stem, which is related to the leaves, is lacking in the root. As in the stem,

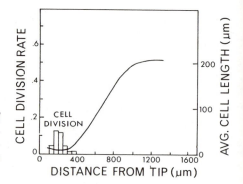

Figure 12.3. *The distribution of cell division and cell elongation of epidermal cells in the root tip of* Phleum pratense *(Timothy grass). (Adapted from Goodwin and Avers, 1956.)*

however, much of the growth that produces the ultimate length of the root is accomplished, not in the meristem itself, but in the subapical region. This growth zone is typically more restricted in extent than the elongating region of a shoot, usually being less than 1 cm long, and often much less. In this growing region both cell division and cell expansion occur, but the balance between these two processes varies along the axis. For example, in studies of epidermal cells of living seedling roots of *Phleum* Goodwin and Avers (1956) showed that cell division frequency is low in the region of the meristem, increases to a maximum at about 150 to 200 μm behind the root cap, and then declines to zero at about 400 μm (Fig. 12.3). As mitotic frequency declines, an increase in net cell length is observed, continuing to a point slightly more than 1 mm from the tip. In *Zea* Erickson and Sax (1956) found the maximum cell division frequency in epidermis at 1.25 mm from the tip and a continuation of division to a total distance of 2 mm. Cell elongation continued to a level 9.5 mm from the tip. The cellular processes involved in growth can be traced basipetally from the root meristem in an unbroken sequence because, as already noted, the axis is not segmented into nodes and internodes. In principle, therefore, the root ought to be a simpler system in which to study growth processes, but this expectation has been realized only in part.

THE ROOT APEX

Structure

In the root where the meristem is surrounded by maturing derivatives, delimitation of the meristem is much more of a problem than in the shoot apex. In the shoot apex the distal surface of the meristem is superficial and thus clearly delimited and its lateral boundaries are indicated by the positions of the leaf primordia. The difficulty of defining

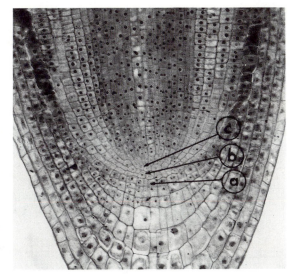

Figure 12.4. *Longitudinal section of the root tip of* Raphanus sativus. *The three tiers of initials in the meristem are labeled a, b, and c. Derivative cells of tier* a *differentiate as root cap and epidermis, those of tier* b *as the cortex, and the derivatives of tier* c *differentiate as the vascular tissue of the root.* ×235.

the meristem becomes evident when median longitudinal sections of root apices are examined. In such preparations longitudinal files of mature and maturing cells may be traced acropetally as they converge upon a small region just below the root cap, cell rows of which also converge upon this region. In many cases these files of cells can be traced into a small number of apparent initiating layers in the root tip.

An example of this pattern is provided by the roots of *Euphorbia esula* (leafy spurge) in which relatively distinct tiers have been found in the meristem (Raju, Steeves, and Naylor, 1964). The most common number of layers is three, but in different roots from one to four have been identified and there is some variation in their relationship to differentiating tissues of the root. These apical tiers, which reflect the regularity of segmentation patterns in the meristem, have been called *meristem layers* and were considered to constitute the promeristem of the root.

In some species the meristem layers are much more distinct and consistent in their relationship to maturing tissues. A typical pattern is illustrated by the root meristem of *Raphanus sativus* (radish) (Fig. 12.4). Here the meristem consists of three superimposed layers. The most distal of these appears to be the initiating layer for the root cap and epidermis. The middle tier has a similar relationship to the cortex, and the innermost layer to the vascular system. Several other layering patterns are found that differ from that of radish. In members of the grass family (Poaceae), for example, *Zea mays*, the root cap is commonly initiated by a separate layer, the epidermis and cortex are both derived from the

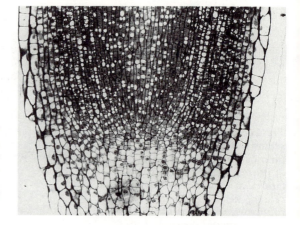

Figure 12.5. *Longitudinal section of the root tip of* Pisum sativum. *Only a single layer of initials can be identified in the meristem, and all the tissues of the root are derived from them.* ×120.

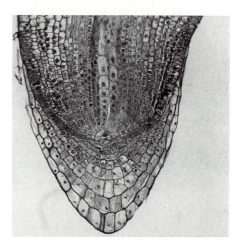

Figure 12.6. *Longitudinal section of the root tip of* Pteridium aquilinum *(bracken). A single enlarged apical cell occupies the center of the meristem. The most recent derivatives of the apical cell are visible as segments around its internal faces. These differentiate as all the tissues of the root.* ×185.

middle layer, and the innermost layer gives rise to the vascular system.

On the other hand, the root apices of some angiosperms and many gymnosperms cannot be described in terms of discrete meristem layers but rather seem to contain a transversely oriented common initiating region for all tissues of the root. In some cases the transverse meristem has been interpreted as consisting of two plates of initials, one specifically for the vascular system and one for the rest of the root. In other cases, such as *Pisum* (Popham, 1955), cell rows can be traced from the central part of the root cap through the meristem to the vascular tissue, and it is argued that there can be only one set of initials (Fig. 12.5). The significance of the difference between root meristems with distinct layers and those without is somewhat diminished by the demonstration of a transition from one to the other during root development (Armstrong and Heimsch, 1976).

The concept of common initials for the entire root has been developed along another line as an alternative interpretation of root apical organization. In many ferns and other lower vascular plants there is an enlarged and conspicuous apical cell present in the root, as in the shoot (Fig. 12.6). Through its segmentation along four cutting faces, the apical cell functions as the ultimate initial of all the tissues of the root. Derivatives of the distal face mature as root cap cells, whereas those from the lateral cutting faces initiate the body of the root. Attempts have been made to describe a similar kind of segmentation pattern based upon a single apical initial in the root apices of several angiosperms (von Guttenberg, 1964). It is held that this central cell serves to renew the initiating layers where these are present. However, in some cases it is believed only to renew certain layers; the layer related to the vascular system apparently functions independently.

Analytical studies

The analysis of root meristem organization has been complicated but greatly enlivened by the discovery that the region of the root apex, which has been interpreted as the initiating center of the root, is relatively, if not totally, inactive mitotically during much of root development (Fig. 12.7). In his analysis of the root of *Zea*, Clowes (1954) encountered great difficulty in interpreting cellular configurations in the apex and concluded that the observed patterns could only exist if the most central cells of the meristem divide infrequently or not at all. This rather startling conclusion was verified when the same investigator (Clowes, 1956a, b) supplied radioactive precursors of nucleic acids to roots of *Zea*, *Vicia*, and *Allium* and used autoradiographic techniques to determine the localization of nucleic acid synthesis in the root meristem. He was able to show conclusively that there is a *quiescent center* in the meristem in which little or no DNA synthesis occurs and presumably little or no mitosis. Other workers have confirmed these results by means of autoradiography as well as by other kinds of observations (Fig. 12.8a). For example, in onion root tips Jensen and Kavaljian (1958) demonstrated by mitotic counts that there is an essentially nondividing region in the center of the apex (Fig. 13.2).

In subsequent studies other characteristics of the quiescent center in certain species have been elaborated. The content of RNA and the rate of synthesis of both RNA and protein are lower in the quiescent center cells than in immediately surrounding regions (Clowes, 1956a, 1958b). Such reports are in agreement with the observed fainter staining of cells

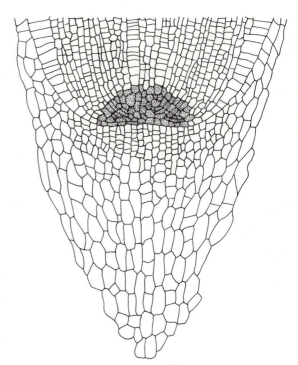

Figure 12.7. *Median longi-tudinal section of the root tip of* Zea mays *(maize). The quiescent center is shown as stippled cells. The relation-ship between differentiating cells and tissues and the mer-istem is very clear in this sec-tion. ×150. (F.A.L. Clowes,* Biol. Rev. *34:501, 1959.)*

in this region with ordinary histological stains. Further distinctive fea-tures of cells of the quiescent center have been revealed by electron microscopy (Clowes and Juniper, 1964; Phillips and Torrey, 1974; Peter-son and Vermeer, 1980). Differences were noted in several cell organ-elles, the most conspicuous being the apparent low activity of Golgi bodies and the relatively poor development of mitochondrial cristae as compared with the cells of the surrounding mitotically active regions. It was also reported that endoplasmic reticulum is poorly developed. All of these features suggest a low metabolic activity.

The phenomenon of mitotic quiescence in the root meristem has stim-ulated great interest among students of development and the resulting investigations have revealed variations in its occurrence in the roots of different species. In the early stages of development, both of the pri-mary roots of seedlings and of lateral roots, the cells of the root meri-stem appear to be uniformly active. Quiescence in the central region appears progressively as the root develops, more rapidly in some spe-cies than in others (Clowes, 1976). As the root grows, more cells are added to the quiescent population until a stable mature size is attained (500 to 600 cells in the meristem of *Zea mays*). There are reports that in

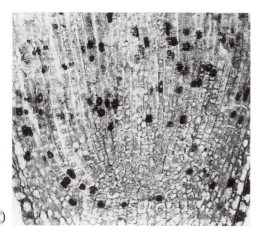

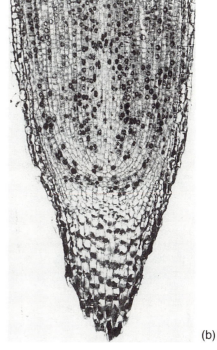

Figure 12.8. *Pattern of thymidine labeling of nuclei in the apex of a perennial long root of* Euphorbia esula *(a) during the height of the growing season and (b) at the beginning of the annual growth period. (a) ×225, (b) ×100. ([a] Raju, Steeves, and Naylor, 1964. [b] Raju, Steeves, and Maze, 1976.)*

(b)

some roots quiescence does not occur at all. For example, in the root system of *E. esula* it was discovered that whereas the vigorous, perennial long roots have a distinctive quiescent center, the determinate laterals that have been designated short roots lack such a center throughout their brief development (Raju, Steeves, and Naylor, 1964).

It has also become clear that quiescence in the root apex may be a variable phenomenon depending upon the circumstances. In *E. esula* there is a seasonal fluctuation in the vigorous long roots where the quiescent center is well developed at the height of the growing season. Early in the growing season when growth is being reactivated after winter dormancy, no quiescent center can be detected (Fig. 12.8) (Raju, Steeves, and Maze, 1976). This indicates that the cells of the quiescent center are intermittently active in cell production. Similarly, Clowes and Stewart (1967) found that the quiescent center participates in the recovery of the root from artificially cold-induced dormancy in seedlings of *Zea mays*. Exposure of root tips to x-irradiation or removal of all or part of the root cap can also lead to an activation of cells of the quiescent center (Clowes, 1976). Finally it has been shown in excised roots in sterile culture, that if growth is arrested by carbohydrate starvation, the quiescent center becomes active when growth is restored by addition of nutrients to the medium (Langenauer, Davis, and Webster, 1974). It thus appears that cells that are predominantly quiescent may have periodic bursts of activity in response to stress or changes in the natural development.

When first recognized, the quiescent center was interpreted as consisting of cells that are completely or almost completely inactive mitotically. The cellular patterns, which appear to indicate derivation of root tissues from this region, were considered to be a reflection of the early stages of development prior to the onset of quiescence. The demonstration of intermittent, or perhaps periodic, cell-generating activity in this region necessitates a reconsideration of its role in the development of the root. It now appears that the cells of the quiescent center do divide slowly, so, as in the shoot apex, quiescence is a relative condition in comparison with the more active state of surrounding cells. It can be expressed in terms of the duration of the cell cycle. Clowes (1961a) has measured the cell cycle times in various parts of the root apex of *Zea* and has found that in the quiescent center it is approximately 170 hours, whereas in the root cap initials, the most rapidly dividing cells, it is only 12 hours. Similarly, Phillips and Torrey (1972) found that in cultured roots of *Convolvulus* the root cap initials had a cycle time of 13 hours, whereas in the quiescent center the cycle duration was an estimated 430 hours. In other regions of the root apex the duration of the cycle was somewhat greater than in the cap initials but much less than in the quiescent center. Moreover, there appears to be variation in the duration of the cell cycle among the cells of the quiescent center (Barlow, 1973), but it is not clear how the cells with different cycle times are distributed. A logical pattern might be one in which there is a gradient of decreasing cycle times from the least active cells at the center to those at the periphery of the region. In fact, Phillips and Torrey (1971) have provided some evidence that this may be the case in cultured roots of *Convolvulus*.

As has been pointed out, the absence of mitotic activity in the center of the root apex can be interpreted to mean that this region, including the meristem layers if present, is not involved in histogenesis, at least in the adult root. Clowes (1961b) has dealt with this problem by locating the initials of the root just outside the quiescent center and completely surrounding it, and he has designated this group of initial cells the promeristem of the root. This analysis of the root meristem is in certain respects reminiscent of the méristème d'attente concept of the shoot apex discussed in Chapter 5.

If, on the other hand, even occasional divisions occur in the quiescent center, serving to renew the more actively divided regions around it, it is difficult to refrain from considering the cells of the center as the initials of the root. The suggestion that the true initials of the shoot need divide only very infrequently in comparison with surrounding cells in order to fulfill their function applies equally to the root. The latter interpretation, in fact, agrees with the views expressed by von Guttenberg (1964) regarding the organization of the root apex. He recognized that cell divisions occur much more frequently in a peripheral zone surrounding a central group of relatively less-active cells, but he regarded the peripheral cells as rapidly dividing derivatives of the true initials. His postulated central cell or cells would, of course, lie within the relatively inactive group.

The divergence of opinions regarding the role of central apical initials in root meristems is brought into clear focus in relation to the root apices of lower vascular plants where, in many cases, a morphologically distinctive apical cell is present. It should be possible here at least to obtain an unequivocal evaluation of the role of that cell in root histogenesis. Even here, however, interpretations vary. A number of investigators have described the apical cell in roots of several species as mitotically inactive beyond a very early stage of development, a kind of one-celled quiescent center (D'Amato, 1975). It is even reported to become polyploid as a result of DNA replication, without subsequent mitosis. Others, however, have found a high level of mitotic activity in the apical cell of actively growing roots (Gifford, 1983). These conflicting claims appear to have been reconciled by Gunning, Hughes, and Hardham (1978) in a meticulous analysis of cell lineages in roots of the aquatic fern *Azolla*. Their documentation of the highly regular pattern of cell divisions in the root leaves little doubt that all cells are ultimately derived from the apical cell. However, divisions in this cell are completed before the root reaches 1 mm in length and the root is determinate in its growth. The occurrence of a similar determinate growth in other species could explain why some workers observe mitotic activity in the apical cell while others do not. In fact a comparable situation appears to occur in the roots of some aquatic angiosperms, such as *Pistia* and *Eichornia*. Clowes (1958a) has observed that all mitosis in the apex of lateral roots of both of these ceases soon after emergence from the tissue of the parent root and subsequent growth is by cell elongation only. Clearly, it is important to understand the growth of an organ when attempting to interpret the functioning of its meristem.

Analytical investigations of the type described thus far have not resolved the question of the location of the cells that give rise to all the tissues of the root. In the analysis of the shoot apical meristem (see Chapter 5) it was shown that the method of clonal analysis, in which the progeny of marked cells can be traced, is a valuable tool in dealing with such problems, but this method has had a very limited application to roots. The most detailed analysis of this type was carried out by Brumfield (1943), who x-irradiated seedling roots of *Vicia faba* and *Crepis capillaris* and after a period of growth sectioned the roots and searched for patterns of distribution of x-ray-induced chromosomal aberrations. Where such aberrations were found in dividing cells they occurred in sectors occupying approximately one-third of the root and including all root tissues from the center to the surface and even the root cap. Such sectorial chimeras led Brumfield to postulate the existence of three initial cells, each of which ultimately gives rise to all the tissues of a sector of the root. This conclusion has been criticized on the grounds that the results might reflect the regeneration of a root apex from a small number of cells left viable after the radiation treatment and might have little relationship to normal apical structure. However, Brumfield has called attention to the occurrence of apparently stable polyploid root chimeras reported by earlier workers whose interest was primarily spontaneous changes in chromosome number. Since these chimeras were naturally occurring, they cannot be ascribed to the results of experimental treatment. In almost all cases these chimeras were sectorial, indicating a central group of initials, each of which gives rise to all of the cells in a particular sector of the root, rather than a large number of initials surrounding an inactive center. Further, it raises questions about the significance of meristem layers where these are found, since each layer would have been expected to have its own initials.

Some revealing observations on the location of the initial cells in root meristems have been reported by Davidson (1961) from his study of colchicine treated seedling roots of *V. faba*. The treatment induces polyploidy in cells that are in division during exposure to this substance. Subsequent to the treatment the apex of the primary root contained many polyploid cells, but these were progressively lost as growth continued and no stable chimeras were formed. If, on the other hand, the colchicine treatment was given following surgical or radiation damage to the root apex, then stable polyploid chimeras were formed. These findings have been interpreted as showing that the true initials of the root are located in the quiescent center. Being relatively inactive, they were not affected by the colchicine in the initial root and gradually repopulated the meristems with normal cells. In the damaged roots, however, some of the quiescent center cells, having been stimulated into mitotic activity, were affected by the treatment and, as altered initial cells, continued to provide altered derivatives.

EXPERIMENTAL INVESTIGATION OF ROOT DEVELOPMENT

As in the shoot apex, it is possible to learn something about functional organization of the root apex using experimental procedures in which normal developmental processes are disturbed. In view of the often stated simplicity of the root, it is surprising to find that relatively little experimental work of this type has been done. It must be pointed out, however, that, in addition to the difficulties of handling roots and root systems, the fact that the root meristem is enclosed makes the use of surgical methods somewhat awkward, because the meristem is not visible and usually cannot be exposed. Nevertheless a few workers have extracted important information about the root by these means.

The growth in sterile culture of excised root tips has been a means of amassing a large body of physiological information about root growth and the technique is essentially a routine one. Since, however, the excised tips are ordinarily at least 0.5 cm in length and include a substantial segment of differentiating tissue, the results tell us very little about the inherent potentialities of the apical meristem. Information of the latter type has been provided by Torrey (1954), who was able to culture excised apices of *Pisum* only 0.5 mm in length including the root cap, which was left in place. The isolating cut passed through a region of the root just proximal to the meristem, at which level the vascular tissue was in an early procambial stage of differentiation. These explants were capable of forming roots, but the nutritional requirements were rather exacting and half-millimeter tips failed to grow on a medium that would support the growth of larger explants. However, the additional requirements for organic supplements, such as vitamins, were entirely ones of concentration and the only new substances required were inorganic micronutrient salts.

These experiments point to the essential autonomy of the root meristem, but they do not establish it conclusively, because the small explants that were capable of organized growth included differentiating tissues of root cap and root axis in which patterns of organization are already established. All doubt, however, has been removed by the more recent experiments of Feldman and Torrey (1976) on seedling roots of *Zea* in which the quiescent center alone, when cultured separately, gave rise to an entire root (Fig. 12.9). The isolated pieces, no more than 0.25 mm^3 in volume, gave rise to entire root apices directly and retained the original polarity of the root from which they had been isolated. Only a culture medium supplemented with organic nitrogen, IAA, and kinetin was adequate to support organized root development from isolated quiescent centers. Nonetheless, this is a remarkable demonstration of the inherent or autonomous organizing capacity of the root meristem, in fact of only a portion of the meristem, a capacity comparable to that of the shoot meristem.

A different kind of experiment has been used to investigate organization within the meristem itself. Longitudinal incisions that split the root meristem have been carried out by Ball (1956) on the mature embryo of

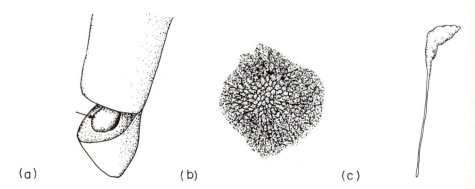

(a)   (b)   (c)

Figure 12.9. *Diagrams illustrating the isolation and growth in culture of the quiescent center of the root of* Zea mays. *(a) Root cap and quiescent center (arrow) partially dissected from root. (b) Surface view of the isolated quiescent center. (c) Root developed from an isolated quiescent center. (a) ×30, (b) ×80, (c) Approximately natural size. (Drawn from Feldman and Torrey, 1976.)*

Ginkgo and by Pellegrini (1957) on *Phaseolus* (bean) seedlings, and the results have been the same in both cases. Each side of the split meristem underwent reorganization and produced a complete root. Ball has reported that where the division was unequal, two roots of unequal size were formed. It therefore must be concluded that, as in the shoot, a portion of the root meristem has the capacity to form a complete meristem.

A similar sort of capability was demonstrated by Clowes (1953, 1954) in surgical experiments on root apices of *Vicia, Fagus, Zea,* and *Triti-*

Figure 12.10. *Surgical experiments that examine the size and organization of the root meristem. (a–c) Diagrams that show three types of experiments performed on the root tip. The incisions remove different regions of the terminal meristem. The incision in (a) removes an oblique flank of the meristem; the incision in (b) removes a vertical wedge; and that in (c) removes a horizontal wedge. (d) Median longitudinal section of a root tip of* Zea mays *fixed eight days after an operation to the left side of the apex of the type shown in (a) or (b). On the right side of the root the normal relationship between cell layers in the meristem and in the differentiating tissues has been maintained. The epidermis and the cortex on this side of the root are shown limited by the dark boundaries that trace their origin into the meristem. The root cap originates from a distinct cell layer in the meristem on this side of the root. On the left side of the root regeneration following the operation has obliterated the normal cell relationships. No distinct root cap initiating cell layer in the meristem is evident, and files of cells extending from the meristem toward the surface provide evidence of extensive cell proliferation. One such file of cells is shaded in the diagram, and others are visible near it. Key: C, cortex; E, epidermis; RC, root cap. (d) ×470. ([a–c] Clowes, 1953. [d] Clowes, 1954.)*

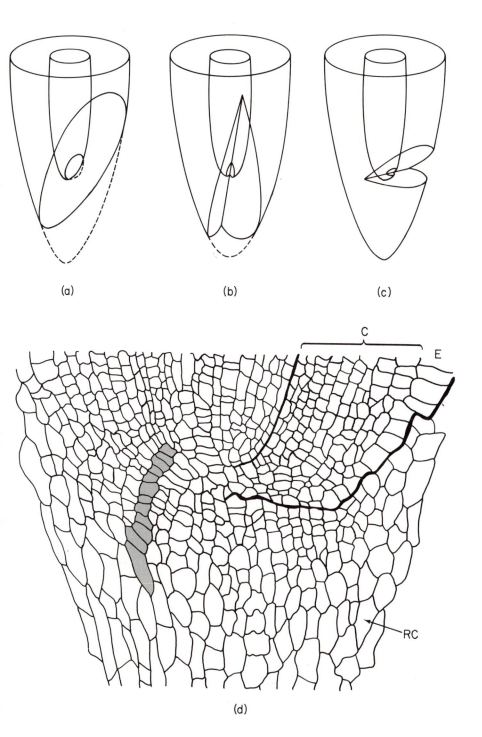

Figure 12.10. *(Caption on facing page).*

cum. He made a series of glancing or wedge-shaped excisions of portions of the meristem to various depths. In all cases, as long as some portion of the meristem remained, a complete apex was reconstituted. A further important observation was made in the roots in which distinct meristem layers could be noted (*Zea* and *Triticum*). In the undamaged portion of the root apex the normal layers could be observed and their relationships to specific mature tissues were as expected. In the regenerated portions of the same apices these relationships had been altered, particularly in the lack of distinctness of a separate layer associated with the root cap, which is a conspicuous feature of root apices of grasses. In effect, the organization was different in different parts of the same root tip yet an essentially normal mature structure was produced (Fig. 12.10).

The capacity of a portion of the root meristem to regenerate a complete apex is also revealed by the experiments previously described in which excised quiescent centers of *Zea* in culture gave rise to functional roots (Feldman and Torrey, 1976). A significant corollary of these experiments was that a root from which the root cap and quiescent center had been removed was also able to regenerate a complete apex with a new quiescent center within three days (Feldman, 1976). The regeneration was accomplished by cells in close proximity to the quiescent center. It thus appears that in the root apex, as in the shoot apex, different regions of the meristem have equivalent capacity to form an entire meristem. By contrast, if an additional 0.3 to 0.35 mm were removed from the root tip, the meristem was not regenerated but precocious lateral roots developed close to the cut surface.

The rapid replacement of the quiescent center in regenerated root apices confirms the conclusion, based upon extensive studies of normal root development, that the quiescent center is an important feature of the organization of the root meristem. Yet, as has been shown in the previous section, it appears to be variable in occurrence, depending upon the stage of root development and conditions to which the root is subjected. There is, naturally, considerable interest in knowing the mechanism by which mitotic activity is restrained in a group of cells in the center of the meristem, and attention has turned to interrelationships among the various regions of the root apex. It has been shown that mitotic activity in the quiescent center is stimulated by partial or complete removal of the root cap, as well as by a number of treatments that cause a cessation of division in surrounding cells, such as x-irradiation or exposure to cold and starvation in cultured roots, when conditions favorable to growth are restored. Two alternative interpretations have resulted from these observations (Clowes, 1976). On the one hand, it is argued that quiescence is maintained as a result of the successful competition for nutrients and hormones by the surrounding active cells, leaving the most central cells inadequately supplied. Alternatively, it is suggested that the active proliferation of surrounding cells places a physical restraint upon the central cells, preventing their expansion and consequently their division. The information presently at hand is inadequate to permit a choice between these two interpretations, each of which has its advocates.

ROOT BRANCHING

With the exception of certain lower vascular plants, chiefly the lycopods, where branching is terminal, branching in the root is a phenomenon that does not directly involve the apical meristem. In the vast majority of vascular plants lateral roots have their origin in the pericycle, one or more layers of cells that constitute the outermost region of the vascular system. The endodermis, the innermost layer of the cortex, may also participate in lateral root formation and in the ferns it is the actual site of origin. Thus, lateral roots do not arise from superficially placed buds as do shoot branches; rather, they have an endogenous origin that necessitates their subsequent penetration of cortical and dermal tissues. Observation of lateral root primordia is thus as difficult as observation of the terminal meristem of the root. Although in a few aquatic angiosperms, such as *Eichornia crassipes* (water hyacinth), lateral root primordia arise very close to the root apex, in most cases their origin is well removed from the apex to the region of the parent root in which elongation has been completed.

A root primordium is formed by divisions in a localized group of pericycle cells that give rise to a hemispherical mound of meristematic tissue (Fig. 12.11). Continued oriented divisions lead to the establishment of a terminal meristem and root cap well before the young root has penetrated to the exterior of the parent root. In most cases the pericycle cells from which the lateral root primordia arise are partially or even highly differentiated, and cytological reorganization precedes cell division. This includes increased cytoplasmic density, reduction in vacuolation, enlargement of nuclei, and wall modifications. Further evidence that cell division is not the initial event is provided by certain cases in which division was blocked by colchicine or other mitotic inhibitors. In these instances the general form of young primordia was attained by enlargement of localized groups of pericycle cells (Fig. 12.12). When the inhibitor was removed, these pseudoprimordia gave rise to lateral roots (Foard, Haber, and Fishman, 1965).

The observation that lateral roots are initiated at some distance from the tip of a growing root suggests either that a certain level of tissue

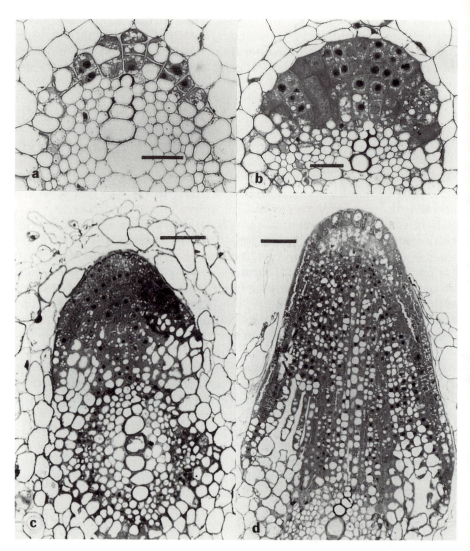

Figure 12.11. *Origin and early growth of lateral root primordia in* Raphanus sativus. *(a) Initial divisions in the pericycle opposite a protoxylem strand. (b,c) Stages in the organization of the root meristem. (d) Lateral root emerging from the parent root cortex. Scale bars (a,b) 25 μm, (c,d) 50 μm. (a) ×440, (b) ×340, (c) ×240, (d) ×190. (L. M. Blakely, M. Durham, T. A. Evans, and R. M. Blakely,* Bot. Gaz. 143:341, 1982.)

maturity is required before initiation can occur or that the parent root exerts some regulatory influence upon the process of initiation. The second of these suggestions has attracted some interest and has been the object of several experimental investigations. A number of workers have shown that removal of the root tip promotes the formation of lateral roots. Other experiments have shown that excision of more mature regions of roots growing in sterile culture reduces the formation of laterals. Thus, it appears that the initiation of laterals is inhibited by the root tip and promoted by more mature regions. Presumably the position at which a new lateral root is initiated is determined by the interaction of stimuli from these two locations (McCully, 1975).

Although all the cells of the pericycle at a particular distance from the root apex presumably are equally capable of initiating a root, it is an interesting observation that lateral roots are spaced with some degree of regularity. The most obvious pattern is one that is related to the internal organization of the parent root. Lateral roots are ordinarily distributed in rows or orthostichies opposite the protoxylem poles of the vascular system, with one or two rows opposite each pole. Other patterns, such as roots situated opposite the phloem poles, are known. There is a small amount of evidence suggesting the existence of distribution patterns beyond that of longitudinal rows. Working with roots of *Musa* (banana) and several other species, Riopel (1966, 1969) observed that the spacing between any particular lateral root and the next to arise in acropetal sequence was greater than would be expected on the basis of random distribution. If the next root was relatively close in the lateral direction (30° divergence or less), it tended to be displaced longitudinally more than the average distance. This suggested that there is an inhibiting influence exerted by existing lateral roots upon the initiation of new ones. However, a somewhat different pattern has been described in other species, notably *Cucurbita maxima* (Mallory et al., 1970). In these species, which had relatively fewer orthostichies than did those described by Riopel, although the lateral roots were well spaced in the rows, they arose in horizontal groups having a clumped appearance. In these cases it seems unlikely that there is an inhibitory influence of lateral roots upon one another. Rather, the interaction between

Figure 12.12. *Longitudinal section of a colchicine-treated root of* Triticum *showing the apparent initiation of a lateral root without cell division. Enlarged pericycle (P) and endodermal (E) cells indicate the site of the pseudoprimordium.* ×35. *(Drawn from Foard, Haber, and Fishman, 1965.)*

an inhibition from the parent root apex and a promoting effect from its more mature regions seems to offer a more plausible interpretation.

ASSOCIATIONS WITH MICROORGANISMS

The roots of vascular plants form a variety of symbiotic relationships with microorganisms in the soil which, in addition to their physiological significance, may have a profound effect upon root development. The root systems of many temperate-zone forest trees and shrubs are distinctively modified by the development of ectomycorrhizae in which fungal hyphae penetrate between the cells of the root epidermis and cortex and also form a dense mantle over the surface of the root (Marks and Kozlowski, 1973). In these root systems it is mainly lateral short roots that are affected. The longitudinal growth of these roots is restricted. They become swollen by the radial extensions of the epidermal or cortical cells, there is no development of root hairs, and the root cap, although present, is greatly reduced in extent. In some species there is extensive branching of the infected short roots that may result in a coralloid or almost nodular appearance. It has been demonstrated that the exudates of mycorrhizal fungi can induce the characteristic mycorrhizal morphology in roots growing in sterile culture or in attached roots in soil, and auxins can duplicate at least some of these effects. Surprisingly, in endomycorrhizae where fungal hyphae actually invade the cells of the host root, there is little or no influence upon root morphology. Both types of mycorrhizae play an important role in the nutritional economy of the host plant in absorbing nutrients from the soil.

Another association involving lateral roots is that formed between a number of mainly woody angiosperms, for example, *Alnus, Casuarina*, and various forms of filamentous bacteria (actinomycetes) of the genus *Frankia* (Becking, 1975). In these actinorrhizal associations, which are important in the fixation of atmospheric nitrogen, filaments of the microorganism gain entry to a host root by way of root hairs and advance to the cortical region, where cells are stimulated to enlarge and divide. When a branch root is initiated, if it grows through the infected cortical region, it itself becomes infected. In one type of development there is a repeated initiation of successive orders of stunted lateral roots resulting in an enlarged coralloid structure sometimes called a nodule. In other species the development is similar except that the successively formed branch roots are only temporarily arrested and each one ultimately grows out in a seemingly normal manner, except that it has no root hairs and lacks the normal response to gravity. These latter roots are determinate and rather short lived but the overall nodular outgrowth in both types is perennial, being maintained by the initiation of successive orders of branch roots, and may attain a diameter of several centimeters.

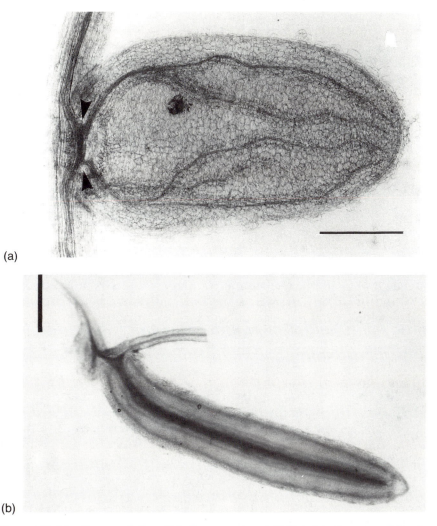

(a)

(b)

Figure 12.13. *Root nodules of* Medicago sativa *(alfalfa). (a) Mature nodule cleared with lactic acid showing the pattern of vascular strands. The attachment of the nodular vascular bundles to the root stele is indicated by arrows. (b) Aberrant nodule formed as a result of inoculations with a mutant strain of* Rhizobium. *There is a single, central vascular strand as in a normal root. (a) ×22, (b) ×15. Scale bars, 1 mm. (M. E. Dudley, T. W. Jacobs, and S. R. Long,* Planta *171:289, 1987.)*

Probably the most familiar root–microorganism associations are those formed by bacteria of the genus *Rhizobium* with the roots of most members of the legume family (Fabaceae). These associations have long been recognized as major contributors to the fixation of atmospheric nitrogen (Dart, 1975). In these, as in the actinorrhizal associations, access to the root is gained through root hairs, and this is followed by the formation of an infection thread that invades the cortex. There localized meristematic activity is stimulated, primarily in cortical cells but in some cases with the participation of the pericycle. As the nodule develops, meristematic activity in it is localized on the side facing the root surface, and the resulting growth leads to the emergence of a nodule that through further growth develops an elongate form. The cells containing bacteria are concentrated in the center of the elongate nodule, and external to this zone are found vascular strands and an endodermis, both of which are in continuity with corresponding tissues in the parent root (Fig. 12.13). In some legume species meristematic activity is more generally distributed around the exterior of the nodule, resulting in a more spherical form. Unlike the two previous types of association, legume nodules are not modified lateral roots, but they do show interesting similarities, particularly those with elongate form. A further point to note is that apparently normal roots sometimes develop from nodules. In essence it appears that, when meristematic activity is induced in the root by the invading bacteria, some portions of the genetically controlled developmental patterns of lateral roots are activated.

GENERAL COMMENT

This treatment of developmental processes in the root has been relatively brief compared to that devoted to the shoot and its component parts. This brevity of treatment is possible largely because the root and shoot systems, in spite of their seemingly great differences, have much in common in their basic organization. Both represent continually expanding systems that carry out a three-dimensional exploration of the environment, although the environments are rather different. In both systems the form depends largely upon the activity of the terminal meristems and of the recent derivatives of these meristems, not only for the continued or indefinite growth pattern but for the kind and degree of branching as well.

Although the apices of root and shoot appear rather dissimilar in organization, the differences are more apparent than real. In the descriptions of many root apices, emphasis is often placed upon a layered or tiered structure. However, a tendency to meristem layering is present in both shoot and root meristems of seed plants, but it is often more striking in roots because regular patterns of cell division are not disturbed there by the periodic initiation of appendages as in the shoot.

Perhaps more striking is the seeming comparability of root and shoot apices in the possession of a central zone that is quiescent, or relatively so, as far as cell division is concerned. Although this feature of organization was more readily accepted for the root than for the shoot, its presence in some form in the shoot apex has gained substantial support. There is no need to suppose that the quiescent region must be identical in shoot and root, but its presence in some form in both would give strong support to the idea that it plays a fundamental role in the functioning of terminal meristems. It may not be too soon to propose that a low rate of division is a requirement for the long-term retention of initial cells within this region. This suggestion is supported by the fact that the analogous stem cells in animals also have a low mitotic frequency.

Finally, it may be noted that experiments have shown that both shoot and root apices are to a large extent autonomous or self-determining in their development. These formative centers are differentiated in the early development of the embryo, and they subsequently appear to produce their respective systems with little more than basic nutritional support from the already mature portions of the plant. Although this last statement is not absolutely proven, experimental evidence increasingly points to its validity.

There are, however, important differences between developmental processes in shoot and root that must be noted. Whereas the shoot meristem is located terminally in the shoot, the root meristem is subterminal as a result of its production of mature derivatives acropetally as well as basipetally. The functional significance of this difference is obvious, but its bearing upon the organization of the root and shoot apex should perhaps be considered. In the shoot apex, the surface is extremely important, and only portions of meristem with the ordinary surface layers present can regenerate whole apices. In the root, on the other hand, this limitation does not apply: Decapitated roots can regenerate new apices. The appendages of the shoot, whether leaves or buds, have their origin at the surface and at the margin of the meristem. In the root, however, there is no appendage formation at the surface or near the meristem, and indeed there could not be because any protuberance would be destroyed as the apex is forced through the soil by subapical elongation or would impede that necessary elongation. Thus, lateral roots are initiated at some distance from the apex, by the activation of cells of the pericycle, and at some distance within the tissue of the root. The absence of appendages at the apex results in a much greater regularity of developmental processes within the root apex and ultimately in the development of a nonsegmented axis in contrast to the nodal–internodal organization of the stem. This has important consequences for the processes of elongation and differentiation.

Thus there are significant differences between the developmental processes in roots and shoots, but reflection upon the even greater similarities leads to the conclusion that these differences are superimposed upon a fundamentally homologous plan of organization. If the original land vascular plants were rootless, as is commonly supposed – that is, if the plant body was not differentiated into root and shoot systems as in more recent vascular plants – then these two systems probably represent evolutionary modifications of an original organizational plan in relation to two different environments. The fundamental similarities of root and shoot apices may well reflect the common origin, and the differences may reflect the evolutionary adaptations to contrasting environments.

REFERENCES

Armstrong, J.E., and C. Heimsch. 1976. Ontogenetic reorganization of the root meristem in the Compositae. *Am. J. Botany* 63:212–19.

Ball, E. 1956. Growth of the embryo of *Ginkgo biloba* under experimental conditions. II. Effects of a longitudinal split in the tip of the hypocotyl. *Am. J. Botany* 43:802–10.

Barlow, P. W. 1973. Mitotic cycles in root meristems. In *The Cell Cycle in Development and Differentiation*, eds. M. Balls and F. S. Billett, 133–65. Cambridge: Cambridge University Press.

Becking, J. H. 1975. Root nodules in non-legumes. In *The Development and Function of Roots*, eds. J. G. Torrey and D. T. Clarkson, 507–66. New York: Academic Press.

Brumfield, R. T. 1943. Cell-lineage studies in root meristems by means of chromosome rearrangements induced by X-rays. *Am. J. Botany* 30:101–10.

Clowes, F.A.L. 1953. The cytogenerative centre in roots with broad columellas. *New Phytol.* 52:48–57.

──────. 1954. The promeristem and the minimal constructional centre in grass root apices. *New Phytol.* 53:108–16.

──────. 1956a. Nucleic acids in root apical meristems of *Zea. New Phytol.* 55:29–34.

──────. 1956b. Localization of nucleic acid synthesis in root meristems. *J. Exp. Botany* 7:308–12.

──────. 1958a. Development of quiescent centres in root meristems. *New Phytol.* 57:85–8.

──────. 1958b. Protein synthesis in root meristems. *J. Exp. Botany* 9:229–38.

──────. 1961a. Duration of the mitotic cycle in a meristem. *J. Exp. Botany* 12:283–93.

──────. 1961b. *Apical Meristems.* Oxford: Blackwell.

──────. 1976. The root apex. In *Cell Division in Higher Plants*, ed. M. M. Yeoman, 254–84. New York: Academic Press.

Clowes, F.A.L., and B. E. Juniper. 1964. The fine structure of the quiescent centre and neighbouring tissues in root meristems. *J. Exp. Botany.* 15:622–30.

Clowes, F.A.L., and H. E. Stewart. 1967. Recovery from dormancy in roots. *New Phytol.* 66:115–23.

D'Amato, F. 1975. Recent findings on the organization of apical meristems with single apical cells. *Giorn. Bot. Ital.* 109:321–34.

Dart, P. J. 1975. Legume root nodule initiation and development. In *The Devel-*

opment and Function of Roots, eds. J. G. Torrey and D. T. Clarkson, 467–506. New York: Academic Press.

Davidson, D. 1961. Mechanisms of reorganization and cell repopulation in meristems of roots of *Vicia faba* following irradiation and colchicine. *Chromosoma* 12:484–504.

Erickson, R. O., and K. B. Sax. 1956. Rates of cell division and cell elongation in the growth of the primary root of *Zea mays. Proc. Am. Phil. Soc.* 100:499–514.

Feldman, L. J. 1976. The *de novo* origin of the quiescent center in regenerating root apices of *Zea mays. Planta* 128:207–12.

Feldman, L. J., and J. G. Torrey. 1976. The isolation and culture *in vitro* of the quiescent center of *Zea mays. Am. J. Botany* 63:345–55.

Foard, D. E., A. N. Haber, and T. N. Fishman. 1965. Initiation of lateral root primordia without completion of mitosis and without cytokinesis in uniseriate pericycle. *Am. J. Botany* 52:580–90.

Gifford, E. M., Jr. 1983. The concept of apical cells in bryophytes and pteridophytes. *Ann. Rev. Plant Physiol.* 34:419–40.

Gill, A. M., and P. B. Tomlinson. 1975. Aerial roots: An array of forms and functions. In *The Development and Function of Roots,* eds. J. G. Torrey and D. T. Clarkson, 237–60. New York: Academic Press.

Goodwin, R. H., and C. J. Avers. 1956. Studies on roots. III. An analysis of root growth in *Phleum pratense* using photomicrographic records. *Am. J. Botany* 43:479–87.

Gunning, B.E.S., J. E. Hughes, and A. R. Hardham. 1978. Formative and proliferative cell divisions, cell differentiation and developmental changes in the meristem of *Azolla* roots. *Planta* 143:125–44.

von Guttenberg, H. 1964. Die Entwicklung der Wurzel. *Phytomorphology* 14:265–87.

Jensen, W. A., and L. G. Kavaljian. 1958. An analysis of cell morphology and the periodicity of division in the root tip of *Allium cepa. Am. J. Botany* 45:365–72.

Langenauer, H. D., E. L. Davis, and P. L. Webster. 1974. Quiescent cell populations in apical meristems of *Helianthus annuus. Can. J. Botany* 52:2195–2201.

Mallory, T. E., S. Chiang, E. G. Cutter, and E. M. Gifford, Jr. 1970. Sequence and pattern of lateral root formation in five selected species. *Am. J. Botany* 57:800–9.

Marks, G. C., and T. T. Kozlowski. 1973. *Ectomycorrhizae: Their Ecology and Physiology.* New York: Academic Press

McCully, M. E. 1975. The development of lateral roots. In *The Development and Function of Roots,* eds. J. G. Torrey and D. T. Clarkson, 105–24. New York: Academic Press.

Pellegrini, O. 1957. Esperimenti chirugici sul comportamento del meristema radicale di *Phaseolus vulgaris* L. *Delpinoa* 10:187–99.

Peterson, R. L., and J. Vermeer. 1980. Root apex structure in *Ephedra monosperma* and *Ephedra chilensis* (Ephedraceae). *Am. J. Botany* 67:815–23.

Phillips, H. L., Jr., and J. G. Torrey. 1971. The quiescent center in cultured roots of *Convolvulus arvensis* L. *Am. J. Botany* 58:665–71.

———. 1972. Duration of cell cycles in cultured roots of *Convolvulus. Am. J. Botany* 59:183–8.

———. 1974. The ultrastructure of the quiescent center in the apex of cultured roots of *Convolvulus arvensis* L. *Am. J. Botany* 61:871–8.

Popham, R. A. 1955. Zonation of primary and lateral root apices of *Pisum sativum*. *Am. J. Botany* 42:267–73.

Raju, M.V.S., T. A. Steeves, and J. Maze. 1976. Developmental studies on *Euphorbia esula:* Seasonal variations in the apices of long roots. *Can. J. Botany* 54:605–10.

Raju, M.V.S., T. A. Steeves, and J. M. Naylor. 1964. Developmental studies of *Euphorbia esula* L: Apices of long and short roots. *Can. J. Botany* 42:1615–28.

Riopel, J. L. 1966. The distribution of lateral roots in *Musa acuminata* "Gros Michel." *Am. J. Botany* 53:403–7.

———. 1969. Regulation of lateral root positions. *Bot. Gaz.* 130:80–3.

Torrey, J. G. 1954. The role of vitamins and micronutrient elements in the nutrition of the apical meristem of pea roots. *Plant Physiol.* 29:279–87.

CHAPTER 13

Differentiation of the plant body: the origin of pattern

The continued growth of the plant body depends upon the production of new cells by mitotic activity in its meristematic regions. One might predict that this would result in a homogenec 's cell population, because mitosis ordinarily leads to the formation of identical sister cells. It is obvious, however, that the plant body does not consist of such a uniform assemblage of cells. Rather, it is composed of diverse specialized cells arranged in patterns having functional significance. If this were not the case, the plant could function, at best, in only a very restricted manner. The phenomenon of *differentiation,* as this production of diverse cell types in definite patterns is called, has been alluded to in earlier chapters because it is almost impossible to consider growth apart from it. Now it is necessary to turn attention specifically to the phenomenon itself, one of the major topics of interest in modern developmental biology.

GENETIC CORRELATES OF DIFFERENTIATION

The diversity of differentiated cell types might suggest that genetic changes must be involved in differentiation. The preponderance of evidence, however, indicates that cellular diversity within the organism is accomplished in a framework of genetic homogeneity. The most striking evidence in support of this principle is to be found in the well-known regeneration phenomena characteristic of plants, which will be discussed fully in Chapter 17. Roots, shoots, and in many cases whole plants are often regenerated from fully differentiated cells either as a normal process or as a result of wounding or some other stimulus. Because the regenerated plants are apparently quite normal and contain all of the cell types characteristic of the species, it must be concluded that the differentiated cells that gave rise to them had not undergone mutation as part of their differentiation. Although it is not possible to observe such regeneration from all tissues of all plants, the phenomenon is of such widespread occurrence that a mutational basis for differentiation is essentially ruled out.

A possible exception of the principle of genetic homogeneity in development has been recognized in the widespread occurrence of quantitative chromosomal changes in differentiated plant cells. Differentiated tissues in many plants have been shown to contain some polyploid cells (Nagl, 1978). In *Scilla decidua* all differentiated tissues and organs examined were found to contain endopolyploid cells, the plant as a whole having endopolyploid nuclei in 70 percent of its somatic cells (Frisch and Nagl, 1979). However, in some plants endopolyploidy is absent (Partanen, 1959). It is unlikely, therefore, that these quantitative changes have a causal role in cellular differentiation. Furthermore, there is usually no regular pattern of distribution of endopolyploid cells that can be correlated with particular patterns of cell types. In a few cases such a relationship has been demonstrated. Root hairs of *Hydrocharis* are initiated by unequal division of a cell in the epidermis. The larger cell divides to produce four or more epidermal cells; the smaller cell develops as a root hair. Cutter and Feldman (1970) found that nuclei of the epidermal cells were diploid but the nucleus of the root hair cell became endopolyploid, reaching an octoploid DNA content. In another example it was shown by Shanks (1965) that in the formation of stomata on leaves of *Galtonia candicans* there is an unequal division of an initial cell into a small cell that subsequently divides to form the two guard cells of the stoma, and a larger cell that enlarges and matures as an epidermal cell. The nuclear DNA content of the guard cell precursor was diploid, but that of the epidermal cell increased about tenfold. As in the case of root-hair formation in *Hydrocharis* there is a precise relationship between nuclear DNA content and cell type in the leaf epidermis of *Galtonia*, but in both cases the initial differentiation involved an unequal division that occurred while the nuclei contained the diploid DNA content, and endopolyploidy occurred during subsequent development. Whether endopolyploidy is required for, or is simply another manifestation of, differentiation in these cases is not known. It is interesting that in both *Hydrocharis* and *Galtonia* it was the cell that enlarged greatly after division that underwent DNA endoreduplication; in both cases the smaller cell and its derivatives remained diploid.

It has been observed in some plants that although the nuclei of differentiated cells contain more than the diploid amount of DNA, the amount does not represent an exact doubling of the genome. It has been suggested that this represents the differentiatial replication or amplification of specific genes or DNA sequences. This has been investigated most fully in *Linum usitatissimum* (flax) by Cullis (1984), who showed that under certain growth conditions several portions of the genome, including the ribosomal genes, were amplified during development. Present methods of analysis do not permit determination of whether there are cell-specific or tissue-specific differences in the degree of amplification, which would be required if gene amplification were the basis for cell differentiation.

If differentiation cannot be equated with mutational or quantitative changes in the genes, it becomes a problem to visualize how so many different cell types can arise under the control of the same genetic matial, yet it is unrealistic to seek mechanisms that operate only in the extranuclear portions of the cell. It has long been suggested that the answer to this dilemma must lie in differential gene expression so that, though a particular cell type during its development contains all of the genes underlying all of the characteristics of all of the cell types, only certain of these are active in a cell at any one time. Thus there *are* differences among the nuclei of diverse cell types but they are of a functional nature and are not necessarily irreversible.

The rapidly developing field of molecular genetics has begun to suggest a mechanism by which differential activation and repression of genes can be accomplished. It has been proposed that other components of chromosomes, for example histones and other nuclear proteins, may interact with the DNA in such ways that certain genes are repressed and others activated, perhaps by physical masking and unmasking that affects the transcription of messenger RNA. The mechanism of gene masking and unmasking is still far from being understood, but it does provide a model for differential gene expression. The individual activation or repression of genes in development, however, would require a maze of stimuli that would seem to be most unlikely to culminate in organized development in a multicellular organism. This problem has been greatly alleviated through the discovery by Jacob and Monod (1963) that groups of genes are linked together into functional units, each under the control of an operator gene that in turn is controlled by a regulator gene. In this system a single stimulus can result in the activation, or derepression, of all the genes in the operon, or functional unit. Thus, a relatively small number of stimuli or inducing substances, working through the regulator genes, could set in motion complex developmental phenomena involving many genes. As development continues, new products formed in the cytoplasm could evoke a sequence of new events, and the whole system could remain continuously responsive to external influences.

To date this operational mechanism has been elucidated primarily in bacteria, and operons have not been found in higher plant genomes. But in plants, as in bacteria, there are occasions when numerous genes are expressed in a coordinated manner. One example is the expression of the approximately 120 genes that code for zein, the storage protein accumulated in the endosperm of maize (Viotti et al., 1979). These genes are located at numerous sites throughout the genome and none is expressed during early development of the endosperm. However, they are expressed in a synchronous manner later in endosperm development. The molecular mechanism by which expression of the zein gene family is regulated is not yet understood, but it is known that expression of these genes and of several other genes that code for endosperm proteins in maize is affected by another gene, opaque-2. Opaque-2 causes alteration in the level of expression of these genes, some increasing and others decreasing in activity. How the opaque-2 gene affects the activity of the genes that are under its control is still not known, but its role may be somewhat similar to that of the regulator gene in the Jacob and Monod operon model of gene expression.

The way in which one gene affects expression of other genes in plants has been partially elucidated by studying still other maize genes that are active during endosperm development. Gene *A* is required for the production of anthocyanin. When it is present and fully active, the aleurone layer of the endosperm has a deep purple color. Expression of *A* is affected by another gene, suppressor-mutator (*Spm*). *Spm* is a member of a class of genes designated controlling elements by McClintock (1965), having the ability to affect expression of other genes in the same nucleus and to transpose from one chromosomal site to another during cell development. More than one copy of *Spm* may be present in the genome. The different members of the *Spm* gene family are located at various chromosomal sites, and different *Spm*s can have different actions on other genes. McClintock found that if an *Spm* that was defective in its action was located on the chromosome adjacent to *A*, there was only a low level of expression of *A* and the aleurone was light purple in color instead of being deeply pigmented. If a copy of nondefective *Spm* was also present in the genome, expression of *A* was completely suppressed and aleurone cells were colorless. However, *Spm* has a second function that McClintock termed mutator, resulting in the occasional transposition of the defective *Spm* away from its chromosomal site adjacent to *A*. In those cells where transposition of the defective *Spm* occurred, *A* resumed its full activity and a clone of deeply pigmented aleurone cells was formed. Maize kernels in which transposition of the defective *Spm* has occurred during development have a background of colorless aleurone cells and a pattern of pigmented spots, each spot being the result of a single transposition event.

Controlling elements have been shown to be active in other parts of the maize plant. Those that are active in the endosperm are especially amenable to detailed study and may represent a model of how genes of this type control the timing of expression of other genes that are important in cell differentiation throughout the organism. Controlling elements have been identified in several different higher plant species, and functionally similar mobile genetic elements occur in animals and in prokaryotic organisms. Their action results in differential gene expression, but to what extent they represent a widespread mechanism for cell differentiation will not be known until further investigations have been carried out.

The concept of differential gene expression seems to provide a reasonable framework for the analysis of the mechanisms of cellular differentiation, although much further work will be needed to establish its validity in higher organisms. There remains, however, the much more difficult problem of the organized patterns within which cellular differentiation occurs. It is important to know why a particular cell in a developing stem becomes a vessel element, but it is perhaps even more important to understand how this particular cell is integrated into the linear sequence of such elements that constitutes the functional vessel and is in turn part of the vascular system continuous throughout the plant body. Only in this context is the individual vessel element of physiological significance to the plant. The occurrence of levels of organization above the cellular level obviously demands further elaboration of the concepts proposed for cellular phenomena, but it will be well to defer this until after some of the phenomena of differentiation have been considered.

METHODS FOR THE STUDY OF DIFFERENTIATION

The study of differentiation can begin in a variety of ways. It can begin with mature structures that are traced back to their origins or with embryonic features that are followed to maturity. It can start with a study of cellular changes later integrated into the overall plan of the organism, or it can explore the origin of the larger plan and then fill in the cellular details. Because the point of view of this book has been a concern for the plan of the organism, it seems appropriate to explore the origin of higher levels of organization first. Because the basic plan emerges in the earliest stages of differentiation, these stages will be dealt with first in this account, and in Chapter 14 the later stages, together with some of the cellular aspects, will be discussed.

It might be expected that the ideal place to begin this consideration of differentiation would be the zygote, and indeed in the earlier discussion of embryology this was done. However, in the plant, developmental phenomena are not concentrated in an embryonic phase early in the life of the organism. Rather, growth and differentiation are continuing processes associated with the activity of the meristems. Much of the work done on differentiation in plants has been concerned with the derivatives of the meristems rather than the embryo, and particularly with the primary meristems of shoot and root. It is of course recognized that the shoot and root apical meristems themselves reflect a rather high level of differentiation, which becomes apparent when they are set off early in the embryogeny of the plant as distinct shoot- and root-producing entities. They are, however, stabilized in a state that permits their continued proliferation. On the other hand the derivatives of these meristems undergo further differentiation, which, although it may involve active cell proliferation for a time, ultimately results in a cessation of mitotic activity as maturity is approached. Many cells, of course, if their structural modification does not preclude it, may be reactivated subsequently by a suitable stimulus. With regard to the stabilization of the meristems in a partially differentiated state, it is interesting to note that the vascular cambium apparently represents such a stabilization at a further level of specialization. Thus, a consideration of differentiation in relation to the primary meristems does not encompass all aspects of the phenomenon, but it does include most of the work done in this field.

There is a very logical basis for the concentration of effort upon this aspect of differentiation rather than upon embryological stages, which should provide closer comparisons with animal studies. The accessibility of the apices of shoot and root for analysis and experimentation has made them attractive objects for investigation. Perhaps more important, however, is the property of repetitive growth that characterizes these regions, so that organs, tissues, and cell types are produced predictably in a continuing sequence. This has led to a methodology in plant studies that is not ordinarily encountered in animal embryology. The animal embryologist must study a sequence of individual organisms at different stages of development in order to observe progressive differentiational changes. A botanist who studies plant embryos must do the same. However, when investigating the shoot or root apex, one may observe stages of differentiation by progressing basipetally from the meristem into the maturing regions of the organ along the axis. This effectively substitutes an axis of space for time. Caution must be exercised in the practice of this method because there are ontogenetic changes that disturb the strictly repetitive pattern as in the onset of flowering, and the leaves of the shoot, being determinate organs, do not show repetitive growth in their own differentiation. These qualifications, however, pose no problem for the perceptive worker.

EARLY STAGES OF DIFFERENTIATION IN ROOTS

The above method of examining the phenomena of differentiation is equally applicable to shoot and root apices, but because of the presence of leaves and the influences they exert, the patterns in the shoot are more complex. Consequently it seems appropriate to begin this consideration with the root. Differentiation in the root of *Pisum sativum* (garden pea) has been described by several workers, and roots of this species have been used in experimental analysis of differentiation. Thus it will provide a good starting point for this discussion (Torrey, 1955).

Immediately proximal to the promeristem of the *Pisum* root may be detected the first indications of tissue differentiation, but as might be expected they are vague and rather generalized (Fig. 13.1a). This initial differentiation leads to the establishment of three tissue systems: dermal, fundamental or ground, and vascular, distinguishable primarily because of the varying planes of cell division and the early vacuolation of ground tissues. In the ground tissues of the future cortex early divisions are predominantly transversely oriented. This leads to the formation of files of cells extending back along the root axis. However, the number of files of cells in the cortex increases basipetally as a result of periclinal divisions. This is most evident in the innermost layer, and in transverse sections a highly regular radial arrangement of cells may often be seen.

The mature vascular system of the root is a solid core of xylem, fluted in outline, with pholem peripheral to it contained in the bays of the xylem. The protostelic nature of the future vascular system is evident immediately behind the meristem, where it is recognizable as a solid core of rather small densely staining cells within the vacuolating ground meristem of the cortex (Fig. 13.1b). In these cells the plane of early divisions is predominantly longitudinal. As a result cell length soon exceeds transverse width since the longitudinal growth of the core keeps pace with that of the cortex in which the cells are dividing transversely. When the cells have achieved this form it is customary to designate the tissue as *procambium*, a generally recognized early stage in vascular differentiation. Separating the procambium from the surrounding ground meristem is a single layer of cells, the *pericycle*, which becomes distinct at a very early stage and which is of significance as the later site of lateral root initiation. In some species the initial uniform appearance of the procambium is rapidly obscured by the early transverse enlargement of certain cells in the future xylem region in close proximity to the promeristem. This phenomenon will be considered in greater detail in the next chapter.

These considerations of initial differentiation lead to the question of the distribution of cell division and cell enlargement in the root tip. Any suggestion that division, enlargement, and differentiation are spatially separated along the axis must obviously be discarded. As pointed out in Chapter 12, divisions in the root are not limited to a restricted apical

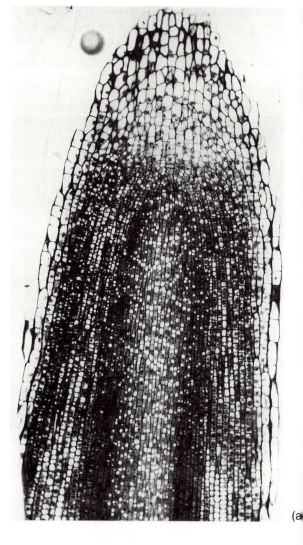

Figure 13.1. *Initial differentiation of tissues behind the root apex of* Pisum sativum. *(a) Longitudinal section of the root tip. The central vascular tissue consists of narrow elongated cells, and the superficial dermal tissue consists of cubical cells. Between the two is the ground tissue in which cell shape is cubical. The root cap covers the root surface at this level. (b) (Figure on next page) Transverse section just behind the terminal meristem. The central vascular tissue consists of very narrow cells. In the vascular tissue the centrally located procambium that will differentiate as metaxylem has started to vacuolate. (a) ×75, (b) ×150.*

(a

zone but continue for some distance along the axis. This was conclusively shown by Jensen and Kavaljian (1958), who undertook an analysis of cell-division frequency and distribution in root tips of *Allium cepa*. These investigators found that the frequency of mitosis in the terminal region of the meristem is low (Fig. 13.2). In fact, they considered their data to support the concept of the quiescent center that had been proposed two years earlier by Clowes on entirely different grounds. The frequency of cell division rose and then declined to zero basipetally along the axis, but the most significant finding was that the peak of mitotic

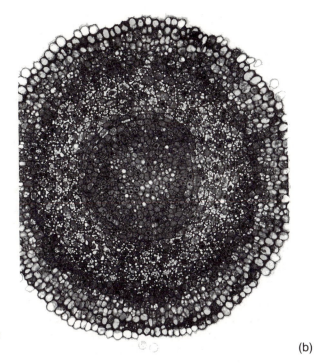

Figure 13.1(b). *(Caption on preceding page).*

(b)

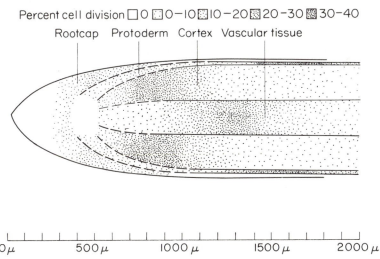

Figure 13.2. *The distribution of cell division in the onion root tip. For each tissue the highest rate of cell division occurs at a different distance behind the tip. The data shown are pooled data from five root tips and were measurements made on noon collections. It was found that the percent of dividing cells varied at different times of the day, but the noon values were high for all tissues. (Jensen and Kavaljian, 1958.)*

frequency occurred at a different position in each of the three tissue systems, being reached first in the ground tissue of the future cortex, then in the vascular core, and finally at about 1 mm from the tip of the meristem, in the dermal tissue. There was also a difference among the tissues in the division frequency achieved at the peaks. Thus, it is apparent that differences in the frequency of cell division form an important part of the initial differentiation of tissues in the root.

The interrelationship between cell division and cell enlargement in a growing organ has been noted previously. The importance of this relationship in the early differentiation of tissues is particularly evident in the root. For example, in a study of the rates of cell division and elongation in roots of *Pisum* and *Hyacinthus,* Webster (1980) found that in the terminal millimeter the rate of division balanced the rate of elongation in the cortex and pericycle, so mean cell length did not change in these tissues. By contrast, in the stele there was a steady increase in cell length resulting from a lower rate of cell division. Further, in *Pisum* cells of the outer region of the stele, which would be the first to complete differentiation, elongated more between successive divisions. Thus, one of the major features of initial differentiation is a consequence of differences in the duration of the cell cycle in different tissues.

It is also evident that the orientation of cell division plays a role in the early delimitation of tissues in the root. Longitudinal divisions increase the number of cell rows extending back from the promeristem while transverse divisions increase the number of cells in each row. Ordinarily divisions having these two orientations are intermingled, although one or the other may predominate in a particular tissue region. Evidence that the orientation of cell division is a highly controlled process that has predictable consequences for subsequent differentiation has come from studies of roots of the water fern *Azolla* (Gunning, 1982). The roots of this plant develop from an apical cell and at any level in the mature root there are usually only fifty-six cells in a cross section. This very small number of cells makes it possible to trace cell lineages with unparalleled accuracy (Fig. 13.3). In each segment cut off from the apical cell there is a complete temporal separation of longitudinal and transverse divisions. The former establish a precise number of cell files

Figure 13.3. *Diagrammatic representation of the production of segments from each of the three internal cutting faces of the apical cell of a root of* Azolla pinnata. *The segments are numbered from 1 to 12 in the order of increasing age. In these twelve segments only longitudinal divisions occur. Numbered arrows indicate the sequence of these divisions. These divisions establish the tissues of the mature root, which are amplified by transverse divisions beyond segment 12. (Adapted from Gunning, 1982.)*

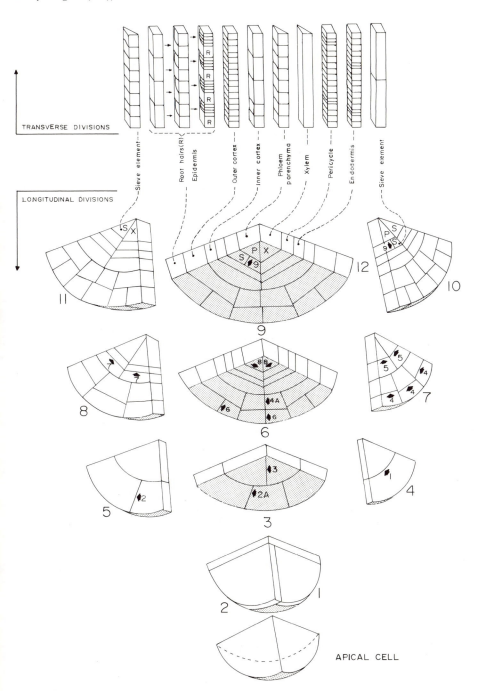

Figure 13.3. *(Caption on facing page).*

before a succession of transverse divisions multiplies the cell number in each file. In effect, the precise destiny of each cell can be predicted. While the *Azolla* root probably represents an extreme example of developmental precision, it does emphasize the significance of the orientation of cell division in the initial differentiation of tissues.

EARLY STAGES OF DIFFERENTIATION IN SHOOTS

Histodifferentiation in the shoot can be studied in the same manner employed for the root. However, the pattern of differentiation is so markedly influenced by the developing leaves that even a morphological description of developmental phenomena in the stem is much more complex than in the root. It cannot be doubted that the physiological basis of differentiation is equivalently complex. In Chapter 11 the segmental nature of the stem with its nodal–internodal organization was discussed in relation to the question of elongation. This pattern has its counterpart in the internal organization of the stem, which must be considered in any treatment of tissue differentiation.

Topography of the Mature Vascular System

Although our concern is the pattern of differentiation of the principal tissues of the shoot, it is appropriate to begin by describing the structure of the mature organ. This is because the course traced by the vascular tissue system in particular is so complex that its differentiation can be comprehended best after the final state is understood.

Examination of the course of the mature vascular system in the stem of a seed plant such as *Linum* or *Helianthus* – either by the study of stems cleared by chemical treatment for translucence, which enables the vascular tissue to be seen internally, or by serial transverse sections – reveals that it consists of a system of discrete strands called vascular bundles (Fig. 13.4a). The bundles, each of which consists of both xylem and phloem, extend along the axis as a system that forms a ring around the central pith (Fig. 13.4b).

An important feature of this vascular system is that at each node one or more bundles turn out from the central cylinder and extend across the cortex into the petiole of the leaf arising at that node. If a vascular bundle extending through a leaf petiole as a leaf trace is followed basipetally into the central cylinder, it may be found to join directly to another bundle, which did not diverge at that node, or more commonly it may be seen to extend downward, sometimes vertically, sometimes obliquely, through one or more internodes before joining another bundle, and other smaller bundles may join it in its basipetal course.

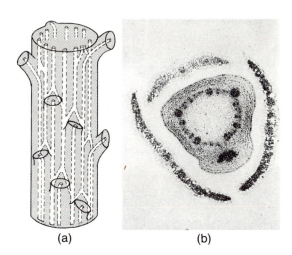

Figure 13.4. The vascular system of Linum. *(a) Three-dimensional diagram showing the vertical extent and interconnection of leaf traces in the stem. (b) Cross section of a stem showing the arrangement of bundles around a central pith and the departure of leaf traces at nodes. (b) ×40.*   (a)   (b)

If the vascular system is traced acropetally along the axis, it may be seen to consist of branching bundles. In some plants these bundles form a system that is completely interconnected, but more commonly the stem contains several independent units that do not interconnect. It has been found that the bundles show a high degree of regularity in their sequence of interconnection. Not surprisingly, these patterns can be correlated with the phyllotaxy of the shoot, and a knowledge of the phyllotaxy can be most useful in interpreting a particular system. If each leaf is served by a single trace, the pattern may be relatively simple, but where multiple traces occur, each with its own course through the stem and ultimate connection with other bundles, the system may be extremely complex. This is very well illustrated in many monocotyledons where large numbers of traces enter the stem from the sheathing bases of the leaves.

Thus, the impression emerges of a unitary vascular system in stem and leaves that could be separated into cauline and foliar components only with great difficulty if at all. In fact, the vascular system in the stem may be described as being largely, if not entirely, leaf oriented. This orientation seems to be eminently sound physiologically, considering that the vascular system is the conducting apparatus of the plant and that the leaf is the major source of elaborated organic compounds and the destination of most of the water that passes through the stem. This vascular pattern may be said to be characteristic of the seed plants generally.

In the lower vascular plants the situation seems to be different. In some, such as *Lycopodium*, there is a central core of vascular tissue to

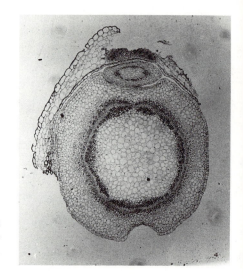

Figure 13.5. *Cross section of the stem of* Dianthus *(carnation) showing an apparent ring of primary vascular tissue.* ×35.

which the leaf traces attach peripherally. Many ferns have a cylinder of vascular tissue surrounding the central pith. Where a leaf trace diverges into a leaf petiole, the region of the vascular cylinder that confronts it is occupied by parenchyma rather than by xylem and phloem. This parenchymatous region, which connects pith and cortex, is called a *leaf gap.* In a great many ferns leaf gaps have a substantial axial dimension, extending through several internodes. Their consequent overlapping causes the vascular system to have the appearance of a ring of bundles when viewed in cross section (Fig. 13.13a), and there may in fact be additional interruptions that are not related to leaf trace departure. It should also be recognized that in some seed plants, such as *Syringa, Dianthus,* and *Hypericum,* the bundles in the stem are extremely broad and may be contiguous, so in transverse section the vascular cylinder seems to be in the form of a continuous ring of xylem and phloem interrupted by leaf gaps as in the ferns (Fig. 13.5). However, it is now believed that, where apparently similar vascular configurations occur in seed plants and lower vascular plants, they have been reached by different evolutionary paths (Beck, Schmid, and Rothwell, 1982).

Differentiation Related to Leaves

The close correlation between leaf and stem vascular systems in the mature state suggests a developmental relationship, and it is to the exploration of this relationship that attention must now be turned.

The study of tissue differentiation in the stem ordinarily is carried

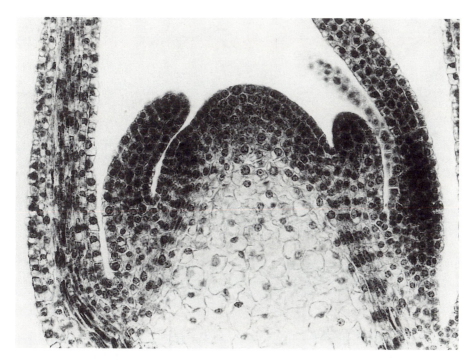

Figure 13.6. *A median longitudinal section of the shoot apex of* Linum *showing early differentiation of the pith and initiation of the procambium in relation to a young leaf primordium on the left side of the meristem.* ×400.

out by examination of both transverse and longitudinal serial sections of the shoot apex together with the subjacent regions in which maturation is occurring. This is essential because the often oblique course of bundles may lead to erroneous interpretations of their course and connections if they are seen only in longitudinal sections. This method and the kind of result it yields are illustrated in a study of tissue differentiation in the shoot of *Linum perenne* carried out by Esau (1942) and generally considered to be the classic description of vascular differentiation in a seed plant. The terminal meristem is a high, domed mound, and leaf primordia are initiated on its sloping flanks (Fig. 13.6). A distinctive feature of differentiation in the stem immediately behind the meristem is that there are differences among the tissues in the rapidity of maturation. Cells of the pith begin vacuolation very close to the summit of the apex and enlarge much more rapidly than do the cells that surround them peripherally. Moreover, at the surface of the apical flanks the protoderm forms a distinctive superficial layer. Between these two distinc-

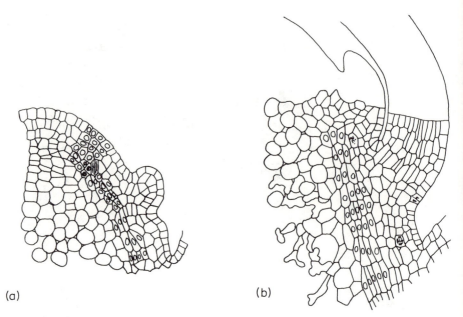

(a)

(b)

Figure 13.7. *Longitudinal sections of the shoot apex of* Linum perenne *showing the correlated initiation of a leaf primordium and its associated procambial strand. Cells with nuclei drawn in are those involved in leaf and procambium development. (a) The upper part of the procambial strand showing the association with the leaf primordium at the time of inception. (b) The lower part of the same strand seen in an adjacent longitudinal section where the strand has diverged around the gap of an older leaf. ×320. (Esau, 1942.)*

tive regions are the relatively unvacuolated cells that will give rise to the cortex and the vascular tissue. It is from this peripheral region also that the leaves arise. The close relationship between the leaves and the vascular tissue of the stem can be seen to have its origin at this level for, from the time of its inception, each leaf primordium has associated with it and extending into it a recognizable strand of procambium (Fig. 13.7a). At its uppermost limits the procambium is identifiable both by the shape of its constituent cells and by their intensified staining reaction to a variety of histological stains. Procambium is initiated by longitudinal divisions that occur with greater frequency than in surrounding cells with the result that procambial cells have an elongate shape, that is, they are longer than they are wide. However, they are initially not necessarily longer than are surrounding cells, although this is clearly true at later stages of development. Thus, the developmental phenomenon that initiates procambium is a preferential orientation of the planes of

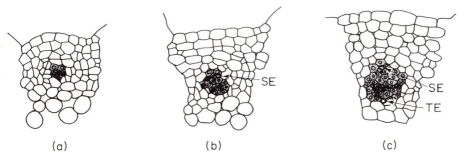

(a)   (b)   (c)

Figure 13.8. *Diagrams of procambial strands associated with successively older leaves of* Linum perenne. *The strands are seen in transverse section. Procambial cells have the nucleus indicated, and it can be seen that the number of cells in the strands increases basipetally. Key:* SE, *sieve element;* TE, *tracheary element.* ×320. *(Esau, 1942.)*

cell division in localized groups of cells. This does not mean that procambial initiation occurs in isolation. Rather, it is clear that this process represents the upper extremity of a sequence of similar divisions occurring in a continuous column of differentiating vascular tissue. Thus, the procambial system differentiates acropetally into the growing apex and in continuity with the existing vascular system. Reports of discontinuity or of basipetal procambial differentiation in shoot apices appear to have been in error and to have resulted from a failure to recognize the oblique course that procambial strands may follow, causing them to appear to end blindly when observed in longitudinal sections (Fig. 13.7a, b). There are, however, some circumstances in which procambium does differentiate basipetally, and examples of these will be discussed in their proper context.

In the procambial system the pattern of the primary vascular tissue of the shoot is established. In the terminal region of the shoot it is much foreshortened as is the stem itself, but it becomes extended in the process of internodal elongation. This is particularly conspicuous in the development of long shoots. During this process there is considerable net cell elongation in the procambium, in contrast to the ground tissues where there is a preponderance of transversely oriented cell divisions. In a series of transverse sections progressing basipetally from the level of initiation, it can be seen that the number of procambial cells in each bundle increases (Fig. 13.8). This is accomplished by longitudinal divisions in the procambial cells themselves and by addition of cells from surrounding regions by the appropriate orientation of cell division planes. This process is of particular importance in those species in which the bundles become very wide and in effect form a continuous primary vascular cylinder.

Differentiation Related to the Terminal Meristem

Before considering the further differentiation of procambium and the final maturation of procambial cells as xylem and phloem, a question concerning initial differentiation must be discussed. This is the question of whether, in fact, the procambium differentiated in relation to leaf primordia is the initial stage of vascular tissue differentiation or whether there is a stage in the process that precedes procambium. Because the evidence pertaining to this question must be obtained in the region immediately subjacent to the terminal meristem where cellular characteristics are at best indistinct, it is not surprising that the question has not been settled. Nonetheless, it is an important issue because it pertains to the interpretation of the relative influences of terminal meristem and leaf primordia in regulating vascular differentiation. If the procambium, initiated in strands associated with the leaves, is the first stage, it might be argued that the terminal meristem plays a passive role in the process and the developing leaves determine the pattern of the vascular system. If, on the other hand, there is a stage preceding procambial initiation, the suggestion might be that this earlier stage, influenced by the terminal meristem, sets out the basic plan of the vascular system, a plan that is subsequently modified by the influence of the developing leaves. Although final answers are not yet available, it may be well to examine the evidence pertaining to this difficult question as an indication of the current status of investigations.

In a description of vascular differentiation to *Geum chiloense* McArthur and Steeves (1972) examined the relationship between leaf primordia and the earliest sign of vascular differentiation. This study confirmed the relationship between leaf primordia and procambial initiation. However, it was found that the procambial strands were initiated within a cylinder of small, densely staining cells that resemble procambial cells in cytological characteristics but not in size or shape (Fig. 13.9). Because the leaf-oriented procambial strands arose as only part of this cylinder, it does not seem likely that the cylinder itself can be interpreted as being related to the leaves in its differentiation. Rather it seems to arise under the influence of the terminal meristem. This cylinder appears to be the forerunner of the vascular system of the stem and as such may be designated *provascular tissue*. Although no procambium could be detected in relation to the youngest leaf primordium, it was present in all others and was recognizable at essentially as high a level

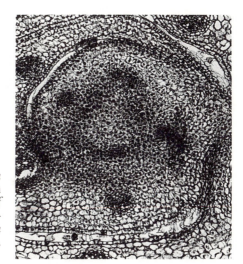

Figure 13.9. *Transverse section 80 μm below the apex of the stem of* Geum chiloense. *There is a distinct ring of provascular tissue within which procambial strands are differentiating in relation to developing leaf primordia.* ×90. (McArthur and Steeves, 1972.)

in the stem tip as any other part of the cylinder. The heterogeneity of the cylinder at its first appearance results from differences in timing in the early differentiation of procambium. In a sector of the apex occupied by a young leaf primoridium, underlying provascular tissue is converted almost immediately to procambium. In interfoliar positions at the stem tip the cylinder consists of typical provascular tissue, some of which will later be converted to procambium in the acropetal differentiation of strands that will form in relation to primordia initiated later.

The concept of a stage of vascular differentiation preceding procambial initiation is not a new one and the tissue here described as provascular tissue has appeared repeatedly in the literature under a variety of names including *prodesmogen, meristem ring, Restmeristem* or *residual meristem,* and *prestelar tissue.* There are two rather different interpretations of the tissue described by these various terms and the difference between the two is not merely one of terminology. On the one hand, the faintly delimited cylinder of meristematic tissue in the axis immediately subjacent to the terminal meristem is interpreted, as in the description of *Geum,* as the first stage of vascular differentiation occurring more or less independently of the leaves, but in relation to the activity of the terminal meristem (Fig. 13.10). Further differentiation to the procambial stage occurs only in relation to the leaves, but the provascular tissue, as it may appropriately be designated in this context, is held to represent a differentiational departure from the promeristem condition in the direction of vascular tissue. On the other hand, another view holds

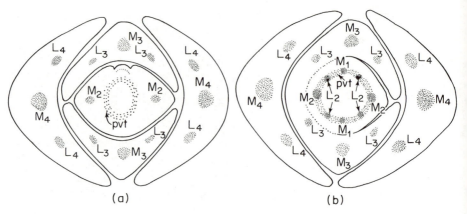

(a)   (b)

Figure 13.10. *The initiation of vascular tissues in the stem of* Garrya elliptica *seen in a series of transverse sections at increasing distances behind the apex. (a) Section in the internode below leaf pair 2. In the stem the vascular tissue is present as a ring of provascular tissue. (b) Section at the level of the third node. The vascular tissue in the stem consists of provascular tissue, part of which has differentiated as median and lateral trace procambium associated with the second leaf pair above. (c) A section lower in the stem showing more of the provascular tissue differentiated as procambium. Key: L, lateral leaf trace; M, median leaf trace; pvt, provascular tissue. (R. M. Reeve, Am. J. Botany 29:697, 1942.)*

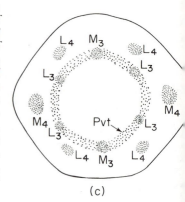

(c)

that the tissue in question has not undergone the first stages in vascular differentiation but rather represents a perpetuation of the uncommitted meristematic state and is recognizable only because of the precocious differentiation of the ground tissue of pith and cortex. In this view the procambium does represent the first step in differentiation toward vascular tissue, and the tissue of the cylinder consists of *residual meristem* – that is, cells in a still uncommitted state – which may or may not become procambium depending upon the relation to developing leaf primordia. It is not difficult to understand why this difference of interpretation is difficult to resolve by observation only. However, there are experiments that have yielded additional information on this point, and these will be discussed later in this chapter.

Another approach to interpretation of vascular differentiation is to examine it in the lower vascular plants. The differences between these

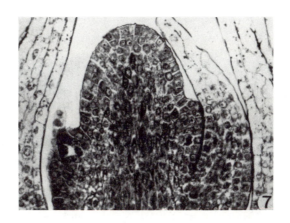

Figure 13.11. *Longitudinal section of shoot apex of Ly-copodium sabinaefolium (club moss). Note the extension of elongate procambial cells into the apex above the youngest visible leaf primoridum. ×210. (R. H. Wetmore.* Torreya *43:16, 1943.)*

plants and the seed plants, both in their meristems and in their vascular systems, have been noted previously, and it is possible that these differences might facilitate the recognition of events that are difficult to observe in seed plants. In *Lycopodium* Freeberg and Wetmore (1967) have described a central column of procambium in the stem tip extending above the level of the youngest leaf primordia, which are themselves initially without procambium (Fig. 13.11). In this plant the leaves appear to exert no influence upon the initiation of the stem vascular system so that in many respects the differentiation pattern is like that of the root. There is, however, some influence of the leaves upon the later stages of xylem and phloem maturation.

In the ferns, because the terminal meristem consists of enlarged, highly vacuolated cells, it may be possible to examine initial differentiation of vascular tissues without the complication of the probable resemblance of the earliest stages to the cells of the terminal meristem. This expectation was realized in *Osmunda cinnamomea*, in which Steeves (1963) recognized, almost immediately beneath the enlarged cells of the surface prismatic layer, cells appearing to be in the early stages of differentiation and quite different in cytological characteristics from the prismatic cells (Fig 13.12). In the center was a cluster of isodiametric cells that, because they appeared to be ontogenetically related to the pith rib meristem and the pith, were considered to be *pith mother cells.* Around these in the form of a flattened, truncated cone was a layer of smaller, more intensely staining cells. Because these were in basal continuity with obvious procambium, they were termed collectively *incipient vascular tissue.* This tissue appears to be comparable to provascular tissue of other stems. In view of the distinctive characteristics of these cells in

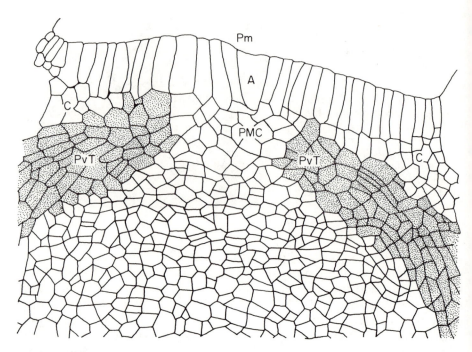

Figure 13.12. *Interpretation of initial vascular differentiation in the shoot of* Os-munda cinnamomea. *Longitudinal section of the shoot apex showing the relation of the provascular tissue to the terminal meristem. (Compare with Figure 4.11.) Key:* A, *apical cell;* C, *cortex,* Pm, *promeristem,* PMC, *pith mother cells;* PvT, *provascular tissue.* ×180. (*Adapted from Steeves, 1963.*)

comparison with those of the prismatic layer in *Osmunda,* it would be difficult to consider them as constituting a mere residual meristem. Essentially similar observations on the initial differentiation of the vascular system have been made in *Osmunda japonica* by Imaichi (1977). In the youngest leaf primordia of *O. cinnamomea* incipient vascular tissue, similar to that of the stem and in continuity with it, could be detected. Procambium was differentiated acropetally into the leaf primordia within the incipient vascular tissue and in continuity with older procambium at lower levels in the stem. Above the level of the fifth or sixth youngest leaf primordium, the layer of incipient vascular tissue in the stem was uninterrupted. In relation to the older leaf primordia, parenchymatous gaps began to develop confronting the procambial leaf traces so that the originally continuous cone of incipient vascular tissue was interrupted by the development of leaf gaps. Thus, the concept of initial differentiation of the vascular system independent of the leaves but subsequently modified in relation to leaf development is supported by the pattern of differentiation in *Osmunda* in which, as in the seed plants, procambium appears to be initiated largely if not entirely in relation to leaf primordia.

A similar interpretation of the initial differentiation of vascular tissues in the fern shoot apex had been advanced earlier by Wardlaw (1944a), based upon observations of *Dryopteris* and several other species. In his descriptions, however, the incipient vascular tissue was considered to constitute a solid core beneath the prismatic layer and the pith was held to arise by the further changes in cells of potentially vascular tissue.

On the other hand, several workers also examining the early stages of differentiation in ferns came to different conclusions. Kaplan (1937), who proposed the *Restmeristem (residual meristem)* interpretation in the seed plants, was able to account for the histological features he observed in fern shoot apices within the same framework. Hagemann (1964) also failed to detect anything that he could call potentially vascular tissue in the fern shoot apex above the level of procambium in the leaf traces. Moreover, the interpretation of shoot apical organization adopted by McAlpin and White (1974) appears to support this conclusion, because the region in which the initial differentiation of provascular tissue would be expected was considered by them to be part of the uncommitted promeristem. It appears that histological investigation alone cannot be relied upon to resolve the question of the identity of the first stage in vascular differentiation. Fortunately there is a substantial body of experimental evidence for ferns that deals with this question and lends support to the existence of a definite incipient vascular or provascular stage preceding the leaf-related procambial stage in the shoot apex.

Experimental Analysis of Early Stages of Vascular Differentiation

In *Dryopteris* and several other ferns Wardlaw (1944b, 1946) sought to eliminate the influence of developing leaf primordia on vascular development by puncturing successive leaves at the time of, or shortly after, their initiation. He reasoned that such an operation ought to reveal the extent to which the terminal meristem alone is capable of initiating a vascular system and the nature of the modifications that the developing leaves exert upon that system. The result of this experiment in several species was the differentiation of a complete ring of mature vascular tissue uninterrupted by leaf gaps (Fig. 13.13). Thus, in the absence of leaves the terminal meristem does produce a vascular system, thereby providing evidence for the validity of the interpretation of provascular or incipient vascular tissue differentiated in the apex independently of leaves. The experiment also reveals the modifying influence of leaves on

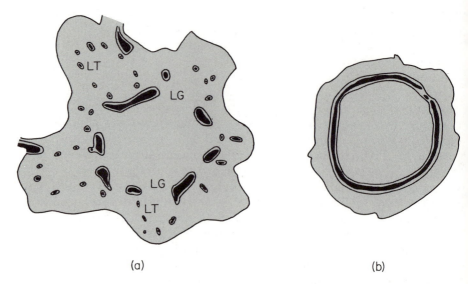

(a)   (b)

Figure 13.13. *The effect on differentiation of vascular tissue in the stem of* Dryopteris dilatata *of systematically puncturing leaf primordia at the time of their initiation. (a) Distribution of vascular tissue in the intact stem seen in transverse section. (b) Transverse section of a stem in which successive leaf primordia had been punctured. The vascular tissue forms an uninterrupted ring. Key: LG, leaf gap; LT, leaf trace. ×4. (Wardlaw, 1944b.)*

the initial vascular system because it appears that they are causal in the differentiation of leaf gaps.

These experiments have been criticized on the grounds that the conclusions were based upon the configuration of the resulting mature stem tissues and did not consider the actual development of the experimental plants. Although it is most unlikely that the mature structures described could have developed in any way other than that suggested, it is reassuring that other experiments of a comparable nature that included a developmental analysis have led to the same conclusion. Soe (1959) punctured young leaf primordia in the fern *Onoclea sensibilis* as they were formed and found that, although the resulting vascular system had very small leaf gaps, they were nonetheless present. However, when the sites of prospective leaf primordia were punctured before the leaf actually protruded from the surface of the meristem, the resulting stem vascular system formed an uninterrupted cylinder of xylem and phloem. In the unoperated apex of *Onoclea*, although procambium initiation in relation to leaf primordia is readily observed, it is difficult to detect incipient vascular tissue as a continuous cylinder. In experimentally treated apices, however, such an uninterrupted cylinder is clearly present and is basally continuous with procambium. Comparison of the two kinds of apices reveals that the difficulty in observing incipient vascular tissue in the normal apex is the result of the very early development of large leaf gaps in relation to the youngest primordia. This also explains why puncturing leaf primordia after they have emerged does not prevent gap formation, the gap already having been initiated at this stage of development. Thus, the experimental study of differentiation in fern apices lends support to the recognition of a provascular stage of differentiation.

In fact, there are ontogenetic changes in the vascular pattern in ferns that could be considered natural experiments having a bearing upon this question. For example, there are cases among the ferns in which both leaf-bearing and leafless shoots occur on the same plant. Both kinds of shoots have a well-organized vascular system that, in the case of the leafless shoots, must have been organized by the terminal meristem alone. In *Onoclea* Soe (1959) has described the development of lateral shoots from detached meristems (see Chapter 8). At a very early stage provascular tissue was clearly recognized as a solid core beneath the promeristem and matured as a core of vascular tissue. As the meristem enlarged the core of provascular tissue gave way to a cylinder surrounding pith mother cells and the mature vascular system became tubular. Up to this point no leaves had been initiated. Leaf gaps of the first formed leaves were small and did not overlap, so that the general cylindrical form of the vascular system was maintained. The large overlapping gaps associated with later formed leaves resulted in the configuration characteristic of the adult plant, both in the mature axis and in the shoot apex.

In view of the usefulness of the experimental approach in elucidating initial vascular differentiation in ferns, it would seem reasonable to apply the same technique to flowering plants. In fact, some experiments of this type have been done, but the results have given answers that are less clear than in the ferns. Wardlaw (1950) and Ball (1952) carried out surgical experiments that had the effect of removing, or at least greatly reducing, leaf influence upon the stem vascular system during its development in the apices of *Primula* and *Lupinus*, respectively. In both cases the terminal meristem was first isolated laterally by vertical incisions. During the ensuing period of reorganization the shoot meristem developed for a time without forming leaf primordia. In the case of *Primula* subsequently formed leaf primordia were punctured as they emerged for a time, but ultimately, in both cases, leaves were allowed to develop on the isolated meristem. Under these conditions in which the leaf influence was reduced, or probably absent for a time, the development of a provascular cylinder was more readily detectable than in unoperated apices; but because leaves were subsequently permitted to develop it is difficult to determine the extent of the leaf influence. The fact that the resulting mature vascular system tended to be in the form of separate bundles suggests that the leaf influence may have been rather large. This last conclusion is supported by the work of Young (1954), also on *Lupinus*, which showed that if all leaves were systematically removed at an early primordial stage, a distinct provascular cylinder developed but there was no further differentiation of this tissue into procambium or mature vascular tissue.

Some of the confusion surrounding the question of initial vascular differentiation in flowering plants has been resolved in a study by McArthur and Steeves (1972), referred to earlier in this chapter, which was designed to examine differentiation of vascular tissues in the stem under conditions where the influence of the leaf could be evaluated accurately. The terminal meristem of *Geum chiloense* shoots was isolated laterally by three vertical incisions and successive leaf primordia were punctured as they appeared. When these apices were fixed and examined histologically, they were found to contain a provascular cylinder throughout the region of new growth. The provascular cylinder was also prolonged basally as an indistinct zone into the supporting pith plug, presumably by the conversion of previously existing immature pith cells. Thus, a provascular ring was produced and maintained in the stem in the absence of leaves, as in Young's experiment on *Lupinus*. It is significant that no further development of provascular tissue, even to the procambial stage, occurred in the absence of leaves.

In further experiments, after several leaves had been punctured on the partially isolated meristem, later leaves were allowed to develop for several weeks. These shoots contained a vascular pattern rather like that of the normal shoot, with procambial traces containing some mature vascular tissue associated with the developing leaves. The procambial strands had also extended basipetally into the supporting pith plug. This suggests that the leaf is essential for further development beyond the provascular stage. Because developing leaves are known to be important sources of auxin in the shoot apex, their influence was replaced in the leafless experimental apices by applying the auxin indoleacetic acid in lanolin as a cap over the apex. Following this treatment the provascular ring became much more distinct and easily recognizable, but there was no differentiation beyond this stage (Fig. 13.14a). Thus, auxin alone did not replace the influence of the leaf primordium in promoting vascular differentiation beyond the provascular stage. From other observations it seemed that carbohydrate nutrition might be a limiting fac-

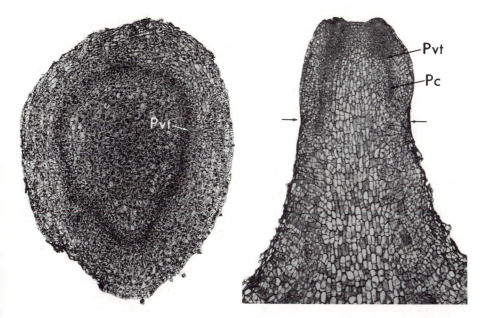

Figure 13.14. *Experimental modification of vascular differentiation in* Geum chiloense. *(a) Transverse section 270 μm below the tip of a surgically isolated apex on which successive leaf primordia had been punctured at the time of their initiation and to which indoleacetic acid had been applied. Note the distinct ring of provascular tissue that has been formed. (b) Longitudinal section of an isolated shoot apex after leaf puncturing, application of indoleacetic acid, and addition of sucrose. Provascular tissue in the apical region has differentiated as procambium at lower levels in the stem. Arrows indicate the junction between the newly formed shoot above and the supporting pith plug below. Key:* Pc, *procambium;* Pvt, *provascular tissue. (a)* ×65, *(b)* ×70. *(McArthur and Steeves, 1972.)*

tor. Auxin treatment was therefore combined with application of a sucrose solution in which the bases of the shoots were immersed. Growth of the isolated apices was considerably enhanced and more complete vascular differentiation occurred. Apart from the terminal portion of the isolated apex, which contained a distinct provascular core or cylinder, all of the other vascular tissue of the isolated shoot had differentiated at least to the procambial stage (Fig. 13.14b). A procambial cylinder without leaf traces extended through most of the experimentally produced shoot. Xylem elements were present in the pith plug below the isolated shoot, but it is not known whether final vascular maturation would have occurred throughout the procambial ring if the experiment has been continued for a longer time. However, the formation of procambium indicated the potentiality of the provascular tissue to differentiate as a vascular cylinder in the absence of leaves if suitable nutritional and hormonal factors are provided artificially.

It now seems possible to suggest a resolution of the apparent disparity between the descriptions of initial differentiation of vascular tissues in ferns and seed plants. In both it appears that the shoot apex directly promotes an initial blocking out of the vascular system in the form of a provascular ring. In the ferns the vascular differentiation seems to proceed to completion without the contribution of leaves, although the leaves normally produce extensive modifications in the pattern of development. In the seed plants, on the other hand, later stages of differentiation, including the procambial stage, seem to be dependent upon the contribution of leaves and in their absence do not occur. The leaf effect can be artificially replaced, but its exact nature is not clear at the present time, although auxin would appear to be implicated. The experiments reported here point out a potentially fruitful approach to the analysis of relative contributions of the terminal meristem and leaf primordia to the complex process of vascular differentiation in the shoot of vascular plants.

GENERAL COMMENT

The view that emerges from this examination of initial stages of differentiation in both roots and shoots is one of blocking out of major tissue systems in relation to the activities of the terminal meristems. These meristems, which in Chapters 6 and 12 were shown to be autonomous with respect to their ability to develop normally when isolated from the rest of the plant, are now seen also to be the organizers of development in the organs to which they give rise. This is particularly clear in roots in which lateral organs do not complicate the initial blocking out of the three tissue systems and in those shoots that develop without the formation of leaves. It is less obvious in shoots in which leaves are initiated in close proximity to the terminal meristem. In these, histological examination alone may not reveal the relationship between the terminal meristem and the blocking out of the tissue systems, but experiments described in this chapter confirm that it exists here also. Thus, in a real sense it can be said that meristems make plants.

On the other hand, the diversity of the mature structure of root and shoot axes indicates that there are significant modifications during the maturation of tissues that follows the initial blocking out of pattern. Since these modifications are mostly concerned with later stages of differentiation, which are the subject of the next chapter, a general commentary on pattern formation in relation to tissue differentiation will be deferred to the end of that chapter.

REFERENCES

Ball, E. 1952. Morphogenesis of shoots after isolation of the shoot apex of *Lupinus albus*. *Am. J. Botany* 39:167–91.

Beck, C. B., R. Schmid, and G. W. Rothwell. 1982. Stelar morphology and the primary vascular system of seed plants. *Bot. Rev.* 48:691–815.

Cullis, C. 1984. Environmentally induced DNA changes. In *Evolutionary Theory: Paths into the Future*, ed. J. W. Pollard, 203–16. New York: Wiley.

Cutter, E. G., and L. J. Feldman. 1970. Trichoblasts in *Hydrocharis* II. Nucleic acids, proteins and a consideration of cell growth in relation to endopolyploidy. *Am. J. Botany* 57:202–11.

Esau, K. 1942. Vascular differentiation in the vegetative shoot of *Linum*. I. The procambium. *Am. J. Botany* 29:738–47.

Freeberg, J. A., and R. H. Wetmore. 1967. The Lycopsida – A study in development. *Phytomorphology* 17:78–91.

Frisch, B., and W. Nagl. 1979. Patterns of endopolyploidy and 2C nuclear DNA content (Fuelgen) in *Scilla* (Liliaceae). *Plant Syst. Evol.* 131:261–76.

Gunning, B.E.S. 1982. The root of the water fern *Azolla*: Cellular basis of development and multiple roles for cortical microtubules. In *Developmental Order: Its Origin and Regulation*, 40th Symp. Soc. Devel. Biology, eds. S. Subtelny and P. B. Green, 379–421. New York: Liss.

Hagemann, W. 1964. Vergleichende Untersuchungen zur Entwicklungsgeschichte des Farnsprosses. I. Morphogenese und Histogenese am Sprossscheitel leptosporangiater Farne. *Beit. Biol. Pflanzen* 40:27–64.

Imaichi, R. 1977. Anatomical study of the shoot apex of *Osmunda japonica* Thunb. *Botan. Mag. Tokyo* 90:129–41.

Jacob, F., and J. Monod. 1963. Genetic repression, allosteric inhibition, and cellular differentiation. In *Cytodifferentiation and Macromolecular Synthesis*, 21st Growth Symp., ed. M. Locke, 30–64. New York: Academic Press.

Jensen, W. A., and L. G. Kavaljian. 1958. An analysis of cell morphology and the periodicity of division in the root of *Allium cepa*. *Am. J. Botany* 45:365–72.

Kaplan, R. 1937. Ueber die Bildung der Stele aus dem Urmeristem von Pteridophyten und Spermatophyten. *Planta* 27:224–68.

McAlpin, B. W., and R. A. White. 1974. Shoot organization in the Filicales: The promeristem. *Am. J. Botany* 61:562–79.

McArthur, I.C.S., and T. A. Steeves. 1972. An experimental study of vascular differentiation in *Geum chiloense* Balbis. *Botan. Gaz.* 133:276–87.

McClintock, B. 1965. The control of gene action in maize. *Brookhaven Symp. Biol.* 18:162–84.

Nagl, W. 1978. *Endopolyploidy and Polyteny in Differentiation and Evolution*. Amsterdam: Elsevier/North-Holland Biomedical.

Partanen, C. R. 1959. Quantitative chromosomal changes and differentiation in plants. In *Developmental Cytology*, 16th Growth Symp., ed. D. Rudnick, 21–45. New York: Ronald.

Shanks, R. 1965. Differentiation in leaf epidermis. *Austral. J. Botany* 13:143–51.

Soe, K. 1959. Morphogenetic studies on *Onoclea sensibilis* L. Ph.D. dissertation, Harvard University.

Steeves, T. A. 1963. Morphogenetic studies on *Osmunda cinnamomea* L: The shoot apex. *J. Indian Bot. Soc.* 42A:225–36.

Torrey, J. G. 1955. On the determination of vascular patterns during tissue differentiation in excised pea roots. *Am. J. Botany* 42:183–98.

Viotti, A., E. Sala, R. Marotta, P. Alberi, C. Balducci and C. Soave, 1979. Genes and mRNAs coding for zein polypeptides in *Zea mays. Eur. J. Biochem.* 102:211–22.

Wardlaw, C. W. 1944a. Experimental and analytical studies of pteridophytes. III. Stelar morphology: The initial differentiation of vascular tissue. *Ann. Botany (London) (NS)* 8:173–88.

———. 1944b. Experimental and analytical studies of pteridophytes. IV. Stelar morphology: Experimental observations on the relation between leaf development and stelar morphology in species of *Dryopteris* and *Onoclea. Ann. Botany (London) (NS)* 8:387–99.

———. 1946. Experimental and analytical studies of pteridophytes. VII. Stelar morphology: The effect of defoliation on the stele of *Osmunda* and *Todea. Ann. Botany (London) (NS)* 10:97–107.

———. 1950. The comparative investigation of apices of vascular plants by experimental methods. *Phil. Trans. Roy. Soc. London* Ser. B 234:583–604.

Webster, P. L. 1980. Analysis of heterogeneity of relative division rates in root apical meristems. *Bot. Gaz.* 141:353–9.

Young, B. S. 1954. The effects of leaf primordia on differentiation in the stem. *New Phytol.* 53:445–60.

CHAPTER 14

Differentiation of the plant body: the elaboration of pattern

The early stages of differentiation, considered in Chapter 13, are characterized by the blocking out of regions within which cells are relatively homogeneous and behave similarly. In contrast, the later stages of differentiation show highly localized specializations, often with adjacent cells differing markedly in the developmental changes they undergo. Therefore, in studying these stages of differentiation it is essential to pay close attention to the events taking place in individual cells, and this will be the first task. When this has been done, it will be necessary to return to the higher levels of organization to consider the interrelations that keep the differentiating cells as part of an organized system.

CELLULAR CHANGES DURING DIFFERENTIATION

In contrast to the body of higher animals with its large number of differentiated cell types, relatively few cell types, possibly no more than twelve, are differentiated in the body of the vascular plant. However, to describe the cellular events related to differentiation of even this small number of cell types would fill many pages. Rather than do this, attention will be concentrated on the differentiation of the tracheary elements of the xylem, the tracheids and vessel members, and the sieve elements and companion cells of the phloem. These differentiate in the primary body from procambium and in the secondary body from cambium, and because the sequence of differentiation is generally similar in both cases, it usually will not be necessary to make distinctions between them in the following description. Both the procambium and the fusiform initials of the cambium appear to be homogeneous in respect to their cellular composition and do not give evidence of the numerous cell types, such as tracheary elements, fibers, sieve tube members, companion cells, or parenchyma, that will subsequently be differentiated from them. Therefore, it is reasonable to think that the events leading to the final differentiation of these cells do not occur much before the first visual evidence of differentiation.

At the level of light microscopic observation the first evidence of dif-

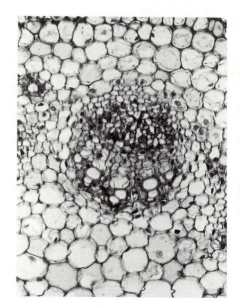

Figure 14.1. *Transverse section of a differentiating vascular bundle in the stem of* Helianthus annuus. *In the center of the bundle can be seen the expansion of procambial cells as they differentiate as xylem and begin to lay down secondary wall material. ×60.*

ferentiation of tracheids and vessel members is the enlargement, often both transverse and longitudinal, of the cells. Subsequently, the wall becomes increasingly thick by the deposition of additional wall substance against the inner face of the preexisting primary wall (Fig. 14.1). Deposition of this secondary wall occurs in highly distinctive patterns and may consist of a number of discrete bands circling the cell, or one or more spirals traversing the long axis in helical fashion (Fig. 14.2). These circular or helical tracheary elements differentiate from procambium, typically in regions that are still elongating. The secondary wall structure of these cells allows them to remain extensible. In other tracheary cells the secondary wall is deposited more extensively and forms a network over the inner surface of the primary wall or, as in some cells in the primary body and all those differentiating from cambium, it may form an essentially continuous layer between the primary wall and the protoplast. In these cells regions where the secondary wall is not formed are called *pits*. They tend to occur over areas of primary wall where there are dense accumulations of submicroscopic cytoplasmic strands called *plasmodesmata* running through the primary wall and maintaining continuity between the protoplasts of contiguous cells. These often thinner regions of primary wall are called *primary pit fields*. The final stages of tracheary differentiation involve chemical specialization of the wall by deposition of lignin within the previously formed cellulose framework. This is followed by lysis and death of the protoplast and, in

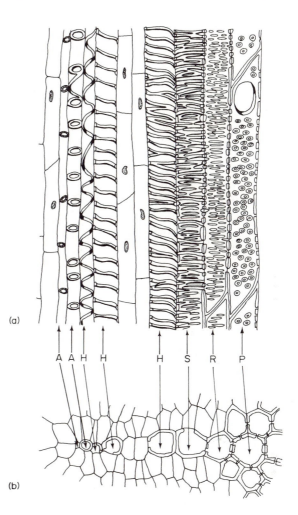

Figure 14.2. *A sequence of tracheary cells in the stem of* Lobelia *showing the variety of patterns of secondary wall deposition. The cells are named according to the secondary wall pattern and from left to right are: (A) annular (rings of secondary wall material); (H) helical (spirals); (S) scalariform (ladderlike); (R) reticulate (netted); (P) pitted. Thin-walled cells in the sections are xylem parenchyma. (a) Longitudinal section. (b) Transverse section. (A. J. Eames and L. H. McDaniels,* An Introduction to Plant Anatomy. *New York: Mc-Graw-Hill, 1947.)*

the case of cells differentiating as vessel members, there is also lysis of parts or all of the end walls to form a continuous multicellular water-conducting tube or vessel (Fig. 14.3). The mature, functioning cells are, therefore, dead, and it is a matter of some interest that death of a cell may be of developmental significance. In animal development morphogenetic cell death contributes to shaping of the extremities of the embryonic limbs, but such cases are rare in plants where cell death seems to be associated more with functional specialization.

The cytological events associated with differentiation of tracheary elements have been studied in more detail by high-resolution examination made possible by electron microscopy. As in other elongating cells,

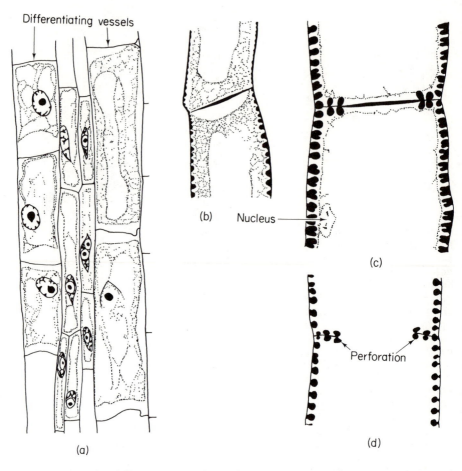

Figure 14.3. *The differentiation of vessel elements in* Apium graveolens *(celery) showing (a) enlargement of the cells, (b) early stage in deposition of the secondary wall, and (c,d) thickening and subsequent dissolution of the end wall to form the perforation and death and lysis of contents of the cell. (a) ×600, (b–d) ×950. (K. Esau,* Hilgardia *10:479, 1936.)*

cellulose microfibrils in the primary wall are deposited predominately at right angles to the axis of elongation and are pulled into a longitudinal orientation as the cell elongates. In contrast the secondary wall layers, laid down after cell elongation is complete, are composed of highly oriented layers of microfibrils that encircle the cell in a helical fashion and are not subsequently reoriented. At the time when secondary walls are being deposited, endoplasmic reticulum is abundant in the cytoplasm and the Golgi system is especially abundant and appears to be

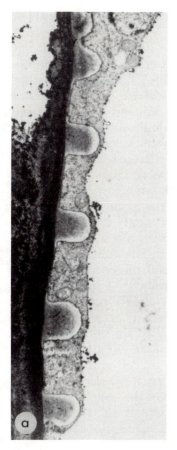

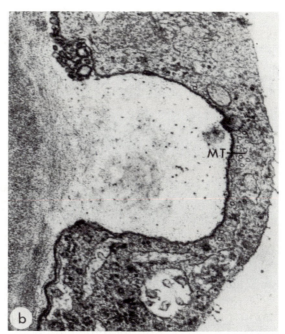

Figure 14.4. *Ultrastructure of xylem differentiation in* Coleus. *(a) Portion of a longisection of a differentiating tracheary element showing bands of secondary wall. (b) Enlarged section of a secondary wall band showing microtubules clustered over secondary wall. Key:* MT, *microtubules. (a) ×6,300, (b) ×54,000. (P. K. Hepler and D. E. Fosket. Protoplasma 72:213, 1971.)*

actively forming vesicles thought to contribute material to the wall (Roberts, 1976). The well-documented relationship between the orientation of cytoplasmic microtubules and that of wall microfibrils is particularly clear in these cells (Fig. 14.4). In cases in which the secondary wall is deposited in bands, microtubules in the peripheral layer of cytoplasm are also grouped into parallel bands underlying the developing wall. The concept that microtubule orientation is functionally related to microfibril orientation is supported by experimental evidence. Treatment with colchicine, a substance known to block microtubule assembly, causes the secondary wall microfibrils to be deposited in an irregular fashion rather than in a highly ordered array (Torrey, Fosket, and Hepler, 1971).

Differentiation of sieve elements in the closely associated phloem tissue is characterized by very different but no less striking changes from those observed in the xylem (Evert, 1984). The protoplast of the mature sieve element, unlike that of a tracheary cell, is alive at functional maturity, but has undergone extensive modifications. These include disappearance of the nucleus, ribosomes, microtubules, and Golgi system; the loss of the tonoplast, as a result of which there is no distinction between vacuole and cytoplasm; the partial degeneration of plastids and mitochondria; and reorganization of the endoplasmic reticulum (Fig. 14.5). In addition, the cell lumen contains an abundance of fibrous protein, which is characteristic of these cells. A particularly distinctive feature of sieve elements is the development of *sieve areas* in the walls. These are areas in the cell wall through which connecting strands, essentially enlarged plasmodesmata, extend. In the angiosperms, particularly well-developed sieve areas occur, either singly or in groups, in the end walls of sieve tube members. These end walls are called *sieve plates* and their presence is what distinguishes a sieve tube. In the gymnosperms and lower vascular plants such specialized end walls do not occur and the individual cells are called sieve cells.

The complexity of sieve element differentiation is illustrated by the development of a typical sieve plate. Early in the process portions of the endoplasmic reticulum become closely applied to the cell wall at the sites of plasmodesmata, and a carbohydrate known as callose is deposited beneath these on each side of the developing cell plate. Gradually the callose replaces the wall material around each plasmodesma. The subsequent dissolution of a substantial portion of the callose allows the plasmodesma to enlarge into a typical connecting strand.

It is clear from this brief summary of the differentiation of tracheary and sieve elements that there are numerous developmental problems to be solved at the cellular level of organization. For example, the patterned deposition of the secondary wall in tracheary cells and the extensive alterations of the protoplast of sieve elements present problems of developmental control that must be answered. However, these questions go beyond the scope of this book and, however interesting they are, we must leave them to the cell biologists for resolution.

LATER STAGES OF DIFFERENTIATION IN ROOTS

After consideration of some of the changes that occur in individual cells during the process of differentiation, it is appropriate now to consider once again the higher levels of organization within which these changes occur. As with the initial stages of differentiation, consideration will be given first to the final stages of differentiation in roots where the events are not complicated by the presence of lateral organs.

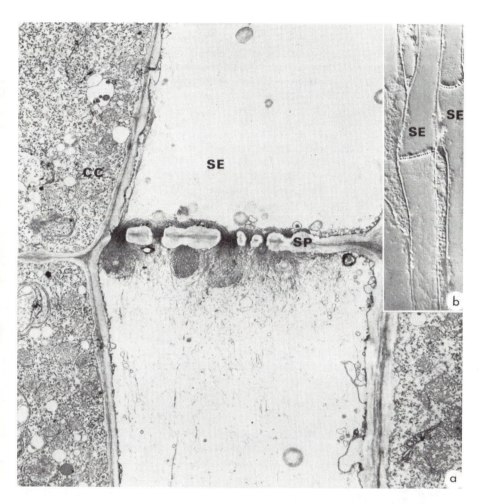

Figure 14.5. *(a) Electron micrographs of sieve tube elements and associated companion cells of* Nicotiana tabacum *in the region of a sieve plate. The striking contrast in cellular organization between sieve tube and companion cell is evident. (b) Nomarski interference contrast photograph of a similar specimen. Key: CC, companion cell; SE, sieve tube element; SP, sieve plate. (a) ×5,300, (b) ×530. (J. Cronshaw and R. Anderson, J. Ultrastruct. Res. 27:134, 1969.)*

The late stages of differentiation, involving the maturation of tissues, have been studied in the roots of numerous species. Torrey's study (1955) of differentiation in the root of *Pisum* is a good example, combining structural description with experimental evidence. Attention will be directed mainly to the events that convert the central core of the procam-

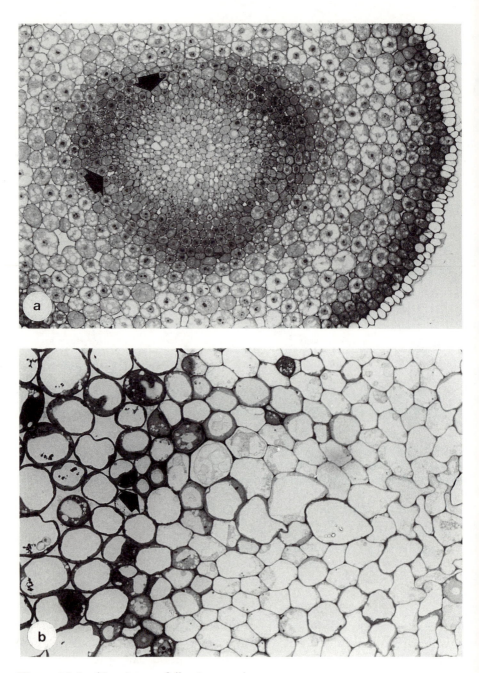

Figure 14.6. *(Caption on following page).*

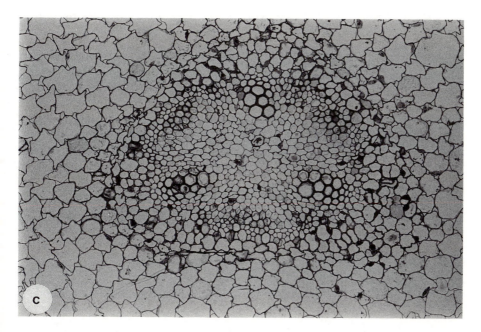

Figure 14.6. *Later stages in the differentiation of vascular tissues in the root of* Pisum sativum. *(a) Cross section of a root showing the central procambial cylinder in which the first protophloem cells have differentiated at the margins (arrows) and in which cells that will differentiate as metaxylem in the center of the cylinder have started to expand. (b) Cross section of part of a pea root showing a later stage of differentiation in which the protoxylem cells (arrow) are maturing but the metaxylem cells (at the center of the section) are still immature. (c) The central cylinder in which protoxylem has differentiated at three points. Protoxylem cells can be identified by their thick lignified walls. Inside each group of protoxylem cells the first metaxylem cells are maturing. These differ from the protoxylem in their larger diameter. The primary phloem lies on radii alternating with the protoxylem. (a) ×170, (b) ×750, (c) ×170.*

bium into the mature xylem and phloem and associated tissues. In the seedling roots of *Pisum* the mature vascular system has a so-called triarch arrangement in which there are three symmetrically arranged ridges of primary xylem in the central core. The establishment of this pattern in the procambial cylinder occurs within 0.2 mm of the promeristem by transverse enlargement of future xylem cells in each of the ridges. The differential enlargement of certain cells occurs first near the center of the root and progresses outward toward the tips of the ridges, or arms as they appear in transverse section (Fig. 14.6). Meanwhile, other procambial cells continue to divide longitudinally and retain a small transverse diameter. In *Pisum* the difference in diameter between cells in the

Figure 14.7. *Longitudinal section of the root apex of* Zea mays *showing enlargement of the future metaxylem vessel elements beginning close behind the promeristem.* ×75.

center and at the margin of the central cylinder is not very great, but in some other species the cells that will differentiate as the central xylem vessels attain a very large diameter immediately behind the promeristem (Fig. 14.7).

The final maturation of procambial cells as functional vascular elements occurs first in regions lying between the arms of the future xylem, that is, in the phloem where the first cells to mature are conducting sieve elements located near the outer margin of the procambial core and adjacent to the pericycle (Fig. 14.6a). The small group of elements that differentiate in this region constitutes the *protophloem*. Subsequently additional sieve elements differentiate deeper in the procambial core. This constitutes the *metaphloem*. The terms *protophloem* and *metaphloem* are used to designate a time sequence in differentiation, and they are most valuable when used in this sense without other histological connotations.

The maturation of procambial cells as xylem in the root of *Pisum,* as in other species, occurs some distance proximal to the earliest mature phloem cells, and it is found first at the tips of the arms or ridges in the procambial core (Fig. 14.6b). This is called the *protoxylem* and ordinarily consists of elements with a relatively small transverse diameter. Subsequently xylem differentiation proceeds toward the center of the procambial core (Fig. 14.6c). The later-formed xylem is designated *me-*

taxylem, and as in the case of the phloem, the prefixes *proto* and *meta* designate a temporal relationship only. The transverse sequence of final xylem differentiation poses an intriguing problem in that it progresses from the periphery to the center whereas the earlier blocking out of future xylem elements by transverse enlargement of procambial cells occurs in the opposite direction. One might expect the first elements that begin the process to be the first to complete it, but in fact they are the last. Undoubtedly the relatively small diameter of the protoxylem elements in comparison with that of the central metaxylem cells is related to the duration of the transverse enlargement phase that precedes the deposition of a secondary wall and the death of the protoplast. In some species, notably among the monocotyledons, the differentiation of xylem does not proceed completely to the center of the root, and a pith of varying dimensions may result from the differentiation of central procambial cells into parenchyma or into nonvascular sclerenchyma. When this occurs, the xylem arms or ridges have the form of separate bundles or strands alternating on separate, equally spaced radii with corresponding bundles of phloem. In large roots of monocotyledons the number of such strands may be 100 or more.

The process of differentiation in the root is a continuous one that proceeds steadily toward the apex without discontinuities (Fig. 14.8). This does not mean, however, that there is a constant distance between the apical meristem and the sites at which cells derived from it differentiate. It has been found in a number of species that the distance at which differentiation occurs varies with the growth rate of the root. This is well illustrated in a study of seedlings of *Sinapis alba* (white mustard) (Peterson, 1967) grown under different conditions of aeration, which strongly affected root growth rate. In more rapidly growing roots the distance from the apex at which specific differentiation events occurred was consistently greater than in more slowly growing roots. A possible explanation for this phenomenon might be that a cell must reach a particular age before differentiation can occur. If this were the case, one would expect to find uniform degrees of differentiation at any level of the root, at least within any particular tissue. Clearly this is not the case, because in both xylem and phloem there is a distinct transverse temporal pattern of differentiation. This point will be important later in considerations of the control of differentiation processes.

The outermost cells of the procambial core differentiate in a highly distinctive fashion, and at a level very close to the meristem they can be recognized by their cuboidal shape. The *pericycle,* as this layer is designated, is composed of cells that appear structurally to be typical parenchyma but possess distinctive developmental features. Whereas other differentiated tissues of the root achieve a relatively stable mature

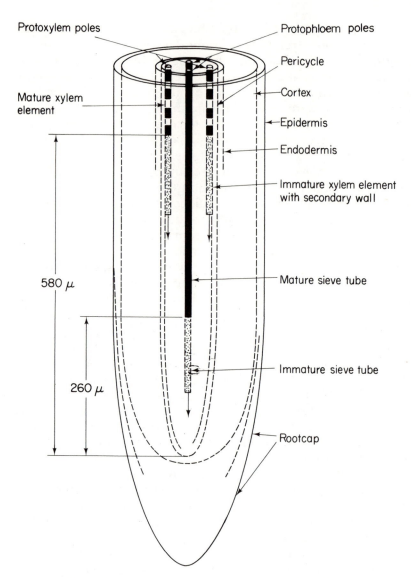

Figure 14.8. *Three-dimensional diagram illustrating the differentiation of primary vascular tissues in the root tip of* Nicotiana tabacum. *Phloem and xylem both differentiate continuously and acropetally with the phloem in advance of the xylem.* *(K. Esau,* Hilgardia *13:437, 1941.)*

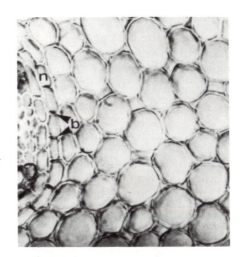

Figure 14.9. *Cross section of a root of* Allium cepa *showing a portion of the endodermus (n) with distinctive Casparian bands (b) visible in the radial cell walls.* ×*135. (C. A. Peterson, M. E. Emanuel, and C. Wilson,* Can. J. Botany *60:1529, 1982.)*

state, the pericycle retains the capacity to initiate lateral roots, to contribute to the origin of the vascular cambium and the cork cambium, and, in some long-lived roots, to give rise to a considerable amount of parenchymatous tissue. It is interesting to note that in species in which polyploidy occurs in some of the mature cells of the root, the increase in chromosome number does not occur in the pericycle, which remains diploid.

One of the most interesting tissues of the root is the *endodermis,* which ordinarily consists of a single layer of cells lying immediately outside the pericycle. It is developmentally related to the cortex, and the interface between the pericycle and the endodermis marks the boundary between the vascular core and the cortex of the root. Mature endodermal cells are characterized by the presence of a *Casparian band,* a narrow strip of suberized material running round the cell on its radial and transverse walls, and extending out through the intercellular cementing substance that attaches adjacent endodermal cells (Fig. 14.9). The overall effect of the Casparian bands is to make the walls impervious to the passage of water and dissolved materials, thus forcing transport through the protoplasts of the endodermal cells, and this layer has figured prominently in considerations of water uptake and transport. The endodermis is relatively late to differentiate and does not appear as a distinctive tissue until after the first xylem has matured. Before any specific morphological feature may be detected, endodermal cells acquire distinctive biochemical characteristics that are the result of the interaction of substances originating in the vascular system with those having their origin in cortical parenchyma cells. Under conditions of high oxidation in the endodermal cells unsaturated fats are oxidized and deposited in the wall as the Casparian strip. This reaction is subject to experimental manipulation, and several environmental factors have been shown to influence the presence or absence of Casparian bands in some species.

LATER STAGES OF DIFFERENTIATION IN SHOOTS

The final differentiation of the vascular system in the shoot consists of the conversion of procambium blocked out behind the apical meristem into mature elements of the xylem and phloem. This process has been studied in a number of species; Esau's investigation of *Linum* (1943), in which the maturation of these tissues has been correlated with the phyllotactic pattern of the shoot, is a particularly good example that seems to be typical. It has already been shown in Chapter 13 that the procambium of *Linum*, as of most seed plants, forms a system of inter-connected strands or leaf traces. Each of these behaves as a unit as far as the timing of its differentiation phenomena is concerned, and it is therefore easier to describe the sequence of events occurring in individual strands than to deal simultaneously with the entire system of the shoot. The overall pattern of final differentiation is illustrated in Fig. 14.10 for tobacco, which is essentially similar to *Linum* in this respect.

The first cells to mature in a leaf trace are sieve elements of the phloem (see Fig. 13.8), and the final maturation of phloem progresses acropetally into and through the strand from its basal connection with other leaf traces. This process does not take the form of a uniform maturation of all of the phloem procambium but initially involves only a single file of cells at the outer face of the procambial strand. Subsequently other files of procambial cells located successively deeper in the bundle undergo final differentiation and become mature. Thus, in addition to the acropetal wave of maturation as seen longitudinally, there is a centripetal progression as viewed transversely at any level in the bundle and the primary phloem forms a continuous system throughout its development. As in the root, the first cells to mature at any level are designated protophloem, the later elements metaphloem.

In a number of seed plants, xylem maturation has been found to be considerably more complex than phloem differentiation. In *Linum*, as in all seed plants so far investigated, xylem maturation is delayed until after the start of phloem maturation. In *Linum*, the first phloem was found in the traces of leaves just over 0.25 mm in length, but the first xylem did not appear in traces of leaves much less than 1 mm in length. Moreover, the first xylem was found in the leaf itself, near its base, and was not connected by mature elements with fully differentiated ele-

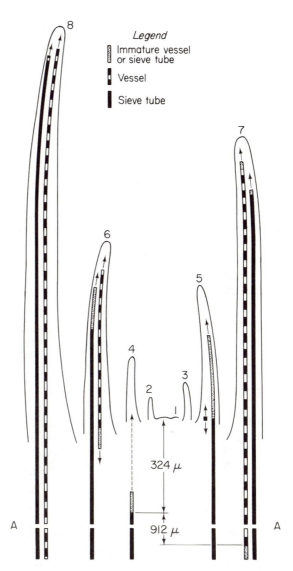

Figure 14.10. *A diagrammatic representation of the differentiation of the first phloem and xylem in the stem and leaves of* Nicotiana tabacum. *The leaves are drawn as if they arose in two ranks rather than in a helical pattern. Distances below the shoot apex are shown on the diagram and a segment of the stem has been omitted at A–A. The procambial stage of differentiation is not shown. Mature vascular elements are absent from the leaf traces to the three youngest leaves. (K. Esau, Hilgardia 11:343, 1939.)*

ments at lower levels in the stem. Subsequently xylem maturation proceeded both acropetally and basipetally in the trace, ultimately establishing a connection with the mature system. Thus, the primary xylem system, although continuous when mature, is discontinuous during its development. In addition, at any level in the trace the first mature xylem elements, the protoxylem, were found at the inner face adjacent to the pith and the subsequent maturation of metaxylem progressed outward through the procambial trace (see Fig. 13.8). The centrifugal differentiation of xylem, together with the centripetal differentiation of phloem, progressively restricts the remaining procambium (Fig. 14.1). In monocotyledons and some dicotyledons all the procambial cells ultimately differentiate as xylem and phloem, but in many dicotyledons this does not occur and the remaining procambial cells give rise to cambium. The transformation of procambial cells into the cambial meristem that forms the secondary body of the plant will be considered in more detail in Chapter 15.

The pattern of vascular maturation in *Linum*, although complicated to describe, is simple because each leaf is served by only one trace. Where several traces enter each leaf, there are usually differences in the timing of differentiation in each of the traces. Ordinarily differentiation in the median trace is in advance of that in the laterals, and where there are several lateral traces there is usually a median-to-lateral progression of maturation. This considerably complicates the pattern of differentiation seen in transverse sections of the stem because of the great variation in the degree of maturation among the bundles. In monocotyledons, which characteristically have many traces entering each leaf, although the median trace differentiates according to the pattern already described, the lateral traces may show basipetal and discontinuous differentiation, not only of protoxylem and protophloem, but even of procambium (Kumazawa, 1961).

The evidence presented thus far puts emphasis upon the early discontinuous and basipetal pattern of primary xylem differentiation. There are, however, descriptions of an acropetal progression of xylem differentiation that extends from the mature system below up into the leaf trace and meets the descending basipetal wave of differentiating xylem. Esau observed such acropetal maturation in *Linum*, and it has also been described in *Coleus* by Jacobs and Morrow (1957) and in *Ginkgo* by Gunckel and Wetmore (1946). In other cases where xylem maturation has been studied, no mention of this acropetal wave is made, but because most of these studies have been concerned with only the earliest stages of vascular differentiation in the vicinity of the nodes of young leaves, an acropetal wave having its origin relatively far down the stem might easily be overlooked. The significance of the acropetal differentiation is not clear. There is certainly a general impression of overall acropetal progression of metaphloem and metaxylem maturation and of cambial initiation where this occurs. This seems to be superimposed upon the protoxylem discontinuities. However, such an impression could be misleading, and despite the numerous studies made of vascular differentiation in the shoot and the considerable amount of factual information now available, it is clear that numerous unanswered questions remain for further study.

In the lower vascular plants that have been investigated a variety of patterns of differentiation has been observed. In the fern *Osmunda cinnamomea,* which has a vascular system consisting of a series of anastomosing bundles surrounding a central pith, a conspicuous acropetal wave of xylem differentiation meets the basipetal wave of protoxylem in the leaf trace. On the other hand, White (1984) has reported on fern species in which some of the protoxylem strands extend acropetally beyond the youngest leaf primordium and therefore must develop in a different way. In *Lycopodium* (Freeberg and Webmore, 1967) there is a solid core of procambium in the stem in which differentiation is acropetal and maturation of xylem precedes that of phloem. However, in the leaf traces, protoxylem differentiates discontinuously. In *Equisetum* both xylem and phloem are initiated discontinuously at the nodes and connections with older mature vascular tissues are established by basipetal differentiation, with no evidence of acropetal differentiation of either tissue (Golub and Wetmore, 1948).

From these considerations, and those of the previous section, it is evident that the complex phenomena that accomplish differentiation at the cellular level occur within the context of orderly patterns leading to the formation of an integrated primary vascular system throughout the plant. In reviewing experimental investigations of vascular differentiation in the following section, this important perspective should be kept in mind.

EXPERIMENTAL STUDIES OF VASCULAR DIFFERENTIATION

The relative roles of the shoot apex and leaf primordia in the delimitation of provascular tissue and the differentiation of procambium have been discussed in Chapter 13. The patterns of final maturation of xylem and phloem that have now been described would suggest that in the later stages also there is ample opportunity for the investigation of controlling mechanisms. The differences in pattern in different organs and in different groups of plants may prove useful in such investigations and could provide the basis for comparative experimental studies.

The apparent relationship between the final maturation of vascular tissues and the development of leaves has been the point of departure for a number of experimental studies of the differentiation of xylem and phloem. It has been known for some time that if a vascular bundle in the stem of *Coleus* is severed by a cut, a connection is reestablished by the differentiation of intervening pith parenchyma as vascular elements (Fig. 14.11). Jacobs (1952) has used this observation as the basis for a

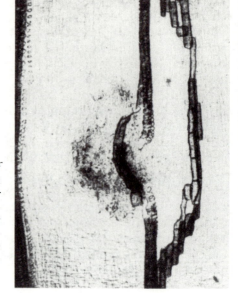

Figure 14.11. *The differentiation of wound tracheary elements around an incision that severed one of the vascular bundles in the stem of* Coleus. *The severed strand is on the left. The cells in it are elongated and aligned. The regenerated cells are short and the course of the regenerated strand through the wound is irregular.* ×78. *(Roberts and Fosket, 1962.)*

series of experiments on the control of xylem differentiation. Removal of the terminal bud and expanding leaves distal to the wound had a marked effect in reducing the amount of xylem differentiated around the wound, whereas removal of leaves and buds from that part of the stem below the wound had little or no effect. The effect of the distal organs could be attributed principally to the expanding leaves, which are the major source of auxin in the plant. That the effect of the leaves was actually due to the auxin they produce was demonstrated by the application of indoleacetic acid through the bases of the petioles of leaves that had been removed. In fact, a quantitative relationship between the amount of auxin applied and the extent of xylem differentiation was demonstrated. Furthermore, most of the xylem differentiation proceeded in a basipetal direction, paralleling the predominantly basipetal direction of polar auxin transport in the stem, but a limited amount of acropetal differentiation that occurred could be correlated with a correspondingly small amount of acropetal auxin transport.

In evaluating these experimental observations, Jacobs has commented upon their possible bearing on the isolated centers of xylem differentiation at the bases of developing leaves in intact plants and the subsequent basipetal differentiation of xylem strands. Such a pattern might be expected if auxin is required for final maturation of xylem and the leaf is the primary source of this substance. On the other hand it is important to remember that the tracheary elements that develop in the experimental plants do so from previously mature parenchyma cells caused to redifferentiate in response to a new stimulus, whereas xylem differentiation in the intact plant is the culmination of normal procambium development. It is therefore pertinent to cite the evidence that auxin plays a role in normal xylem differentiation from procambium. Jacobs and Morrow (1957) have demonstrated that there is a direct correlation between the rate of xylem maturation from procambial cells and auxin production by the leaf associated with a particular vascular strand. They have also calculated that approximately ten times as much auxin is required to convert a parenchyma cell to a wound-vessel member as is needed to bring about the maturation of a procambial cell into a normal vessel element. This fact, together with the demonstration by Wangermann (1967) that externally applied radioactive indoleacetic acid is transported principally in the vascular bundles of *Coleus*, may explain why pith parenchyma in the intact stem does not differentiate as xylem. It is only after the severing of a vascular strand allows the local accumulation of auxin in excess of normal levels in the region above the wound that the concentration reaches sufficiently high levels to initiate redifferentiation of parenchyma.

In the experiments described thus far the emphasis has been on xylem differentiation, but the severance of a vascular bundle is followed by the differentiation of phloem sieve elements as well. Furthermore, it has been shown by LaMotte and Jacobs (1963) in *Coleus* that this differentiation of sieve elements, like that of wound tracheary elements, is auxin dependent. In fact the experiments that documented this dependence were essentially the same as those that demonstrated the auxin dependence of xylem differentiation. In individual plants phloem differentiation slightly preceded xylem differentiation and the pathways of differentiating xylem strands closely paralleled those established by the phloem (Jacobs, 1970).

From these experiments it is suggested that hormones of the auxin type are implicated in the final maturation of xylem and phloem, but it is not clear to what extent auxin interacts with other factors within a complex system of control. Because of the difficulties of isolating the system in the whole plant, several attempts have been made to examine the phenomena of vascular differentiation in excised stem internodes, tissue plugs, or tissue cultures in vitro. It is possible to maintain callus cultures of a number of species on a medium that will support their growth without the differentiation of vascular tissues. Wetmore and his associates (Wetmore and Sorokin, 1955; Wetmore and Rier, 1963) have experimented extensively with such tissues, particularly of *Syringa* (lilac). When vegetative buds were grafted into callus cultures of the same

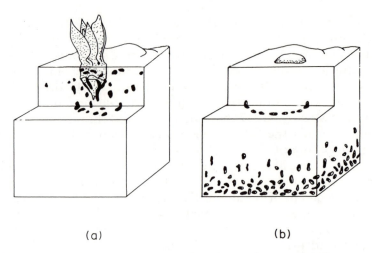

(a)   (b)

Figure 14.12. *Stereodiagrams showing induced vascular tissue differentiation in callus of* Syringa vulgaris. *(a) Differentiation of xylem in nodules and short strands induced by a grafted bud. Vascular tissue occurs in the wedge-shaped base of the bud and in continuity with this in the adjacent callus tissue. (b) Differentiation of vascular tissue induced by chemical treatment of callus. The culture medium and the agar applied to the top of the callus tissue contain the auxin naphthalene acetic acid and sucrose. Vascular nodules containing both phloem and xylem differentiated in relation to both sources of these substances. ([a] Wetmore and Sorokin, 1955. [b] Modified from Wetmore and Rier, 1963.)*

species, nodules and short strands of xylem differentiated, often in a ringlike pattern, in the callus below the grafted shoot (Fig. 14.12a). This suggested the transmission of a stimulus – probably chemical – from the bud into the callus, which had the effect of promoting xylem differentiation. Subsequently it was found that the bud could be dispensed with if in its place agar containing the auxins indoleacetic acid or naphthaleneacetic acid was inserted into the callus (Fig. 14.12b). The distribution of the vascular nodules in the callus could be regulated by the concentration of the applied auxin. When low concentrations were supplied the nodules differentiated close to the auxin source and were progressively further removed with higher concentrations.

Although the vascular tissues differentiated in tissue cultures include both xylem and phloem, their organization in nodules and sometimes in short strands is very different from that of the system of vascular strands in a stem or even from that of strands regenerated around a wound. Wetmore and Rier (1963) have noted that the calluses they studied show no polar auxin transport, so the auxin movement in them must be by diffusion. Interestingly, Clutter (1960) found that plugs of tobacco pith, a tissue in which auxin transport is not highly polarized, when supplied with auxin through micropipettes inserted into the tissue, also differentiated nodules of vascular tissue rather than vascular strands. Further, Roberts and Fosket (1962) found that xylem regeneration in flowering shoots of *Coleus* and in isolated stem segments of the same species was notably less oriented than in vegetative shoots. This was correlated with reduced polar auxin transport. The evidence from these studies suggests that polar transport may be an important factor in auxin-mediated differentiation in intact shoots.

Isolated plant parts and tissues have been used to investigate the possible role of substances other than auxin in vascular differentiation. In these studies such substances as sugar, cytokinins, and gibberellins have been implicated in vascular differentiation, but no consistent picture has emerged. In addition, concentration effects have been investigated, as in the study by Aloni (1980), which showed that in callus of lilac the minimal auxin concentration for phloem differentiation is lower than that for xylem development. This observation, if it is generally applicable, may explain the initial differentiations of phloem in advance of xylem in wound-response studies.

An inherent difficulty in these experimental systems is that it is not possible to separate direct influences upon target cells from indirect effects mediated through other cells and tissues. This problem has been partially overcome by the use of cultures consisting of populations of isolated cells. Cells isolated from the leaf mesophyll of *Zinnia elegans* have shown a capacity to redifferentiate into tracheary elements in liquid medium, and several important points appear to have been established (Fukuda and Komamine, 1980). There was shown to be a requirement for both an auxin and a cytokinin for tracheary differentiation. Because the cells differentiated in isolation it was evident that there is no dependence upon prior phloem differentiation. Finally, many of the isolated mesophyll cells differentiated directly into tracheary elements without preceding DNA synthesis or mitosis. Earlier experiments with whole plants or isolated parts had suggested that one or both of these events might be a necessary step in the process of redifferentiation (Fosket, 1968). Other workers, using the *Zinnia* culture system, have observed the redifferentiation of tracheary elements from more than half of the viable cells in a culture. Of particular interest is the report by Falconer and Seagull (1985) that these elements duplicate the patterns of wall thickening found in the plant (Fig. 14.13). Thus, the method of isolated cell culture appears to hold considerable promise of increasing our understanding of differentiation at the cellular level.

The experiments thus far considered have dealt largely with the actual processes of tracheary and sieve element differentiation and the factors

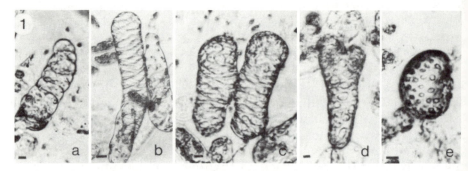

Figure 14.13. *Tracheary elements that have differentiated from isolated mesophyll cells of* Zinnia elegans *in sterile culture. (a) annular, (b) helical, (c) reticulate, (d) scalariform, (e) pitted. Scale bars, 10 μm. (Falconer and Seagull, 1985.)*

Figure 14.14. *The control of vascular strand differentiation in the epicotyl of* Pisum. *In (a) auxin (black spot) has been applied laterally and has induced the differentiation of vascular strands that connect with the central vascular column. In (b) this connection has been prevented by the simultaneous application of auxin to the cut end of the axial column. (T. Sachs, in* Pattern Formation, *eds. G. M. Malacinski and S. V. Bryant. New York: Macmillan, 1984.)*

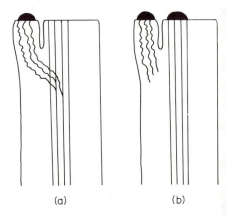

that control them. There have been other experiments in which the objective has been to explore the integrated process of pattern formation in the vascular system. In redirecting attention to normal, orderly development, it must be remembered that this does not occur by the redifferentiation of mature parenchyma cells as tracheary or sieve elements but rather through a sequence of changes from meristem derivatives through provascular and procambial stages. In fact, as was shown in Chapter 13, it is at these early stages of differentiation that the basic pattern of the vascular system is established. Therefore, it is reassuring to find that the main conclusion derived from wound-response and tissue-culture experiments, namely the overriding importance of auxin in vascular differentiation, is probably applicable to developmental events in the whole plant. Bruck and Paolillo (1984) have carried out experiments on *Coleus* in which the effects on vascular differentiation of the excision of young leaf primordia and of their replacement by auxin were assessed. They removed individual leaf primordia, either P_1 or P_2, and observed that the differentiation of the traces associated with the excised primordium did not continue. However, replacement of the excised primordium by a resin bead containing indoleacetic acid at close to physiological concentrations allowed normal development of the traces to be completed in the absence of the primordium.

In a series of experiments designed to explore the patterned differentiation of vascular tissues, Sachs (1981) has shown that pathways of auxin transport can account for some features of integrated vascular differentiation. For example, he found that when auxin was applied laterally to decapitated epicotyls of pea seedlings, a vascular strand differentiated through the cortex from the site of auxin application and established connection with the vascular cylinder of the seedling axis (Fig. 14.14). Application of auxin to the severed end of the cylinder, replacing the excised shoot apex, prevented the establishment of a connection between the induced and the preexisting strands, but by applying a higher concentration of auxin laterally it was possible to bring about the connection. From this and other similar experiments, Sachs has deduced that the interconnection of vascular strands is regulated by the flow of auxin in specific pathways. Although these experiments, like those previously discussed, dealt with the redifferentiation of already mature cells, they offer a basis for consideration of the way in which an interconnected vascular system could be formed. This interpretation is strengthened by an experiment that more nearly approximated normal development. Removal of a young leaf primordium from a lateral bud apex of a pea seedling after its trace had been formed resulted in the joining to that trace of bundles from later-formed leaves that would not have attached to the trace of the intact leaf.

GENERAL COMMENT

A striking conclusion that emerges from this survey of vascular development is the central role of auxin, in effect indoleacetic acid, in the differentiation of both xylem and phloem. In a way that may be unique in plant development, auxin has been shown to influence directly several individual steps in a multistage differentiation process. Not only can it promote procambium differentiation (McArthur and Steeves, 1972; Bruck and Paolillo, 1984), but it can also, in wounding experiments and in cell cultures, bring about the redifferentiation of mature parenchyma cells into tracheary elements completely by-passing the normal procambial stage (Jacobs, 1952; Fukuda and Komamine, 1980). Thus, auxin seems to be involved in the regulation of very generalized changes like orientation of the plane of cell division and the promotion of cell elongation in procambium, as well as in highly specific events of cytodifferentiation in tracheary and sieve elements.

By focusing attention upon the role of auxin in vascular differentiation, it is not intended to minimize the possible importance of other substances. In a variety of experimental systems several different factors, including other hormones and nutrients, have been implicated in vascular differentiation at one time or another (Roberts, 1976). However, as noted earlier, the requirement for these has not been demonstrated consistently. No doubt many factors are involved in normal differentiation, and in particular experimental circumstances any one or more may become limiting. The universal requirement for auxin, on the other hand, suggests that it plays a regulatory role in the plant.

The ability of auxin to influence cellular changes at different stages of vascular differentiation, while satisfying, does not necessarily explain how the pattern characteristic of the plant is established. Therefore, it is of interest to consider a theory of patterned vascular differentiation based largely upon the production and distribution of auxin in the plant. On the basis of evidence from experiments of the type described previously, Sachs (1981) has proposed that the pattern of the vascular system results from initial local differences in the flow of auxin through cells, which leads to the establishment of preferred channels of auxin transport. These channels become progressively improved pathways of auxin movement and drain the surrounding regions at the same time that their cells are induced to undergo differentiation as vascular elements. Connections are established with preexisting channels that are highly preferred pathways but whose auxin supply has been depleted. In this way, Sachs visualizes the building up of the integrated conducting system of the plant body. The value of this theory is that it provides a framework for the interpretation of vascular differentiation and in fact it accounts very well for the reestablishment of vascular connections in wounding experiments and for the basipetal differentiation of procambium in surgically isolated meristems and adventitous buds. It remains to be seen, however, how well it can be applied to events in the apical regions where the normal patterns are established. In these regions differentiation is predominantly acropetal and, in the shoot apex, procambial strands in some cases appear to precede the leaf primordia that would be their most probable source of hormone (Larson, 1983). It should be noted that continuous acropetal differentiation may be expected in response to a basipetal stimulus if that stimulus is also more or less continuous. Only if it is interrupted is basipetal differentiation a necessary consequence. However, the identification of hormones in specific regions of a shoot or root apex, and particularly the recognition of transport channels, will require that refined techniques of analysis be perfected so that this aspect of the theory can be tested.

REFERENCES

Aloni, R. 1980. Role of auxin and sucrose in the differentiation of sieve and tracheary elements in plant tissue cultures. *Planta* 150:255–63.

Bruck, D. K., and D. J. Paolillo, Jr. 1984. Replacement of leaf primordia with IAA in the induction of vascular differentiation in the stem of *Coleus. New Phytol.* 96:353–70.

Clutter, M. E. 1960. Hormonal induction of vascular tissue in tobacco pith *in vitro. Science* 132:548–9.

Esau, K. 1943. Vascular differentiation in the vegetative shoot of *Linum.* II. The first phloem and xylem. *Am. J. Botany* 30:248–55.

Evert, R. F. 1984. Comparative structure of phloem. In *Contemporary Problems in Plant Anatomy*, eds. R. A. White and W. C. Dickison, 145–234. New York: Academic Press.

Falconer, M. M., and R. W. Seagull. 1985. Immunofluorescent and calcofluor white staining of developing tracheary elements in *Zinnia elegans* L. suspension cultures. *Protoplasma* 125:190–8.

Fosket, D. E. 1968. Cell division and the differentiation of wound-vessel members in cultured stem segments of *Coleus. Proc. Nat. Acad. Sci. U.S.* 59:1089–96.

Freeberg, J. A., and R. H. Wetmore, 1967. The Lycopsida – A study in development. *Phytomorphology* 17:78–91.

Fukuda, H., and A. Komamine, 1980. Establishment of an experimental system for the study of tracheary element differentiation from single cells isolated from the mesophyll of *Zinnia elegans. Plant Physiol.* 65:57–60.

Golub, S. J., and R. H. Wetmore, 1948. Studies of development in the vegetative shoot of *Equisetum arvense* L. II. The mature shoot. *Am. J. Botany* 35:767–81.

Gunckel, J. E., and R. H. Wetmore. 1946. Studies of development in long shoots and short shoots of *Ginkgo biloba* L. II. Phyllotaxis and the organization of the primary vascular system; primary phloem and primary xylem. *Am. J. Botany* 33:532–43.

Jacobs, W. P. 1952. The role of auxin in differentiation of xylem around a wound. *Am. J. Botany* 39:301–9.

———. 1970. Regeneration and differentiation of sieve tube elements. *Internat. Rev. Cytol.* 28:239–73.

Jacobs, W. P., and I. B. Morrow. 1957. A quantitative study of xylem development in the vegetative shoot apex of *Coleus. Am. J. Botany* 44:823–42.

Kumazawa, M. 1961. Studies on the vascular course in maize plants. *Phytomorphology* 11:128–39.

LaMotte, C. E., and W. P. Jacobs. 1963. A role of auxin in phloem regeneration in *Coleus* internodes. *Devel. Biol.* 8:80–98.

Larson, P. R. 1983. Primary vascularization and the siting of primordia. In *The Growth and Functioning of Leaves*, eds. J. E. Dale and F. L. Milthorpe, 25–51. Cambridge: Cambridge University Press.

McArthur, I.C.S., and T. A. Steeves. 1972. An experimental study of vascular differentiation in *Geum chiloense* Balbis. *Bot. Gaz.* 133:276–87.

Peterson, R. L. 1967. Differentiation and maturation of primary tissues in white mustard root tips. *Can. J. Botany* 45:319–31.

Roberts, L. W. 1976. *Cytodifferentiation in Plants: Xylogenesis as a Model System.* Cambridge: Cambridge University Press.

Roberts, L. W., and D. E. Fosket. 1962. Further experiments on wound-vessel formation in stem wounds of *Coleus. Bot. Gaz.* 123:247–54.

Sachs, T. 1981. The control of the patterned differentiation of vascular tissues. *Adv. Bot. Res.* 9:151–262.

Torrey, J. G. 1955. On the determination of vascular patterns during tissue differentiation in excised pea roots. *Am. J. Botany* 42:183–198.

Torrey, J. G., D. E. Fosket, and P. K. Hepler. 1971. Xylem formation: A paradigm of cytodifferentiation in higher plants. *Am. Sci.* 59:338–52.

Wangermann, E. 1967. The effect of the leaf on differentiation of primary xylem in the internode of *Coleus. New Phytol.* 66:747–54.

Wetmore, R. H., and J. P. Rier. 1963. Experimental induction of vascular tissues in callus of angiosperms. *Am. J. Botany* 50:418–30.

Wetmore, R. H., and S. Sorokin. 1955. On the differentiation of xylem. *J. Arnold Arboretum* 36:305–17.

White, R. A. 1984. Comparative development of vascular tissue patterns in the shoot apex of ferns. In *Contemporary Problems in Plant Anatomy*, eds. R. A. White and W. C. Dickison, 53–107. New York: Academic Press.

CHAPTER 15

Secondary growth:
the vascular cambium

In considering the apical or primary meristems of the plant body, one of the most perplexing problems is the permanently meristematic condition of these regions, which are somehow spared from the processes of maturation occurring in their derivatives. One might be tempted to relate this property to their terminal position, their three-dimensional mass, or their organization, which is distinct from that of the mature structures they produce. However, the lateral meristems, which share the capacity for continued growth but are strikingly different in every other respect, prevent an easy acquiescence to this temptation. The *vascular cambium* and the *cork cambium*, or *phellogen*, are lateral in position, have the form of cylindrical sheets encircling the plant axis, and are organized in close conformity with the tissues to which they give rise. They initiate only specific tissues rather than whole organs as in the case of the terminal meristems. Furthermore, it must be borne in mind that whereas every vascular plant body must have terminal meristems in order to exist at all, the lateral meristems have a supplemental role and are by no means universal (Barghoorn, 1964).

THE INITIATION OF CAMBIAL ACTIVITY

Nothing emphasizes the differences between primary and secondary meristems more effectively than a consideration of the origin of the vascular cambium. Whereas the shoot and root apical meristems are initiated among the cells of the embryo early in the development of the plant, the cambium has its origin from a partially differentiated vascular tissue, the procambium. Although the major significance of the procambium is that it is a stage in the differentiation of the primary vascular system, certain cells may become stabilized at this level of maturation, retaining the capacity to proliferate indefinitely, and thus give rise to a vascular cambium. The details of cambial origin are best presented as they occur at one level in a stem or root at some distance behind the growing tip; however, it must be kept constantly in mind that such a picture is misleading because it tends to obscure the continuous acropetal extension of cambial activity. The events to be described, in fact, occur at the distal extremity of the already existing cambium and in continuity with it.

In the stem of a typical dicotyledon, the development of procambium in a bundle is accompanied by numerous longitudinal divisions. As these proceed, they ordinarily tend to become oriented parallel to the surface of the stem, that is, periclinally, with the result that the procambial cells tend to occur in radial rows. The final maturation of xylem elements begins, at any level, at the inner face of the bundle adjacent to the pith, and phloem maturation begins opposite to it at the outer face. Subsequently the final maturation of these two tissues proceeds toward the center of the bundle, gradually restricting the procambium to a band of dividing cells. Ultimately a stage is reached, by no means easily recognizable, at which it may be deduced that a common meristem is producing elements that will ultimately differentiate as both xylem and phloem (Fig. 15.1). This normally occurs at about the time that stem elongation is completed at a particular level. At this point it is stated that secondary growth has begun or that cambium has been initiated. Clearly, however, cambial activity represents a continuation of processes begun as part of primary development; there is no sharp demarcation between primary and secondary growth.

Cambial activity is restricted to the bundles in which it is initiated in cases in which there is a very limited amount of secondary growth, but more commonly there is a further development that results in a continuous ring, or more properly a cylinder, of cambium. After the initiation of fascicular cambium – that in the bundles – at any level, partially differentiated parenchyma cells between the bundles begin to divide periclinally and the interfascicular cambium is initiated (Fig. 15.1). Where the bundles are widely spaced, this activity can be observed to begin adjacent to the bundles and to advance across the interfascicular region, suggesting that the stimulus for the renewal of meristematic activity in the interfascicular regions has its origin in the dividing cells of the bundle (Phillips, 1976).

This sequence of events in cambial initiation in stems is generally found in the gymnosperms as well as in the dicotyledons, but in the relatively few monocotyledons that have secondary growth the origin is very different. Here the cambium arises outside the primary vascular tissues and produces both xylem and phloem, in bundles, to the interior. In roots also, the differences from stems in organization and in the pattern of differentiation of primary vascular tissues impose corresponding differences in the manner of cambial initiation. Both primary xylem and phloem differentiate centripetally, that is, toward the center of the root, and along alternating radii rather than in the same bundles.

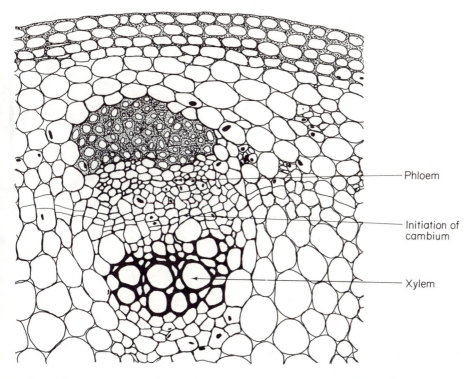

Phloem

Initiation of cambium

Xylem

Figure 15.1. *Initiation of vascular cambium in the stem of* Helianthus annuus. *Cambial activity that starts in the vascular bundles is beginning to expand into the interfascicular regions as a series of tangential divisions in partially differentiated parenchyma cells.* ×125.

The initiation of cambium occurs in procambium situated on the same radii as the phloem strands and lateral to the position of the xylem. The cambial ring is completed by the extension of divisions into the pericycle opposite the protoxylem poles.

THE ORGANIZATION OF THE VASCULAR CAMBIUM

In view of what has been said, it is evident that the methods of study of the vascular cambium, both analytical and experimental, must be quite different from those applied to terminal meristems, and the difficulties are even more acute than in the case of the root apex. The cambium may be observed effectively only in sections, which makes the study of living material difficult; and because of the extended, sheetlike form of the cambium and its relationship to its derivative tissues, the sections

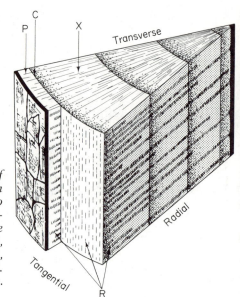

Figure 15.2. *Wedge from the stem of* Pinus *sp. (pine) illustrating the position of the vascular cambium in relation to its derivative tissues, and the orientation of sections used to examine the cambial meristem. Key: C, cambium; P, secondary phloem; R, vascular ray; X, secondary xylem. (Adapted from E. Strasburger, A Textbook of Botany. London: Macmillan, 1921.)*

must be oriented very precisely in order to be meaningful (Fig. 15.2). The general organization of the cambium, as well as the techniques for its investigation, may be illustrated by an examination of this meristem in the stem of *Pinus strobus* (white pine), which was the object of a series of investigations by Bailey (1954) now regarded as classics in the field.

The cambium of *Pinus strobus*

The cambium may be observed in face view by cutting sections that are tangential to the surface of the stem after removing the outer tissues overlying it (Fig. 15.2). Because the cambium is a thin layer sandwiched between the secondary xylem and the secondary phloem to which it gives rise, a section of the meristem ordinarily includes some areas of the derivative tissues on either side of the cambium itself. This is due in part to obliquity in the sections and in part to the curvature of the organ under investigation. Bailey has provided good evidence that if sections are cut in this way from living blocks of tissue, the cambium may be observed in a relatively normal condition. Such tissues slices were kept alive, as evidenced by cytoplasmic streaming, for periods of more than two months in ordinary tap water, but no cell division was noted. For the study of cellular detail tissues blocks may be fixed and sections cut and stained by standard histological methods.

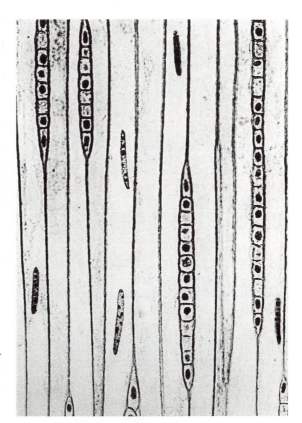

Figure 15.3. *Cambium of* Pinus strobus *(white pine) viewed in tangential section. ×250. (I. W. Bailey, Am. J. Botany 7:417, 1920.)*

When a tangential section of cambium is examined, one is immediately struck by the difference in organization between this tissue and the primary meristems (Fig. 15.3). The tissue is made up of two fundamentally different kinds of cells. The *fusiform initials* are elongate, pointed cells, up to 4 mm in length and 0.04 mm in width. The visible cell walls are thick and are beaded in appearance because of the occurrence of numerous primary pit fields. The cytoplasm is highly vacuolated and, if living sections are examined, protoplasmic streaming can often be observed. In contrast to the fusiform initials, which have no conspicuous vertical or horizontal pattern of alignment, the *ray initials* are arranged in short, uniseriate vertical rows. These cells are only 0.02 mm in height and their width is approximately the same. Comparison with the mature xylem and phloem, cut in the same plane, often visible in the same preparation, reveals a close similarity in histology between the mature and the meristematic tissues. The fusiform initials clearly correspond to the tracheids in the xylem and to the sieve cells of the

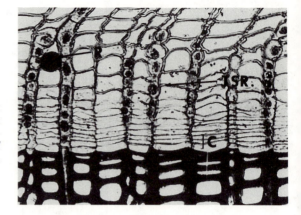

Figure 15.4. *The cambium and its derivative tissues in* Pinus strobus *seen in transverse section. Key:* C, *cambial zone;* R, *ray;* TSR, *tracheid-sieve cell row.* ×250. *(I. W. Bailey,* Am. J. Botany 7:417, 1920.)

phloem, whereas the groups of ray initials are related to the horizontal rays that extend through both xylem and phloem.

The details of the spatial relationship between the cambium and its derivative tissues may be examined more fully in transverse sections (Fig. 15.4). Clearly defined radial rows of cells extend through both xylem and phloem, and these two tissues are separated by a zone several cells in width in which, although the rows are distinct, the cells are immature and are apparently dividing. As in the tangential sections, two kinds of cells can be identified but here their aspect is much different. In the tracheid-sieve cell rows, which must contain the fusiform initials, the narrowest cells have a radial diameter of only 0.006 mm and the corresponding cells in the rays have a radial dimension of 0.03 mm. It is clear from the orientation of cell walls that cell divisions must occur primarily in a longitudinal, periclinal plane, that is, in such a way as to increase the number of cells in the rows. This division process is not difficult to visualize in the rays, but in the case of the fusiform initials, whose length is more than 600 times the radial width, such a division plane raises some interesting questions. A periclinal longitudinal division in a fusiform initial partitions the cell at right angles to its narrowest dimension and forms a wall having the maximum possible surface area (Fig. 15.5b).

In the foregoing discussion descriptions were given of what were called fusiform and ray initials. In transverse sections taken during periods of active growth, however, no single cell may be picked out in each row as an obvious initial. Rather, in each row there are several cells, any one of which on the basis of position and structure could be regarded as an initial. Some observers have maintained that in each row there must be a permanently meristematic initial cell and that the similar cells on

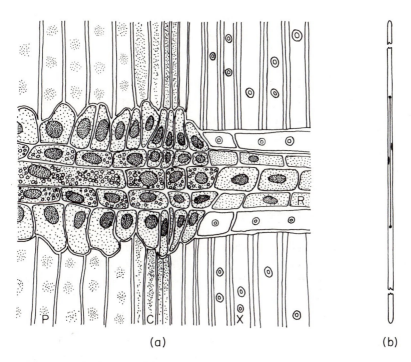

(a)   (b)

Figure 15.5.   *Radial view of the cambium and derivative tissues in* Pinus *sp. (a) The cambium and derivative tissues in radial section. The cambium has elongated fusiform initials and short ray initials.* ×235. *(b) A single fusiform initial undergoing division in the periclinal plane. The nucleus has divided and two daughter nuclei can be seen separated by the new wall, which has not yet grown to the tips of the cell. Key: C, cambial zone; P, secondary phloem; R, ray; X, secondary xylem. ([a] Adapted from E. Strasburger,* A Textbook of Botany. *London: Macmillan, 1921. [b] I. W. Bailey,* Am. J. Botany 7:417, 1920.)

either side of it, although still dividing, are xylem or phloem mother cells. Others have argued that there is merely a zone of dividing cells and that no individuals have any permanent status as initials. A discussion of the merits of these two points of view will be deferred until some aspects of cambial growth have been considered. This problem must be kept in mind, however, in the examination of cambial histology, in that what have been designated fusiform and ray initials are not necessarily permanent initials.

Longitudinal sections cut along a radius of the stem complete the picture of the shape of cambial initials (Fig. 15.5). In such radial sections, the long initials and their immediate derivatives are seen to have blunt ends, which facilitates an understanding of the periclinal division process. Furthermore, the marked disparity in radial width of fusiform and ray initials is very evident in such sections.

Cambium in other species

The cambium has been investigated histologically in a relatively large number of species most of which, however, are trees. Other conifers differ little from white pine in the structure of the cambium, but there is considerable variation in the length of fusiform initials, from less than 1 mm to nearly 9 mm. In woody dicotyledons, the cambium usually contains multiseriate groups of ray initials of various heights and widths in keeping with the occurrence of multiseriate rays in the secondary xylem and phloem. There is also a much greater range in the length of fusiform initials, and in the vast majority of species these are much shorter than in the conifers (Fig. 15.6a). In some primitive, vesselless dicotyledons, however, the length is as great as it is in the conifers. Bailey has shown that there has been a phylogenetic reduction in the length of fusiform initials in dicotyledons that is correlated with a reduction in the length of vessel elements. In some very advanced species, as judged by other criteria, the very short fusiform initials occur in horizontal tiers in what is designated as *storied* (*stratified*) *cambium* (Fig. 15.6b).

Reference has already been made to the distinctive origin of the cambium in the relatively few monocotyledons that have secondary growth. The structure is also very different from that found in conifers and dicotyledons. The cambial initials are all of one type and vary greatly in shape even in the same plant. They may be either fusiform or rectangular, and in some cases the same cell may have one tapering end and one blunt end. Because of its distinctive structure and activity and its developmental relationship with the meristematic activity that increases the thickness of the primary body, there is some doubt as to whether this meristem is truly homologous with the cambium of conifers and dicotyledons.

THE DYNAMIC STATE OF THE CAMBIUM

The picture of the cambium presented thus far is a rather static one, although the extensive cell-producing activity has been emphasized. The continued formation of secondary xylem, however, imposes upon the cambium the necessity of covering an ever-increasing surface. A picture of the cambium as a constantly changing dynamic population of meristematic cells has emerged from several intensive investigations.

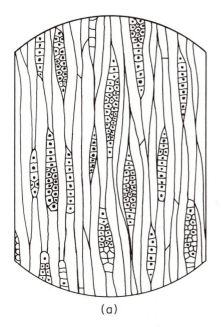 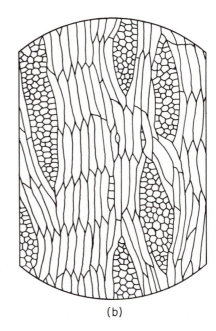

(a)   (b)

Figure 15.6.   *Tangential view of the cambium of* Juglans cinerea *(walnut) (a) and* Robinia pseudoacacia *(locust) (b). There are multiseriate and uniseriate rays in both species, and in* Robinia *the fusiform initials are in a storied arrangement.* ×*100. (Redrawn from A. J. Eames and L. H. MacDaniels,* An Introduction to Plant Anatomy. *New York: McGraw-Hill, 1947.)*

Fusiform initials in conifers

The mechanism by which increase in girth is achieved in the cambium of conifers has been studied revealingly in *P. strobus* by Bailey who has shown that a surprising number of different processes are involved. During a period of approximately sixty years from its initiation, at any level in a tree of this species the cambium is characterized by an increase in the size of its fusiform initials, particularly in length and in tangential width. Increase in length in fusiform initials is accomplished by tip growth, and the pointed ends of the cells intrude between other initials. This process, therefore, leads to an increase in the number of fusiform initials intersected in any transverse section and, together with an increase in width, contributes to the enlarged circumference of the cambial layer. However, Bailey has shown that these processes cannot explain the total increase in circumference and, because they do not continue beyond approximately sixty years, they cannot explain subse-

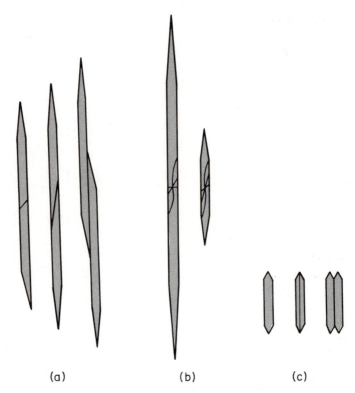

(a)   (b)   (c)

Figure 15.7. *Anticlinal divisions of cambial fusiform initials. (a) Pseudotransverse division and subsequent elongation of the two daughter cells in a conifer. (b) Fusiform initials from a series in dicotyledons showing the increasingly vertical orientation of the anticlinal division associated with decreasing length of the initial. (c) Vertical anticlinal division that results in a storied cambium. (I. W. Bailey, Am. J. Botany 10:499, 1923.)*

quent increase in girth. Because the number of cambial initials in cross sections does continue to increase, it is evident that there must be anticlinal divisions in the fusiform initials. Tangential sections, however, do not show the expected evidence of such divisions in that the fusiform initials are not arranged in pairs or horizontal groups, which would suggest a common origin. The surprising resolution of this difficulty is found in the observation that anticlinal divisions in the fusiform initials are nearly transverse in orientation (Fig. 15.7a). Bailey has termed such divisions *pseudotransverse*. The result of such a division is the production of two initials, each approximately half the length of the original. Following division, intrusive growth occurs at the tips of each of the initials, both of which finally achieve the length of the original. Thus,

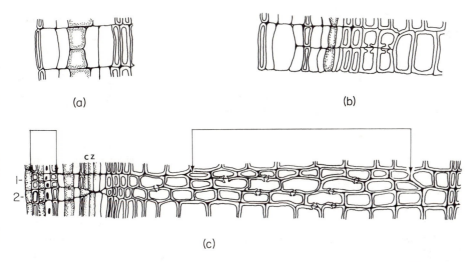

(a)

(b)

(c)

Figure 15.8. *Addition and loss of fusiform initials in the cambium of* Thuja occidentalis *(white cedar). (a) As the result of a pseudotransverse anticlinal division
a fusiform initial of the cambium has doubled. The cambial layer is shown stippled.
The secondary phloem (left) and secondary xylem (right) occur as a single row of
cells that were produced by the activity of a single initial. The divided fusiform
initials have not yet produced derivative cells. (b) A developmental stage in the cambium later than that shown in (a). The two cambial initials have produced a doubled row of phloem on the left and xylem on the right enclosed in the single row
produced before the initial doubled. (c) In the row of cells marked "1" the fusiform
initial doubled producing a double row of phloem (left) and xylem (right). The doubled parts of the phloem and xylem are indicated by the connected arrows. One of
the initials then ceased to function and the xylem and phloem rows became single
again. In the cell row marked "2" the fusiform initial has just ceased to function.
It is not present in the cambial zone (cz). Its derivative xylem and phloem row of
cells will now terminate. (Bannan, 1955.)*

the effective number of cambial cells in cross section is increased (Fig.
15.8).

Bannan (1956) has provided information on the frequency of anticlinal divisions in *Thuja occidentalis* and several other conifers. Because
periclinal divisions that lead to the formation of xylem and phloem derivatives continue during these developmental changes, and because the
derivatives elongate only slightly during differentiation, a record is preserved in the secondary xylem in which the time sequence can be determined. A considerable interval occurs between successive anticlinal
divisions in the same row, but this is highly variable. In *Thuja* Bannan
found that in mature trees the average interval in different trees ranged
from one to eight years and that the overall average for all trees studied

was 3.7 years. He has found that in the first few years after initiation at any level in the stem the cambium is characterized by a high frequency of pseudotransverse divisions, and that this frequency declines in successive years until a relatively stable value is reached. This is correlated with the previously noted trend in the increase in length of fusiform initials during the early years of cambial activity (Bannan, 1960). That the inverse correlation between the rate of anticlinal division and length of fusiform initials is a real one is further demonstrated by Bannan's (1957a) studies on fluted trunks, in which he has shown that in the depressions of such trunks the rate of anticlinal division is higher than elsewhere and the length of fusiform initials is less. Increase in girth, however, is not directly correlated with the rate of anticlinal division because fusiform initials drop out of the cambium by becoming differentiated and thus cease to divide. The evidence for such cessation of activity is the termination of a radial row in both xylem and phloem (Fig. 15.8c). Thus, the increase in circumference is governed by a balance between the rate of anticlinal division and the rate of dropout of initials, and in the depressions of fluted stems there may be a net loss of initials even though the rate of anticlinal division is high. The dropout of an initial can occur at any time in its life history, but it occurs most often in the period shortly after an anticlinal division.

In any consideration of the possible regulation of cell division in the cambium, it is necessary to distinguish between periclinal divisions that are overwhelmingly predominant in frequency and anticlinal divisions that, although much less frequent, are essential in maintaining the integrity of the meristem. Bannan has obtained evidence indicating that although anticlinal division can occur throughout the annual growth period in conifers, there is a tendency for it to occur most often toward the end of the period when periclinal divisions are greatly reduced in frequency. This observation might be extremely significant in any experimental attempt to separate the controls of the two patterns of division.

Fusiform initials in dicotyledons

In the woody dicotyledons, Bailey has found interesting differences in the mechanism of girth increase in the cambium that are related to the phylogenetic shortening of fusiform initials. In the primitive, vesselless species the process is apparently similar to that in conifers. In a series of species having progressively shorter fusiform initials, he has shown that the anticlinal divisions become increasingly oblique, approaching progressively nearer to the vertical orientation (Fig. 15.7b). As the divisions become more nearly vertical, there is a decreasing amount of intrusive elongation following the divisions. The ultimate stage in this series is one in which the shortest fusiform initials are found, and in these the anticlinal divisions are completely vertical and are not followed by elongation (Fig. 15.7c). This process leads to the formation of horizontal tiers, the storied cambium previously mentioned. As in the conifers, there is also a loss of fusiform initials from the cambium so that increase in circumference is the result of a balance between gain and loss of initials. Limited information about herbaceous dicotyledons with restricted cambial activity indicates that in these plants there is no loss of initials and the frequency of anticlinal division is just about adequate to account for increase of girth (Cumbie, 1969).

Reference has been made to an ontogenetic trend in the cambium of conifers that leads to an increase in the length of fusiform initials in the early years of cambial activity. In the dicotyledons, if the series of species with decreasing length of fusiform initials is examined, it is found that this ontogenetic trend is correspondingly suppressed and is essentially absent in species having storied cambium. The suppression of this trend is, of course, an important factor in bringing about the overall reduction in length of the fusiform initials. Because this trend is apparently explained by a reduction in the frequency of anticlinal divisions, with a corresponding enhancement of the elongation of the fusiform initials, it is not surprising that its disappearance is correlated with a shift in anticlinal divisions from the pseudotransverse to the vertical orientation and a loss of the ability to elongate following such divisions. It is significant to note, however, that these changes cannot be explained simply as a result of a loss of the ability of fusiform cells to elongate intrusively because, even in those cases in which there is no elongation at all after anticlinal division, derivatives of the cambium that develop as fibers undergo extensive elongation during their differentiation. Furthermore, in certain circumstances the fusiform initials themselves can elongate intrusively, as will be described subsequently.

Changes in the rays

The foregoing discussion has emphasized that the fusiform initials of the cambium are in a state of constant change. As the secondary body enlarges and the cambium expands, it is evident that the number of rays and consequently the number of groups of ray initials also increase. In fact, Bailey has noted the increase in the number of ray initials as a factor contributing to increase of girth in the cambium in both conifers and dicotyledons. In a study of ray ontogeny in conifers and dicotyledons, Barghoorn (1940a, b, 1941) has investigated the methods by which new rays arise and the subsequent developmental changes that they

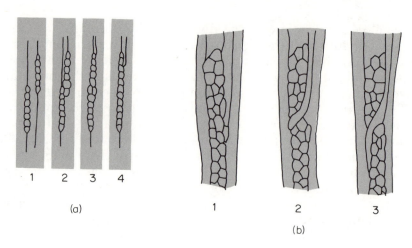

(a)

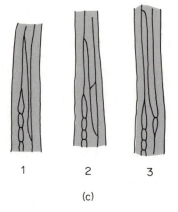

Figure 15.9. *Ontogenetic changes in rays of conifers and dicotyledons revealed by study of the secondary xylem in which the changes are preserved. (a) Fusion of two rays in* Taxus baccata, *which results from dropping out of a fusiform initial. (b) Division of a ray in* Trochodendron aralioides *by intrusive elongation of a fusiform initial. (c) Formation of a new ray from the tip of a fusiform initial in* Viburnum odoratissimus. *([a] Barghoorn, 1940a. [b,c] Barghoorn, 1940b.)*

undergo as a result of changes in the cambium (Fig. 15.9). In the conifers a new ray may arise from a fusiform initial by the cutting off of a cell from the tip, which then begins to function as a ray initial, by the cutting out of a cell from the side of a fusiform initial by means of an anticlinal division in which the new wall curves to intersect the side wall, or by the septation of an entire fusiform initial to form a vertical series of ray initials. The uniseriate rays of conifers do not increase in width but they increase in height as a result of horizontal anticlinal divisions in ray initials or by the fusion of two rays as the result of the dropping out of a fusiform initial. The number of rays may also be increased by the splitting of existing rays as a result of penetration by the elongating tips of fusiform initials.

In dicotyledons new rays have their origin from fusiform initials as in the conifers, but there are two differences to be noted. Ray initials ordinarily are not cut out of the side of fusiform initials, and occasionally a multiseriate ray may arise directly by septations involving several fusiform initials. The developmental changes that occur in rays of dicotyledons are numerous and complex and only a few will be noted here. Rays increase in both height and width by division of fusiform and ray initials, and uniseriate rays are often converted into multiseriate rays. Fusion of rays leads to an increase in size, and ray dissection leads to a decrease in size and an increase in the number of rays. Dissection may be accomplished by intrusion of elongating fusiform initials and by the conversion of ray initials to fusiform initials. In view of the highly dynamic state of rays, Barghoorn has pointed out that it is essential to understand their ontogenetic changes before using them in the systematic description of secondary xylem or attempting to establish phylogenetic trends in ray structure.

CHANGING PATTERNS IN THE CAMBIUM

The dynamic state of the cambium was originally investigated in relation to the phenomenon of increase in circumference. However, it has more recently been recognized that cellular changes in the cambium may occur in patterns that suggest a broader significance. Attempts to interpret situations in which secondary xylem cells are oriented in different ways relative to the longitudinal axis of the trunk led to the discovery of what are called *cambial domains* (Berlyn, 1982). A domain is an area in the cambium, viewed tangentially, within which events such as pseudotransverse divisions and apical intrusive growth of fusiform initials and the splitting and fusion of rays are not randomly oriented but rather show a predominance of left-handed or right-handed inclination (Fig. 15.10) (Hejnowicz, 1975). If these events are occurring with a high frequency, they may have an influence upon the orientation of mature elements in the secondary xylem, that is, the *grain* of the wood. It has also been shown that the cambial domains are not static. They may expand or contract as the number of cells affected increases or decreases. Furthermore, the domains may appear to be displaced vertically, although no cells actually move, as the orientation of cellular processes oscillates between left and right (Hejnowicz and Romberger, 1973; Hejnowicz, 1975). This has led to suggestions that the factors that control these processes are propagated in a wavelike pattern within the cambium, but in the absence of any clear understanding of the morphogenetic controls these suggestions remain speculative.

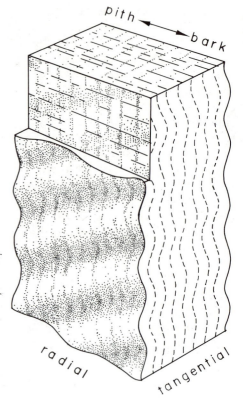

Figure 15.10. *Diagram of a block of secondary xylem showing wavy grain. Looking at the tangential face, it is possible to interpret the grain in terms of the preferred orientation of cells in the xylem, which, in turn, is related to the orientation in cambial domains. Reference to the radial face shows that these have been propagated upward as growth proceeded. (Hejnowicz and Romberger, 1973.)*

THE QUESTION OF CAMBIAL INITIALS

Up to this point the cambium has been treated as if it consisted of a single layer of initial cells, although such a layer is not ordinarily distinguishable from its immediate derivatives. It would be desirable now to examine the evidence favoring this interpretation over the alternative view that the cambium is a multilayered meristem in which there are no cells having any permanent value as initials. The information presented dealing with developmental changes in the cambium provides strong support for the concept of initial cells. It would be difficult to interpret the simultaneous changes occurring in individual radial rows on the two sides of the cambium in terms other than that these changes have occurred in single initial cells that are relatively permanent. Doubling of a row on both sides of the cambium, the loss of a row on both sides, and the various changes in rays occurring in both xylem and phloem are examples of the sort of changes that must be dependent on the existence of single initials (Figs. 15.8, 15.9). Occasionally temporary changes occur on one side only. For example, a xylem row may become double over the extent of several cells and then become single again. Such a phenomenon could result from an anticlinal division in a derivative of the fusiform initial in that particular row, a xylem mother cell, which still had considerable division potential remaining and strongly supports the idea that the more common permanent changes are those that occur in the true initials. The functional distinctiveness of the single layer of initials is emphasized by Bannan's (1957b) observations on twelve species of conifers in which 98 percent of the pseudotransverse divisions noted were of the permanent type.

In view of this convincing but indirect evidence for the existence of functionally distinctive initial cells, it is appropriate to examine the cambial region in detail to determine whether any direct evidence may be obtained. Such examination of transverse sections reveals immediately that the periclinal, tissue-generating divisions are not restricted to a single layer of cells but rather are distributed through a multicellular zone. Wilson (1964, 1966) has described this zone in *P. strobus* as consisting of six to fifteen cells in each radial row that are actively dividing periclinally, and Bannan (1955) has found even wider zones in rapidly growing trees of *Thuja*. It is generally agreed that there is no single cell in each row of elongate cambial cells that stands out structurally as the initial of that row. Newman (1956) has pointed out, however, that in *Pinus radiata* the ray initials can be distinguished on the basis of size from maturing ray cells on either side, and this may be true in many species. This same worker has argued that close scrutiny of cell wall structure in the cambial zone permits the identification of the fusiform initial, at least in some rows (Fig. 15.11). His analysis depends upon the fact that each time a cell divides a complete wall layer is deposited around the protoplast. Thus, the relative age of walls can be determined to a considerable extent by their thickness, and packets of cells within a common wall can be observed. In *P. radiata* the fusiform initials alternately produce groups of derivatives on xylem and phloem sides and one may speak of xylem-forming and phloem-forming phases of activity. By locating the general area of the initial from the position of the ray initials and looking for a particularly thick tangential wall, Newman felt that he could identify the initial cell as the one bounded by that wall and could tell whether it was in a xylem- or a phloem-forming phase according to the side on which the thicker wall was located. Later detailed studies by Mahmood (1968) and by Murmanis (1971), also on conifers, have provided confirmation of Newman's interpretation. On

Figure 15.11. *Diagram illustrating Newman's interpretation of cambial activity in* Pinus radiata *(pine). The approximate location of the initials in the cambium is first identified by the position of the smallest cells in the rays assumed to be the ray initials. The fusiform initials will then be approximately next to the ray initials. The fusiform initials can be recognized by the wide band of wall material on their inner or outer tangential face. When the band is on the outer face of the fusiform initial, as in cell row A, the initial is in its xylem-producing phase (xylem is at the bottom of the diagram and phloem at the top), and when the band is on the inner face of the fusiform initial it is in its phloem-producing phase as is the case in cell row B. Key: I, fusiform initial; R, ray. (Diagram based on data from Newman, 1956.)*

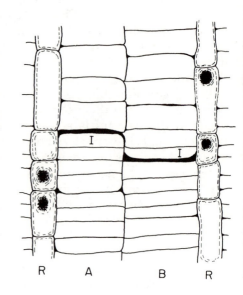

the other hand, Catesson (1974) was not able, even at the ultrastructural level, to find any distinctive cytological characteristics that would identify the initials in the dicotyledon *Acer pseudoplatanus*.

From Newman's study it was also apparent that the individual radial rows were not coordinated in their activity, so of two adjacent rows, at any particular moment one might be forming xylem and the other phloem. Following periclinal division in an initial, either one of the sister cells may retain the initial function while the other becomes a tissue mother cell. Because adjacent radial rows are not necessarily coordinated in this respect, the concept of a single layer of initials is somewhat misleading if it gives a three-dimentional picture of a smooth cylinder. In fact, except during dormant periods, the layer must be rather irregular because the initials of adjacent rows may be considerably offset.

In plants with marked periodicity of cambial activity, during the dormant period the cambial zone is reduced in width and in some cases the presence of an initial in each radial row becomes more obvious. Esau (1948) has described the cambial zone in *Vitis* as being reduced to a single layer of undifferentiated cells, presumably the initials, during the period of winter dormancy. A more typical situation is that described by Bannan (1955) in *T. occidentalis*. In this species the cambial zone in dormant trees is usually three cells wide. The outermost of these cells, adjacent to recognizable phloem, is regarded as the initial and is distinguished from the inner xylem mother cells, which are slightly longer. When growth is renewed in the spring, cell division most commonly begins in the xylem mother cell next to the mature xylem, and somewhat later division occurs in the initial cell. For a time, cell division proceeds more rapidly than cell differentiation so that a broad zone of dividing cells is established. Within this broad zone, however, there is a nonuniform distribution of mitotic activity. The peak of mitotic activity is not found in what Bannan interprets to be the initial cells; rather, it occurs near the center of the group of xylem mother cells. Cell-division frequency in the initial cells and phloem mother cells is approximately one-half the maximum value.

TERMINOLOGY

Now that consideration has been given to the organization and functioning of the lateral vascular meristem, it may be appropriate to comment upon the inconsistencies that are apparent in the terminology applied to this region. In some cases the term *cambium* is used specifically to designate the single layer of initial cells, even though these are extremely difficult if not impossible to identify. The entire band of dividing cells is then designated the *cambial zone* or *cambial region*. An alternative approach has been to use the term *cambium* to describe the whole dividing region and to speak of *cambial initials* or *fusiform* and *ray initials* when these are intended. The latter terminology, which has been more widely followed, seems to be preferable for several reasons. In the first place, the restriction of the term *cambium* to the layer of initials probably would have little influence upon nonspecialists who ordinarily use it in reference to the whole lateral meristem. Furthermore the broader usage of the term serves to make it more nearly comparable to the term *apical meristem* which is considered to include not just the apical initials but their derivatives as well.

GENERAL COMMENT

One of the most significant aspects of the study of the vascular cambium is the bearing it has upon our understanding of the mechanism of continued meristematic activity. If one considers origin, position, histological organization, and cytological characteristics, there is essentially nothing common between the vascular cambium and either of the apical meristems. So great are the differences in fact, it is tempting to speculate that the long retention of meristematic potentialities may have a different physiological basis in different parts of the plant body, but so long as the actual mechanism remains unknown in all cases, such speculation has little value. The one exception to the general dissimilarity between the primary meristems and the cambium could be the occurrence of initial cells in both cases. This may be significant in suggesting that initial cells are a basic requirement for long-continuing meristems, and their similarity to stem cells in animal development may point to an even broader biological role. Although the initial cells of the cambium may appear to function in a manner different from that of root and shoot apical initials, careful comparison will show that they, like the initials of the terminal meristems, conform to Newman's definition (1965) of a "continuing meristematic residue" (see Chapter 5). In fact, it is easier to comprehend this concept in relation to the cambium than with respect to shoot and root apical meristems. Study of this secondary meristem, then, may enhance our understanding of the primary meristems.

One of the particularly striking features of the vascular cambium is the close correspondence in structural organization between the meristem and the tissues that it produces, the secondary xylem and phloem. This is in marked contrast to the terminal meristems, particularly that of the shoot, in which the meristem has its own distinctive organization quite apart from the initial differentiation of tissues. This, of course, is correlated with the fact that the cambium produces certain tissues only, whereas the terminal meristems initiate an entire shoot or root. There seems to be a question here of levels of differentiation. The shoot and root apices, as has been pointed out, are differentiated in the embryo as shoot- and root-producing centers and are seemingly stabilized at this level, retaining for prolonged periods the capacity for cell production while the derivatives they produce undergo maturation. It may be suggested that something of a similar sort occurs in the process of cambial initiation, but at a relatively advanced stage of vascular tissue differentiation, the procambial stage. The resulting meristem is thus very strongly committed to the initiation of vascular tissues, while retaining the capacity for indefinite cell production. It is, in one sense, a histogen, although probably not in the sense in which this term has been used for shoot and root apices. Whether the mechanism of stabilization is the same, or even comparable, in the two cases is, of course, another way of posing the question with which this general comment began.

The close correspondence between the cambium and the mature tissues it produces greatly facilitates the study of this meristem because a record of the changes that occur in the cambium with the passage of time is left behind, particularly in the relatively permanent secondary xylem. This record reveals that the cambium is a surprisingly dynamic population of meristematic cells and one in which the changing patterns of cell division and cell elongation suggest a rather precise control. The discovery of cambial domains and the changes they undergo has focused renewed attention on this question. At present there is very little information about the factors that regulate the frequency and the orientation of cell divisions and the extent and direction of cell elongation, but a limited amount of experimental work has begun to point out ways in which controlling mechanisms may be explored. This is the subject of the following chapter.

REFERENCES

Bailey, I. W. 1954. *Contributions to Plant Anatomy*. Waltham: Chronica Botanica.

Bannan, M. W. 1955. The vascular cambium and radial growth in *Thuja occidentalis* L. *Can. J. Botany* 33:113–38.

———. 1956. Some aspects of the elongation of fusiform cambial cells in *Thuja occidentalis* L. *Can. J. Botany* 34:175–96.

———. 1957a. Girth increase in white cedar stems of irregular form. *Can. J. Botany* 35:425–34.

———. 1957b. The relative frequency of the different types of anticlinal divisions in conifer cambium. *Can. J. Botany* 35:875–84.

———. 1960. Ontogenetic trends in conifer cambium with respect to frequency of anticlinal division and cell length. *Can. J. Botany* 38:795–802.

Barghoorn, E. S., Jr. 1940a. Origin and development of the uniseriate ray in the Coniferae. *Bull. Torrey Botan. Club* 67:303–28.

———. 1940b. The ontogenetic development and phylogenetic specialization of rays in the xylem of dicotyledons. I. The primitive ray structure. *Am. J. Botany* 27:918–28.

———. 1941. The ontogenetic development and phylogenetic specialization of rays in the xylem of dicotyledons. II. Modification of the multiseriate and uniseriate rays. *Am. J. Botany* 28:273–82.

———. 1964. Evolution of cambium in geologic time. In *The Formation of Wood in Forest Trees*, ed. M. Zimmerman, 3–17. New York: Academic Press.

Berlyn, G. P. 1982. Morphogenetic factors in wood formation and differentiation. In *New Perspectives in Wood Anatomy*, ed. P. Baas, 123–50. The Hague: Martinus Nijhoff.

Catesson, A. M. 1974. Cambial cells. In *Dynamic Aspects of Plant Ultrastructure*, ed. A. W. Robards, 358–90. New York: McGraw-Hill.

Cumbie, B. G. 1969. Developmental changes in the vascular cambium of *Polygonum lapathifolium*. *Am. J. Botany* 56:139–46.

Esau, K. 1948. Phloem structure in the grapevine and its seasonal changes. *Hilgardia* 18:217–96.

Hejnowicz, Z. 1975. A model for morphogenetic map and clock. *J. Theor. Biology* 54:345–62.

Hejnowicz, Z., and J. A. Romberger. 1973. Migrating cambial domains and the origin of wavy grain in xylem of broad leaved trees. *Am. J. Botany* 60:209–22.

Mahmood, A. 1968. Cell grouping and primary wall generations in the cambial zone, xylem, and phloem in *Pinus*. *Austral. J. Botany* 16:177–95.

Murmanis, L. 1971. Structural changes in the vascular cambium of *Pinus strobus* L. during an annual cycle. *Ann. Botany* 35:133–41.

Newman, I. V. 1956. Pattern in meristems of vascular plants. I. Cell partition in living apices and in the cambial zone in relation to the concepts of initial cells and apical cells. *Phytomorphology* 6:1–19.

———. 1965. Pattern in the meristems of vascular plants. III. Pursuing the patterns in the apical meristem where no cell is a permanent cell. *J. Linn. Soc. (Bot.)* 59:185–214.

Phillips, I.D.J. 1976. The cambium. In *Cell Division in Higher Plants*, ed. M. M. Yeoman, 348–90. New York: Academic Press.

Wilson, B. F. 1964. A model for cell production by the cambium of conifers. In *The Formation of Wood in Forest Trees*, ed. M. Zimmermann, 19–36. New York: Academic Press.

———. 1966. Mitotic activity in the cambial zone of *Pinus strobus*. *Am. J. Botany* 53:364–72.

16 *Secondary growth: experimental studies on the cambium*

The fact that the activity of the vascular cambium can be traced very precisely in its mature derivatives suggests that this meristem ought to be a useful system in which to study the control of developmental phenomena. On the other hand, its relative inaccessibility to direct manipulation and to observation are obstacles to the kind of experimentation that has been so profitable in the case of the shoot apex. It is perhaps not surprising, therefore, that there has been very little experimental work dealing with the control of developmental patterns in the cambium, while at the same time such phenomena as seasonal activation have received considerable attention. There has also been a great deal of interest in the participation of the cambium in wound reactions and healing and in the establishment of tissue unions in grafts of various types, both areas of considerable applied significance. The relatively few studies that have dealt with fundamental aspects of development in the cambium, together with information that may be extracted from a number of applied investigations, give strong indications that the experimental approach could be as valuable a tool in the understanding of cambial problems as it has been in the case of the terminal meristems.

CAMBIAL INITIATION

The initiation of cambium from the primary procambial tissues, although intensively studied from the structural viewpoint, has not been investigated experimentally to the same extent. The transition from primary to secondary growth is difficult to identify and hence is not readily amenable to manipulation. Rather, most studies of cambial origin have dealt with initiation in parenchymatous tissues between vascular bundles or in the callus tissues formed in response to wounding.

Commonly the initiation of cambial activity in interfascicular regions has been described as spreading laterally from adjacent bundles in which cambium is already functioning (see Chapter 15). Thus, it has been assumed that partially differentiated parenchyma responds to a chemical stimulus propagated from functional cambium in the bundles. Considerable interest is therefore attached to experiments performed by Siebers (1971a, b) on seedlings of *Ricinus communis* (castor oil plant) that indicate that the interfascicular cells from which the cambium arises are highly determined in advance of any overt differentiation.

These experiments were carried out on the hypocotyl of seedlings in which the primary vascular bundles are separated by broad zones of interfascicular parenchyma. In one experiment small pieces of stainless steel razor blade were inserted in pairs into the stem in such a way as to isolate blocks of interfascicular tissue from neighboring vascular strands (Siebers, 1971b). This treatment was later shown by histological study to have had no effect whatever upon the initiation of cambium in the laterally isolated blocks. Siebers, therefore, concluded that cambial initiation is not brought about by a stimulus from the adjacent vascular tissue. An even more surprising result was obtained when blocks of interfascicular tissue were removed and reinserted in a radially reversed orientation (Fig. 16.1) (Siebers, 1971a). To do this, a punch fashioned from four razor blade fragments was used to remove a horizontal plug from the stem and an interfascicular segment was cut out and reversed prior to reinsertion of the plug. Subsequent examination revealed that cambium in the reversed segment had differentiated in the expected location whether or not that location matched the position of the cambium in the adjacent bundles. The startling observation, however, was that the pattern of xylem and phloem formation followed that of the original orientation of the interfascicular segment so that xylem was formed to the outside and phloem to the inside in that segment. Comparable plugs of pith tissue inserted into the hypocotyl as a control were not induced to form cambium at all. Finally it was shown that isolated blocks of interfascicular tissue explanted to a nutrient medium formed a cambium in the expected location. From these experiments, admittedly not yet extended to other species, it may be concluded that the tissue from which the interfascicular cambium will differentiate becomes determined for cambial initiation well in advance of any structural expression, and that determination is independent of local inductive influences acting at the time of cambial initiation.

Before concluding, however, that cambium arises only in such a highly predetermined pattern, it must be recalled that there are numerous accounts of cambial initiation in callus tissue that has proliferated in response to wounding. In fact, cambium initiation is an inherent part of the wound-healing mechanism. In these cases the cambium also arises in predictable locations, specifically at a certain distance below an exposed surface, and in continuity with existing cambium in the wounded organ. The extensive experiments of P. and J. Warren Wilson (summa-

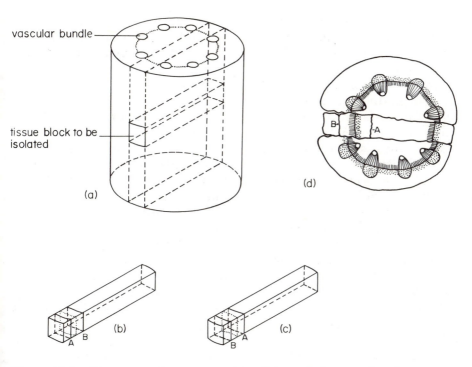

Figure 16.1. *Diagrammatic representation of the technique used to demonstrate the autonomy of radial polarity in the initiation of interfascicular cambium in the hypocotyl of* Ricinus. *(a) Location of tissue block to be isolated. (b) Tissue block with experimental segment (AB) indicated. (c) Tissue block with the experimental segment reversed prior to reinsertion. (d) Result of experiment after 10 days of growth. Secondary phloem (stippled) has differentiated inside the cambium in the experimental segment. (Siebers, 1971a.)*

rized in Warren Wilson, 1978) have indicated that this positioning is very much in response to localized conditions. They have proposed that the position of cambial initiation is determined by gradients of at least two morphogenetically active substances and suggest that auxin and sucrose are probable participants in this induction.

CAMBIUM CULTURES

One of the most effective techniques employed in evaluating the potentialities and the regulation of the terminal meristems of shoot and root has been the excision of these regions with a minimum of their differentiated derivatives and their culture on defined nutrient medium. The application of this method to the cambium – although beset with inherent difficulties because only small pieces can be excised and because removal involves extensive injury – should, if successful, answer the same kinds of questions concerning the degree of developmental autonomy of the meristem as opposed to external controls. Interestingly enough, the fragmentation and injury involved in the removal of pieces of cambium, including the intials and their most recent derivatives, do not appear to be fatal to these cells. Bailey and Zirkle (1931) have shown that tissue slices cut from the cambium of diverse conifers and dicotyledons with a sledge microtome and containing cambial initials and recent derivatives may be kept alive and under intermittent observation for periods exceeding two months in a medium consisting solely of tap water. The vitality of the contained cells was revealed by cytoplasmic streaming and by responsiveness to treatments particularly, in these experiments, changes in external pH. The cells, however, did not undergo division under these conditions, and the addition of various nutrients, such as mineral salts or sucrose, or the substitution of distilled water merely reduced the duration of viability of the explants.

The absence of cell division in these cultures does, however, seem to be a matter of the lack of the proper nutrients and growth factors. It is now well known that tissue slices including the cambium, if removed aseptically from a large number and variety of species and explanted to a nutrient medium containing mineral salts, sugar, certain vitamins, and an auxin, will proliferate freely and indefinitely. In fact, historically such tissues were among the first from which true plant tissue cultures were obtained, that is, cultures capable of unlimited growth (Gautheret, 1942). One of the most interesting characteristics of these cultures is the fact that the cambium, although its cells divide freely, does not continue divisions according to the pattern in the intact plant and does not initiate xylem and phloem. Rather, all the cells capable of division segment into small, approximately isodiametric cells, and the result is a callus culture consisting of relatively homogeneous parenchyma in which the cambium loses its identity. The only cambium present in callus cultures is one that may differentiate later. Thus, a portion of the cambium, even if excised with a considerable number of its derivatives, appears to be unable to maintain a normal pattern of division and differentiation when isolated from the plant.

THE PHYSICAL ENVIRONMENT

The difficulty with isolated pieces of cambium, then, is not that they cannot undergo cell division, but rather that the cambial layer cannot maintain the orderly division pattern so essential to its organization. It is widely believed that physical forces may play an important part in

(a)

Figure 16.2. *The influence of pressure on cambial activity in* Populus trichocarpa. *(a) Diagram showing the manner in which a bark strip was cut and lifted. (b) Organization of the tissues that differentiated in a bark strip that was left free for sixty days. (c) Organization of differentiated tissues in a bark strip that was reinserted and maintained under pressure for sixty days. Key: B, bark; C, original cambium; C', new cambium; CP, callus pad; P, phloem formed by original cambium; P', phloem formed by new cambium; Pd, newly formed periderm; X, xylem produced by original cambium; X', xylem produced by new cambium. In (b) and (c) the original outer surface of the bark strip is at the bottom of the diagram. (b,c) ×5. (Adapted from Brown and Sax, 1962.)*

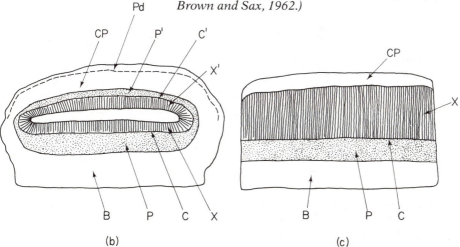

(b)

(c)

maintaining the organized development of the cambium, at least in trees and shrubs. The expanding core of secondary xylem is thought to compress the cambium against the outer tissues of the bark and the same expansion might be expected to cause tensions in the cambium and its derivatives. By making incisions inwardly through the cambium of actively growing trees and observing the gaping of the incision, Hejnowicz (1980) provided evidence for the existence of tensions in the cambium and its differentiating derivatives and suggested that they are not uniformly distributed. There is, however, surprisingly little direct evidence linking physical stress to cambial morphogenesis.

One experiment that has attempted to obtain concrete evidence on this question was that reported by Brown and Sax (1962). These workers investigated cambial development in partially isolated strips of bark containing cambium that, since they remained attached to the plant at one end, could presumably obtain all substances necessary for their continued development. In *Populus trichocarpa* and *Pinus strobus* bark strips 8.0 cm long and 1.5 cm wide, severed at their basal ends but left attached at the tops, were pulled away from the trunk and encased in polyethylene bags to reduce desiccation (Fig. 16.2a). Following this, the enclosed strips either were left free or were reinserted into the slots from which they had been separated with varying amounts of pressure applied. The separation of the strips during a period of intense cambial activity was shown to occur in the zone of recently formed and as yet incompletely differentiated secondary xylem elements so that the cambium was intact within the strip.

In the free strips there was a very rapid proliferation on the inner face, primarily by the immature ray cells but also to some extent in other immature xylary elements, which led to the establishment of a layer of callus up to 5 mm thick by the end of three or four weeks. During this callus development the cambium itself continued to proliferate in a more or less orderly fashion, but its early xylary products were shorter than normal and somewhat irregular in shape. Whether this indicates abnormalities in the cambial initials or in their derivatives is not clear, but because the formation of relatively normal xylem elements was soon reestablished, it would seem that the cambial initials were little affected. After the callus pad had reached the dimensions just noted, a periderm formed near its surface and beneath this, cambial activity, which appeared to be quite normal, gradually extended across the pad from the margins in continuity with the original or outer cambium of the strip. Cross sections of such a bark strip at this time revealed a complete ring of cambium with its associated secondary vascular tissues (Fig. 16.2b).

In a parallel series of plants, the separated bark strips, encased in polyethylene, were forced back into the slots on the trunks and either bound with an elastic grafting band or maintained under measured pressure by the application of a mercury column of a given height with a basal reservoir having one flexible face appressed to the bark strip. These bark strips developed in a strikingly different way from those that were not subjected to pressure. In the first few days callus began to form as in the previous case, but this continued only until irregularities between the two opposing surfaces had been filled. At this time the callus cells differentiated as lignified elements of the tracheary type. The cambium itself functioned quite normally throughout this period and the immature xylem elements present at the beginning of the experiment differentiated in a nearly normal fashion. The range of pressures applied to the strip in this experiment was 0.25 to 1.0 atmosphere, but

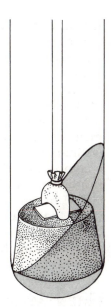

Figure 16.3. *Diagram illustrating the experimental procedure used in applying pressure to tissue slices of* Populus deltoides *containing cambium in sterile culture.* ×⅔. *(Adapted from Brown, 1964.)*

all were equally effective in promoting normal development within the strip, as was the grafting rubber band.

This experiment has been interpreted by Brown and Sax as demonstrating the important role of physical pressure, or perhaps restraint, in bringing about the normal differentiation of tissues derived from the cambium. However, although in the free strips not subjected to pressure there was extensive proliferation of recent cambial derivatives, the cambium itself appeared to remain normal or nearly so. Thus, although this experiment shows that physical factors are important in the differentiation of secondary vascular tissues, in part at least by preventing excessive tissue proliferation, it is difficult to relate these forces directly to the orderly division patterns of the cambium itself.

There is, however, one experimental study that has attempted to deal with physical influences on the cambium. Brown (1964) investigated the development of explants of *Populus deltoides* consisting of inner phloem and cambium on a nutrient medium adequate to support proliferation. In these cultures each explant was supported on a neoprene stopper with its apical end inserted into the nutrient medium (Fig. 16.3). In this way it was possible to apply pressure to the tissue slices by means of a flexible rubber reservoir attached to a mercury column. In the absence of applied pressure there was extensive proliferation that involved parenchyma throughout the explant and the cambial initials themselves, resulting in a complete loss of identity of the cambium and the development of an unorganized callus. When a pressure of 0.1 or 0.05 atmosphere was applied to the explants, the development of callus was greatly reduced and the cambial initials retained their identity. Moreover, there was evidence that the cambial initials had given rise to derivatives that differentiated into xylem and possibly phloem elements. This experiment suggests that physical influences, and particularly pressure, may be important in maintaining the orderly development of the cambium and helps to explain why cambial explants in ordinary tissue culture conditions do not continue to function normally.

Support for the role of physical pressure in the orientation of division in plant cells is provided in an experiment by Lintilhac and Vesecky (1984) in which tobacco pith segments were compressed in an apparatus that maintained constant pressure while they were growing in culture. Cells subjected to compression divided in an orderly fashion to produce radial files resembling the products of a cambium whereas cells in noncompressed regions divided in an irregular fashion.

ORIENTATION IN THE CAMBIUM

The descriptions presented in the previous chapter have shown that the cambium consists of cells and cell groups that are highly oriented. The elongate fusiform initials and the fusiform groups of ray initials lie with their long axes parallel to one another and generally parallel to the long axis of the organ in which they are found. The growth activities of this meristem are intimately connected with this orientation because it has been shown that cell divisions are precisely oriented in these cells. Just as it is important to seek an understanding of the factors that maintain orderly patterns of cell division in the cambium, so it is significant to consider what it is that controls the orientation of cells and cell groups. Several possibilities immediately suggest themselves for consideration. The orientation of cambial initials might be relatively labile, being imposed by the polarity of the axis in which they are located, or more specifically by the polarized movement of nutrients, water, or hormones along the axis. Similarly, one might imagine that the orientation is established when the cambium is initiated and is then maintained by the influence of mature derivatives that, of course, have the same orientation. Because these derivatives are also the conducting cells of the axis, this kind of regulation would be difficult to distinguish from the previous suggestion. Alternatively, the cambial initials could be to a large extent insensitive to stimuli external to themselves in the matter of orientation.

In attempting to evaluate these possibilities, one is faced with a large body of literature dealing with observations incidental to the study of

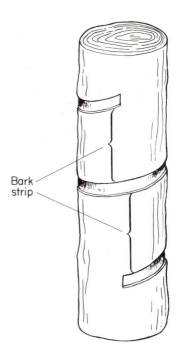

Figure 16.4. *Diagram showing the technique used by MacDaniels and Curtis (1930) to produce spiral ringing of young apple trees.*

grafting and wound healing, which dates back to the early nineteenth century. Much of this information is difficult to evaluate in terms of the behavior of cambial initials, but collectively it provides good evidence for the orientation of cambial initials along the lines of major transport in the axis and for their change in orientation when the direction of transport is altered. In some cases the orientations are extremely complex. One well-documented example of this kind of result illustrates the evidence upon which this conclusion is based. MacDaniels and Curtis (1930) carried out experiments on young apple trees in which helical strips of bark, including the cambium, that extended twice around the trunk were removed and the gap was filled with wax to prevent regeneration (Fig. 16.4). This operation left a continuous helical bark strip several inches in width connecting the upper and lower intact regions of the trunk, its phloem providing the sole connection between the two for the conduction of organic nutrients. In this strip the direction of movement of both nutrients and water must have been changed by about 45°, the pitch of the helix of the bark strip. The first few layers of xylem elements that differentiated after the operation were vertically oriented, but they departed from normal in that the perforation plates were on the lateral walls rather than on the end walls. Progressively, however, the orientation of the elements changed, presumably indicating a change in orientation of the cambial initials, until the long axis of the tracheary cells and of the rays was parallel to the axis of the helical strip, that is, about 45° from the vertical. The elements of the newly formed phloem were correspondingly changed in their orientation, further indication that the cambium itself, which was not examined, had undergone a change in orientation of its initials.

Several interesting points arise from this experiment. First, it is clear that although the cambium reoriented within the first season, its reaction was not immediate. This was not because of a delay in the appearance of the stimulus to reorientation, because an immediate effect upon the differentiation of tracheary cells was noted. Rather, the initials of the cambium must require some time to effect the change, or perhaps they tend to resist it. This raises the question of how the reorientation is accomplished. A recent study by Savidge and Farrar (1984) shows how this process occurs in a conifer. Stems of *Pinus contorta* (lodgepole pine) were girdled leaving either a diagonal or a vertical bridge of bark in each case. In the diagonal bridges there was a progressive reorientation of the fusiform cambial cells so that their long axes became parallel to the direction of the bridge. This was accomplished by apical intrusive growth along the new axis following divisions that reduced the length of the fusiform cells. These divisions were of two types: The majority were of the pseudotransverse type (see Chapter 15), but there were some oblique divisions oriented at right angles to these. Thus, the reorientation of cambial initials appears to be achieved through a modification of normal developmental processes.

In view of the importance of apical intrusive growth in cambial reorientation in some cases, it is of interest to discover how it is accomplished in species with storied cambium where intrusive growth is not part of normal development. This question has been examined by Kirschner, Sachs, and Fahn (1971) in *Robinia*, which has a storied cambium. Here stems were girdled and each girdle was spanned by a bridge, the central portion of which was transversely oriented. Reorientation occurred in the horizontal segment of the bridge. In this case it is reported that the fusiform cambial cells became septated to form short cells. These subsequently elongated intrusively at right angles to the long axes of the parent cell, thus forming elongate cells in the imposed direction. It is not clear whether this should be considered a distinctive mechanism or whether it may be regarded as an extreme case of fusiform initial reorientation evoked by the necessity to turn 90°. The latter seems a reasonable interpretation since Zagórska-Marek (1984) has shown that in the storied cambium of *Tilia* reorientation in diagonal bridges is accomplished by pseudotransverse division and apical intrusive growth.

Although the change in orientation that resulted in these experiments certainly had the effect of improving transport of materials through the newly formed xylem and phloem, there is no direct proof that transport of materials was the causal factor involved in the shift or indication of which materials might be considered most important in producing the change. However, even in a completely ringed tree water can be conducted through the xylem that remains. Therefore, it seems most likely that materials transported in the phloem are implicated. There is experimental evidence that points to growth regulating substances as a factor in the orientation or reorientation of cambial initials. For example, Zagórska-Marek and Little (1986) showed that the intrusive growth that results in reorientation in a helial bridge in a girdled stem of *Abies balsamea* (balsam fir) is inhibited by the application of auxin to the upper edge of the bridge. The movement of auxin from the point of application perpendicular to the bridge angle was thought to interfere with the influence of endogenous auxin transported along the bridge. This conclusion was supported by the observation of some realignment immediately beneath the applied auxin at right angles to the bridge angle. The orientation of fusiform initials parallel to the direction of auxin transport is indicated by these experiments.

At the same time experimental evidence of another type suggests that the cambial initials are relatively autonomous with regard to their orientation. Thair and Steeves (1976) have investigated this question in several species, including *Malus* spp. (ornamental crabapples), *Thuja occidentalis, Cornus stolonifera,* and *Sorbus aucuparia,* by removing small, square patches of bark and either replacing them or exchanging them with comparable pieces from other trees, reorientated by 90° or 180° (Fig. 16.5a). As in experiments mentioned earlier, when bark is lifted the separation occurs in the region of immature xylem elements so that the cambium is removed with the bark piece and is rotated with it. Perhaps the greatest interest in these experiments is attached to the pieces that were rotated 90° so that the cambial initials lay at right angles to the axis of the stem and to the major direction of the transport. In this case the rather surprising result was that the rotated cambium continued to function normally in the altered orientation and that secondary xylem and phloem were formed in a small block oriented at right angles to that of the host plant (Fig. 16.5b). In these experiments, of course, there was no question of releasing pressure because this was supplied by a grafting rubber band until the graft union had formed about two weeks after the operation was made. What is important here is that the cambial initials did not undergo reorientation, and in certain cases in which the experiment was allowed to continue, this development was maintained through several growing seasons as revealed by

Figure 16.5. *Experimental rotation of the cambium. (a) Rotation of a square patch of bark through 90°. (b) Transverse section of a stem cut through such a patch in* Thuja occidentalis *(white cedar) 100 days after the rotation operation. Key:* C, *cambium;* P, *phloem;* R, *ray in original xylem;* R', *ray in new xylem;* X, *original xylem;* X', *xylem formed after the operation. (b) ×45. (Thair and Steeves, 1976.)*

(a)   (b)

the development of growth rings in the secondary xylem matching those of the host plant. Moreover, Thair (1968) has shown that the same result can be obtained with circular patches of bark oriented at angles of 30° intervals from 0 to 180°. With almost incredible accuracy the cambium retained the orientation given to it and laid down secondary conducting elements in this orientation.

These results argue forcefully for the conclusion that the orientation of initials in the cambium is insensitive to external influences, but several additional observations argue against acceptance of this extreme view. First, in the same trees in which the grafted cambium retained its new orientation independently of the direction of transport in the axis, the cambium of the host or stock showed considerable reorientation in the vicinity of the graft, with the result that the graft was largely bypassed in the transport system. Thus, reorientation was found in the same experiments but under different circumstances. Second, various abnormalities were noted in the secondary xylem elements formed in the graft in the period immediately after the operations were made, notably the occurrence of perforation plates on the side walls. Unfortunately, the phloem was not examined in the same detail, so it is not known whether comparable modifications occurred there as well. It does appear, however, that the problems created by an orientation at right angles to the lines of transport had an effect upon the graft, although not necessarily directly on the cambium. However, before the end of the first growing season these abnormalities ceased, although there was no reorientation. It was then discovered that both in the square and in the circular grafts, although all of the edges of the grafts had united with the stock by the formation of callus, only at the original top and bottom of the grafted pieces had a true continuity been established in the cambium with a resulting continuity in derivative xylem and phloem. Furthermore, it appeared that the original upper ends of grafted cambial initials had made contact with the bases of stock initials and the bases of graft initials with the tops of stock initials, with the result that, in the case of a 90° graft, for example, it was possible to tell by later examination in which direction the pieces has been rotated. Thus, it appears that the grafted pieces were incorporated into the transport system in a special way such that they were supplied with materials and were not subsequently subjected to the kind of stress that, in other cases, has resulted in reorientation. It also appears that the cells of the cambium have an inherent polarity that causes them to make the correct end-to-end contacts in graft unions.

GENERAL COMMENT

In the previous chapter emphasis was placed upon the major differences in structure and activity between the vascular cambium and the primary meristems of shoot and root. From the experimental point of view, one of the most striking differences is the relative ease with which shoot and root meristems may be cultured in isolation in contrast to the cambium which, if it develops at all in culture, ordinarily does so in a disorganized fashion. This could be interpreted to mean that the cambium, unlike the shoot and root meristems, is not self-regulating. On the other hand, small patches of cambium grafted into stems in a drastically altered orientation retain their inherent structure and function in a manner normal with respect to their origin but totally aberrant in their new situation. It would be difficult to think of these cambial transplants as being anything other than highly autonomous. In fact, the method of transplantation is a standard approach in animal embryological studies for the demonstration of autonomous function or determination.

It is, as has been pointed out, difficult to analyze the origin of cambium from procambium in such a way as to reveal any determination process that may occur. However, the elegant experiments of Siebers have shown that, in at least one species, the cambial precursor tissue of the interfascicular regions is functionally determined well in advance of any structural changes. It seems important to obtain evidence regarding this phenomenon in a wider range of species. On the other hand, the relative ease with which cambium arises de novo in wound-induced callus as part of the healing process might seem to argue against the significance of cambial determination. But this phenomenon should be given no greater weight than is accorded to the de novo origin of shoot or root apices (or even embryos) from proliferating tissues in a variety of circumstances.

If, then, the cambium is highly determined, why will portions of it not continue to function normally in a growth-supporting culture environment? One possibility may be that nutrient media used in attempts to culture cambial explants are inappropriate for normal morphogenesis since they have generally been devised with the sole objective of supporting cell proliferation. Another serious possibility is that the physical environment of cambial explants may be drastically unsuitable. Since the cambium is an internal tissue it may have specific requirements in terms of pressure, tension, concentration of atmospheric gases, or other factors. One of the rare attempts to deal with this problem, Brown's application of pressure to cambial explants, was strikingly successful in maintaining normal development. This experiment clearly indicated the involvement of physical factors, but it is difficult to ascribe the result to pressure alone. What, for example, was the effect of enclosing the tissue between impermeable layers, which would drastically alter gas exchange? However, the success of this experiment is most encouraging and emphasizes the urgent need for further studies of this type.

In attempting to make some general statement about the control of developmental processes in the cambium, it is evident that the major problem is a lack of information. Of the numerous experiments that suggest themselves as obvious necessities, so few have been done that one can do little else than talk about problems for future research.

REFERENCES

Bailey, I. W., and C. Zirkle. 1931. The cambium and its derivative tissues. VI. The effects of hydrogen ion concentration in vital staining. *J. Gen. Physiol.* 14:363–83.

Brown, C. L. 1964. The influence of external pressure on the differentiation of cells and tissues cultured *in vitro*. In *The Formation of Wood in Forest Trees*, ed. M. Zimmerman, 389–404. New York: Academic Press.

Brown, C. L., and K. Sax. 1962. The influence of pressure on the differentiation of secondary tissues. *Am. J. Botany* 49:683–91.

Gautheret, R. J. 1942. *Manuel technique de culture de tissus végétaux*. Paris: Masson et Cie.

Hejnowicz, Z. 1980. Tensional stress in the cambium and its developmental significance. *Am. J. Botany* 67:1–5.

Kirschner, H., T. Sachs, and A. Fahn. 1971. Secondary xylem reorientation as a special case of vascular tissue differentiation. *Israel J. Botany* 20:184–98.

Lintilhac, P. M., and T. B. Vesecky. 1984. Stress-induced alignment of division plane in plant tissues grown *in vitro. Nature* 307:363–4.

MacDaniels, L. H., and O. F. Curtis. 1930. The effect of spiral ringing on solute translocation and the structure of the regenerated tissues of the apple. Cornell Univ. Ag. Expt. Station Memoir #133.

Savidge, R. A., and J. L. Farrar. 1984. Cellular adjustments in the vascular cambium leading to spiral grain formation in conifers. *Can. J. Botany* 62:2872–9.

Siebers, A. M. 1971a. Initiation of radial polarity in the interfascicular cambium of *Ricinus communis* L. *Acta Bot. Neerl.* 20:211–20.

———. 1971b. Differentiation of isolated interfascicular tissue of *Ricinus communis* L. *Acta Bot. Neerl.* 20:343–55.

Thair, B. W. 1968. Cambial polarity as revealed by grafting experiments. Thesis, University of Saskatchewan.

Thair, B. W., and T. A. Steeves. 1976. Response of the vascular cambium to reorientation in patch grafts. *Can. J. Botany* 54:361–73.

Warren Wilson, J. 1978. The position of regenerating cambia: Auxin/sucrose ratio and the gradient induction hypothesis. *Proc. Roy. Soc. London* Series B 203:153–76.

Zagórska-Marek, B. 1984. Pseudotransverse divisions and intrusive elongation of fusiform initials in the storied cambium of *Tilia. Can. J. Botany* 62:20–7.

Zagórska-Marek, B., and C.H.A. Little. 1986. Control of fusiform initial orientation in the vascular cambium of *Abies balsamea* stems by indol-3-ylacetic acid. *Can. J. Botany* 64:1120–8.

17

Alternative patterns of development

In the development of the plant body, one cell, the zygote, is able to express the full genetic potentialities of the organism; that is, it normally gives rise to the whole plant. All other cells express their potentialities less completely, and such expression is progressively restricted as one proceeds through the later stages of tissue and cell differentiation. Ultimately an individual cell differentiates as a highly specific entity such as a tracheid or a sieve element, which clearly expresses only a small portion of the total genetic capacity of the organism. Where differentiation brings about a drastic change in the morphology of the cell, such as the death of the protoplast in a tracheid, or the loss of the nucleus in a sieve element, no further expression of genetic potentiality is possible and differentiation is irreversible. In most cells, however, no such irrevocable loss occurs, and if differentiation is a manifestation of differential gene action rather than of mutational changes in the nucleus, reactivation of these cells by the appropriate stimuli might be expected to result in further and perhaps different expressions of their potentialities. This expectation is amply realized under many conditions, both natural and artificial, in which differentiated plant cells undergo further development and realize different and more complete expresssions of their potentialities than in their original differentiation. In fact, there is today such widespread exploitation of this phenomenon in applied research, and even commercially, that its fundamental significance for the understanding of plant development may be insufficiently appreciated. It is to a consideration of this fundamental significance that this chapter is addressed.

DEVELOPMENTAL POTENTIALITIES OF ISOLATED CELLS

Proliferation of single cells

Around the turn of the century, the German botanist Gottlieb Haberlandt expressed a conviction that isolated plant cells, if properly nurtured, ought to be capable of developing into whole plants as the zygote

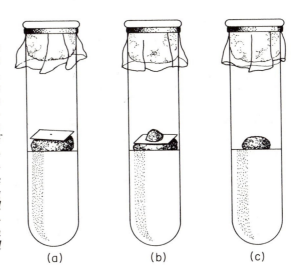

Figure 17.1. *Diagrams of the procedure used to establish tissue clones from single cells. (a) A single cell taken from the surface of a piece of callus tissue or from a liquid suspension culture is placed on filter paper lying on the top of a large tissue culture. (b) Single cells proliferate to produce macroscopically visible masses of tissue. (c) Growing tissue masses are removed from the filter papers to separate culture. (Based on data in Muir, Hildebrandt, and Riker, 1958.)*

(a)   (b)   (c)

does (Krikorian and Berquam, 1969). Although Haberlandt's own experiments failed to achieve this objective, as did those of all other investigators for more than 50 years, modern tissue culture methods have vindicated his view that the zygote is not unique in its ability to give rise to an entire plant. In order to test the potentialities of single cells it was necessary first to culture them under conditions in which they would divide repeatedly, and this proved to be a refractory technical problem. Repeated attempts to isolate single cells from callus tissue and to culture them on media that supported the growth of larger tissue masses failed even when the cells were placed in small volumes of medium such as in hanging drops. Today, when modern culture methods have made the proliferation of isolated cells commonplace, it is easy to forget the effort that was required to achieve this objective.

The successful solution was provided by Muir, Hilderbrandt, and Riker, (1958) working with callus tissue cultures of *Nicotiana tabacum* (tobacco) and several other species. Single cells were carefully picked from the surface of friable tissue masses or were removed from liquid cultures in which tissue dissociation had occurred and were placed separately on pieces of filter paper resting on the surface of nurse cultures of the same or a different species (Fig. 17.1). This method was adopted in the expectation that the nurse culture might provide substances necessary to initiate division in the isolated cell and that would be lacking in an otherwise complete nutrient medium. Under these conditions single cells divided repeatedly and produced callus masses that could then be subcultured indefinitely. Thus, tissue clones were established that had their origin as single cells.

Although this method established the ability of a single cell to proliferate, it did not permit microscopic observation of the stages of development. For this purpose attempts were made to grow isolated cells in microchambers where they could be observed microscopically. Jones et al. (1960) were successful in obtaining cultures from single cells of *N. tabacum* isolated in small drops of a nutrient medium that had been conditioned by prior growth of the callus. Subsequently, Vasil and Hildebrandt (1965a) eliminated the necessity for conditioning the medium by selecting only relatively small cells from liquid shake culture at the time of maximum tissue dissociation and by slightly altering the composition of the medium. The divisions of the isolated cell and its derivatives could be followed until a cluster of cells had been produced that was capable of continued growth. Thus, it was demonstrated by these experiments that an isolated cell can proliferate apart from the influence of other cells.

Morphogenetic potential of single cells

While these studies were being carried out, Steward and his coworkers (1958) were developing a different approach to the study of potency of isolated cells (Fig. 17.2). These workers explanted small pieces of secondary phloem tissue from the root of *Daucus carota* (carrot) and cultured them in large rotating flasks of liquid medium. The tissues proliferated rapidly and numerous individual cells and groups of cells separated from the developing masses and remained freely suspended in the liquid medium, producing a suspension culture. Such suspensions could be maintained through repeated transfers by inoculation of fresh culture medium with small amounts of material from a previous culture. Although under these conditions the proliferation of individual cells could not be followed, careful study of the range of cells and cell aggregates led these workers to the conclusion that isolated cells do divide, following a variety of patterns, and produce small nodules of tissue.

A surprising development in the continuing suspension cultures of carrot root tissue was that many of the nodules that presumably had arisen from single cells gave rise to roots. When rooted nodules were transferred to a medium solidified with agar, many gave rise to shoots. The result, therefore, was the production of complete plantlets that, as they grew, developed tap roots characteristic of the carrot plant. Ultimately, some of these were transferred to soil, where they completed the life cycle by flowering and producing seeds.

The general conclusion from these studies is that isolated plant cells other than the zygote are able to express the full genetic potentialities of the organism by giving rise to entire plants. However, in the suspen-

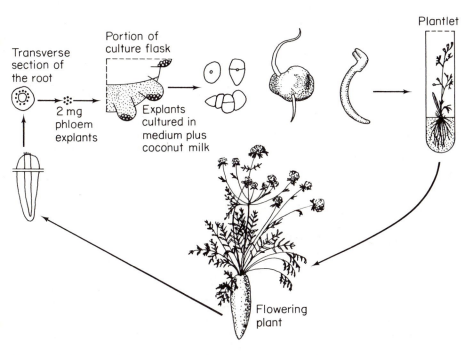

Figure 17.2. *Development of carrot plants from root cells in tissue culture. A slice of carrot tap root, shown at the top left, is used to obtain explants of tissue. Phloem explants cut from the disk are inoculated into liquid culture medium in specially constructed flasks that rotate slowly. Single cells and cell clumps separate from the tissue masses and form a suspension. Some of the varied single cells and clumps are shown in the middle of the top row of the diagram. Some of the small clumps produce roots. Rooted nodules are transferred to agar-hardened nutrient medium where they produce shoots. When transferred to soil these plants form a typical tap root and produce flowers. Tissue from the tap root can be used to repeat the cycle. (Adapted from Steward et al., 1964.)*

sion culture method it is not possible to follow development of an individual cell from its first division to the establishment of a shoot and root. The conclusion that single cells were the source of the plantlets was derived from study of numerous entities in different stages of development and the deduction that these revealed the steps through which an individual entity proceeds. Although this conclusion seems justified on the basis of the evidence presented, it is reassuring to have it confirmed in detail. Vasil and Hildebrandt (1965b), in a continuation of their earlier studies on tobacco cells grown in microculture, found that if the isolated cells were taken from recently established tissue cultures, the calluses that resulted from their proliferation gave rise to roots

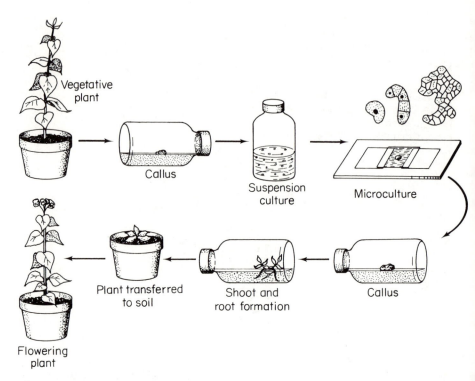

Figure 17.3. *Development of tobacco plants from single cultured cells. A hybrid vegetative plant of the cross* Nicotiana glutinosa × N. tabacum *is used as the source of pith cells, which are transferred aseptically to establish a callus culture on an agar-hardened medium. Transfer of the tissue to liquid medium results in development of a suspension culture. Single cells removed from the suspension culture to microslide culture can be observed directly or by time-lapse photography as they divide and form small tissue masses. These tissue masses are removed to agar medium when they are large enough to be manipulated. On the agar medium shoots and roots are initiated. Plants are then transferred to soil, where they develop vegetatively, flower, and set seed. (Adapted from Vasil and Hildebrandt, 1965b.)*

and shoots and ultimately to plantlets, as in the case of the nodules of carrot tissue from suspension culture (Fig. 17.3). As with carrot, these plants could be transplanted to soil where they flowered.

Although these studies demonstrate that individual cells isolated from callus cultures are capable of fulfilling the role normally reserved for the zygote – that is, the production of a whole plant – the fact remains that the means by which this is accomplished are different from those exhibited in zygote development. The initiation of shoot and root apices in a small nodule derived from a single cell is not the same as the direct

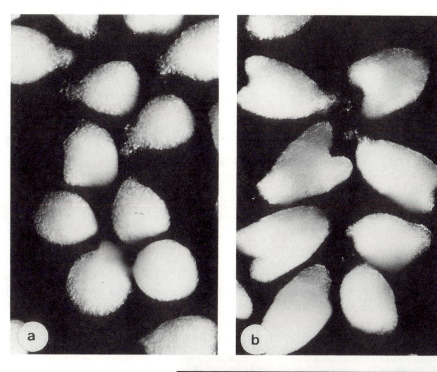

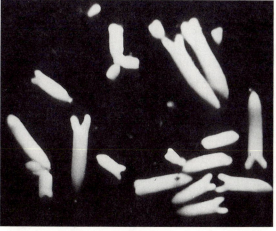

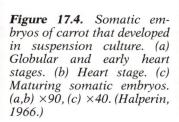

Figure 17.4. *Somatic embryos of carrot that developed in suspension culture. (a) Globular and early heart stages. (b) Heart stage. (c) Maturing somatic embryos. (a,b) ×90, (c) ×40. (Halperin, 1966.)*

formation of an organized embryo from a zygote. However, it soon became apparent that embryolike structures were being formed in cell suspension cultures and also on the surface of callus cultures (Fig. 17.4). For example, somatic embryos were observed in large numbers in suspen-

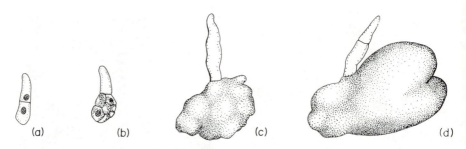

Figure 17.5. *Development of a somatic embryo from a single tissue culture cell of carrot observed by successive microscopic observations of the same unit. (a) First division of the cultured cell. (b) Three-dimensional cell mass formed from one of the first two cells. (c) Continued growth of tissue mass. (d) Somatic embryo formed from tissue mass showing initiation of cotyledons. (a–c) ×115, (d) ×90. (Drawn from Backs-Hüsemann and Reinert, 1970.)*

sion cultures derived from embryos of carrot (Steward et al., 1964) or from root or petiole tissues of wild carrot (Halperin and Wetherell, 1964), which is the species from which cultivated carrots were derived. In succeeding years cultured cells from a large number of diverse species have been shown to have a potentiality for somatic embryogenesis (Ammirato, 1983).

Although it seemed reasonable to assume that somatic embryos that form in cultures have their origin from single cells, it is useful to have direct evidence for this in at least one case. Backs-Hüsemann and Reinert (1970) have followed microscopically the development of somatic embryos from single cells isolated from tissue cultures of carrot. These vacuolated cells first divided unequally and one of the daughter cells gave rise to a small cluster of cells from which one or more embryos developed (Fig. 17.5). Although this does not constitute a direct development of an individual somatic cell into an embryo, because not all cells of the cluster were incorporated into the embryo and because more than one embryo could arise from a single cluster, nevertheless the evidence is convincing that an isolated cell can give rise to an embryo.

Taking advantage of the formation of large numbers of somatic embryos in wild carrot suspension cultures, Halperin (1966) compared the development of these with that of normal zygotic embryos of the same species. Globular embryos arise as three-dimensional, budlike developments at the surface of small cell aggregates that are without apparent organization, and more than one may arise from a single aggregate (Fig. 17.6). These primordia develop directly into globular embryos, and the cells of the original aggregate may remain attached to them as a suspensorlike appendage. This pattern of early development resembles that

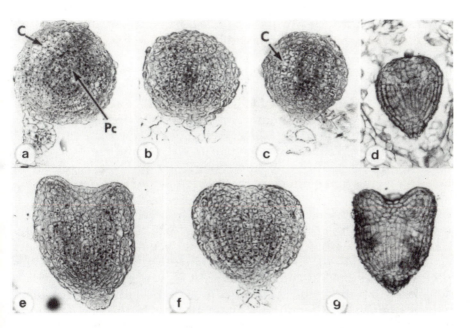

Figure 17.6. *Somatic and zygotic embryos of* Daucus carota *(wild carrot) in longitudinal section. (a–c) Globular stage of somatic embryos showing precocious vacuolation of the cortex and darkly stained procambium. (d) Globular stage zygotic embryo. (e–f) Early heart stage somatic embryos. (g) Heart stage zygotic embryo. Key: C, cortex; Pc, procambium. ×210. (Halperin, 1966.)*

which Backs-Hüsemann and Reinert were later able to trace back to individual cells but contrasts sharply with the normal embryogeny of carrot, in which the zygote gives rise to an elongated filament from the terminal few cells of which the globular stage arises, the remainder of the filament producing the suspensor. On the other hand, Halperin found that the later stages of growth, from the globular or heart stage, are remarkably like those of normal embryogeny. There appear to be some minor but characteristic departures in somatic embryos from the normal pattern, such as larger size at comparable stages, a delay in the initiation of cotyledons, and a precocious vacuolation of the future parenchymatous tissues, which is suggestive of premature germination; however, the end point, under optimal conditions, is a cylindrical, dicotyledonous embryo with well-organized shoot and root apices and the three tissue systems delimited as in zygotic embryos. Many of these develop into functional plantlets.

The differences that occur between somatic embryos and zygotic embryos in the early stages of development should not be allowed to obscure the main result of these studies, namely that the overall sequence of development is remarkably similar to that of a zygote expressing its potentialities in an embryo sac. This similarity has been shown to extend even to the biochemical level (Crouch, 1982). It is characteristic of embryos that they accumulate storage proteins not synthesized at other stages of the life cycle (see Chapter 3). It is, therefore, very significant that somatic embryos of *Brassica napus* (rapeseed) synthesize the major storage protein specific to zygotic embryos, although in a lower concentration than in the normal embryo. It thus begins to appear that the phenomenon of somatic embryogenesis involves a reactivation of much of the developmental program of normal embryogeny.

The results described thus far come close to achieving Haberlandt's goal of stimulating differentiated plant cells to function like zygotes. Even though the embryogenic cells were derived from cultured tissues, it is significant that the cultures were originally established from differentiated tissues of the plant. However, it is now possible to advance even closer to Haberlandt's objective through the culture of protoplasts derived from fully differentiated plant tissues. In this method protoplasts are released from their enclosing walls by exposure to enzymes that selectively hydrolyze wall carbohydrates. Under appropriate conditions a large proportion of such naked protoplasts not only survive, but quickly synthesize a new wall and begin to divide. Surprisingly, in a number of cases, protoplasts derived from leaf mesophyll of such plants as tobacco, rapeseed, and alfalfa give rise rather directly to somatic embryos (Kohlenbach, 1985). As in the case of isolated whole cells, the early divisions produce a cluster to cells within or upon which the somatic embryo is formed (Fig. 17.7). What is different from the whole cell studies is that the individual embryogenic cells are obtained directly from the differentiated plant body rather than from tissues proliferating in a culture. It appears that, at least in some cases, removal of the wall facilitates the expression of latent developmental potentialities. It should also be noted that protoplasts derived from cultured cells can develop in the same way.

The embryogenic capacity of protoplasts is also important in another context: the genetic modification of plant cells. It has frequently been demonstrated that protoplasts can take up foreign DNA and organelles and can be induced to fuse with other protoplasts of the same or different genetic constitution (Giles, 1983). Thus, it is possible to bring about genetic transformations in somatic cells including those that would not be possible through the normal sexual process because of reproductive isolating mechanisms. However, if transformation is to be effective it must be followed by the recovery of viable whole plants bearing the genetic modifications. Thus, the embryogenic capacity demonstrated by

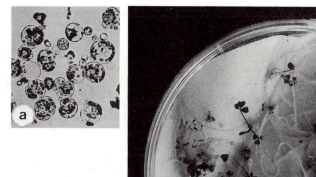

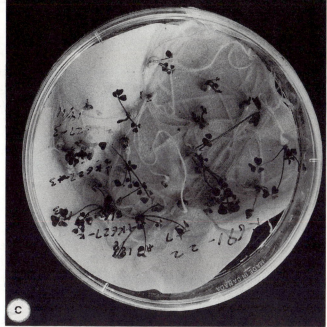

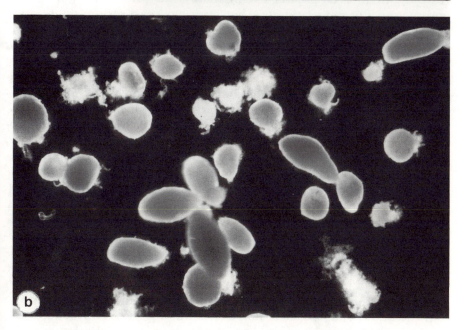

Figure 17.7. *Development of somatic embryos from isolated mesophyll proto-plasts of* Medicago sativa *(alfalfa). (a) Divisions in cells derived from protoplasts 6 days after isolation. (b) Somatic embryos and calluses 42 days after isolation of protoplasts. (c) Plants derived from somatic embryos. (K. N. Kao and M. R. Michyluk,* Z. Pflanzenphysiol. *96:135, 1980.)*

protoplasts of a number of species has significantly enhanced research in this area.

The inescapable conclusion drawn from the experimental studies that have been outlined is that differentiated plant cells retain their full genetic potentialities, and under the appropriate conditions can express them. Thus, the question that stimulated this investigation, "Do differentiated cells retain their full potentialities and can they express them?" has now been replaced by another question, "Since differentiated cells retain their full potentialities and can express them, under what conditions is this new expression possible?"

Requirements for somatic embryogenesis

When entire plants were first produced from single cells in culture, it was thought by many that isolation of a cell is a necessary prerequisite for the expression of its latent zygotic potentialities. In suspension culture, of course, the cells are not truly isolated from the chemical influences of other cells but they are removed from any physical contact or restraint to an even greater extent than is the attached zygote in the embryo sac. Later investigators, however, have cast serious doubt upon this assumption. A number of reports have appeared in which typical embryos are described as arising at the surface of coherent tissues, where the cells clearly are not physically isolated. Even in carrot, callus cultures can produce numerous embryos on their surfaces without dissociation of tissue. One of the most interesting examples of this phenomenon was described by Konar and Nataraja (1965), who observed the initiation of typical embryos from the epidermis of the stem of *Ranunculus sceleratus* plantlets growing in sterile culture (Fig. 17.8). Clearly, then, a cell need not be physically isolated before it can function as a zygote; in fact, there can be no requirement that the tissue of which it is a part be without organization.

The original suspension cultures in which somatic embryos and other manifestations of totipotency were observed were established in nutrient media that included the liquid endosperm coconut milk. It is not surprising that this rich nutrient broth that normally nurtures the development of embryos should have come to be regarded as an essential, or at least optimal, medium for the expression of cellular totipotency. In fact, other similar materials, such as the female gametophyte of *Ginkgo* and the liquid endosperm of horse chestnut and several other species, were ascribed corresponding significance. It now appears that somatic embryogenesis in a large number of species is greatly enhanced by the presence of reduced nitrogen compounds, either organic, for example, amino acids, or inorganic in the form of ammonium salts, and by a low concentration of an auxin. Both of these are components of coconut milk.

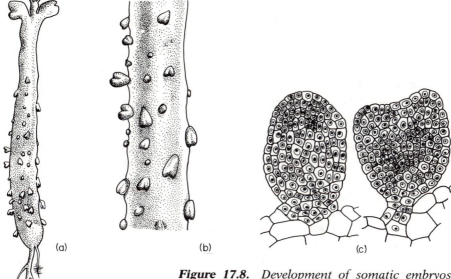

Figure 17.8. *Development of somatic embryos from the stem epidermis of* Ranunculus sceleratus. *(a) One-month-old plantlet grown in sterile culture showing distribution of somatic embryos along the stem. (b) Enlarged view of (a). (c) Longitudinal sections of globular (left) and heart stage (right) somatic embryos showing their development from the epidermis of the stem. (a) ×3, (b) ×6, (c) ×195. (Konar and Nataraja, 1965.)*

Cytokinins, also present in coconut milk, are often included in embryogenic culture media but it is not clear that they have a role more specific than promotion of the necessary cell division. These refinements of the culture medium have largely obviated the need for coconut milk or other natural additions. Furthermore, there is evidence that there may be different requirements for the induction of embryogenesis and for the subsequent development of the embryos. In particular, the hormones required for induction may actually inhibit later stages of development (Kohlenbach, 1978).

In considering the conditions under which cellular totipotency can be expressed, it is important to remember that, as in so many other developmental phenomena, differences among species can be striking. Although somatic embryo formation has been described in a number of species, there are others that have failed to respond to any treatment. For example, Halperin and Wetherell, at the time of their success with wild carrot, failed to obtain any organized development in cultures derived from five other species of the same family (Apiaceae) treated in the same way. Similarly, Steward failed to obtain results with tissues of several other species comparable to those obtained with carrot cultures.

Clearly, the conditions that permit totipotency to be expressed by differentiated cells are difficult to define precisely and are almost certainly diverse. It does not appear that any specific substance or group of substances is universely required. The cells must be induced to divide, often by a division-promoting factor, and they must be provided with adequate nutrients to sustain their own synthetic mechanisms. It is not surprising that there are variations in the ease with which division can be stimulated and in the raw materials appropriate to the necessary metabolic processes within the cells. What does emerge clearly is that the organized pattern of development that leads to the production of a whole plant cannot be ascribed to the external medium. Rather, it has its origin in the intact genetic complement of the cells. The external milieu of the cell must trigger the expression of this pattern and, having triggered it, must sustain its development. It is doubtful that anything different can be said of the zygote and its milieu.

PLANT REGENERATION

When considering the conditions under which differentiated cells are able to express other potentialities, it must be emphasized that this ability is not confined to the artificial conditions of nutrient culture where the cells that produce whole plants are part of a callus mass or are freely suspended in culture. In fact, events that can be interpreted only as manifestations of the totipotency of differentiated cells are encountered with surprising frequency in plants as regeneration phenomena, and in some plants are of such importance that they constitute mechanisms of vegetative reproduction. This strongly suggests that the conditions that permit differentiated cells to express other potentialities do exist in many plants under certain circumstances, and it gives point to the enquiry as to why all cells do not reveal their latent capacities. Although this question cannot be answered on the basis of information presently available, some light may be shed upon it by an examination of the phenomena. From the extensive literature dealing with regeneration and vegetative reproduction in plants, only a few well-documented examples can be selected for discussion.

Patterns of Regeneration

An investigation on the method of vegetative reproduction in the orchid *Malaxis paludosa* by Taylor (1967) has revealed an excellent example of

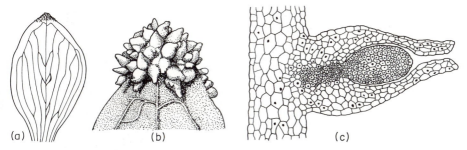

Figure 17.9. *Foliar embryos formed on the leaf of* Malaxis paludosa. *(a) A leaf of* Malaxis *showing the development of a cluster of foliar embryos at the extreme tip. (b) An enlarged view of the leaf tip, with a well-developed group of embryos, each enclosed in a sheathing jacket of cells. (c) Longitudinal section of a partially developed embryo showing its relation to the surrounding jacket and to the tissues of the leaf tip. (a)* ×3, *(b)* ×12, *(c)* ×70. *(Taylor, 1967.)*

the developmental potentialities that may be retained by differentiated cells. In this plant, terminal cells of the mature leaf regain meristematic activity and develop into tiny, egg-shaped embryos essentially like those produced in ovules, each surrounded by a sheathing jacket (Fig. 17.9). After detachment from the leaf, these foliar embryos develop into new plants, thus functioning essentially as does a normal seed embryo. This production of foliar embryos is a remarkable example of the retention of full zygotic potentialities by differentiated cells and of the ability of such cells to express these potentialities within the organization of the intact plant. There are, moreover, other well-known instances in which entire plantlets are initiated on leaves.

Many members of the Crassulaceae, a family of succulent plants, provide examples of this phenomenon. In *Bryophyllum calycinum* Naylor (1932) has described the origin of complete plantlets from residual groups of meristematic cells that persist in the notches along the leaf margin. These meristematic groups first become evident when the leaf is a few millimeters in length. While other cells around them expand and mature, those in the meristematic groups continue to divide and give rise to plantlets with a pair of leaves, a shoot apex, and two roots. At this stage, plantlet development ceases and is not resumed until removal from the leaf, or until the leaf is detached. The development of an entire plant from the partially differentiated cells of the leaf margin does not represent a reversion of fully mature cells as in the case of the orchid already discussed, but it does show that cells committed to the formation of a leaf lamina, and thus highly determined, retain the full zygotic potentialities and can express them in certain localized regions.

Figure 17.10. *Plantlets developing from the marginal leaf notches of* Kalanchoe daigremontiana *while the leaf was still attached to the plant.* ×35.

Other members of the same family show considerable variation in the extent of development of plantlets on leaves still attached to the plant. In *Kalanchoe daigremontiana* they produce several pairs of leaves and numerous roots while still attached to the parent plant (Fig. 17.10). In other species, such as *Brynesia weinbergii*, the meristem remains dormant without the formation of any organ rudiments until the leaf is detached from the parent plant. Finally, in *Crassula multicava* there is no evidence of persistent meristematic areas in the mature leaf, but following detachment of the leaf fully differentiated epidermal cells resume division and give rise to meristematic outgrowths that differentiate as plantlets. There are, in fact, many instances in which fully differentiated cells are reactivated and give rise to whole plants following injury or some other interruption of whole plant integrity. In *Saintpaulia ionantha* (African violet), for example, if a mature leaf is detached and placed on a moist substrate, epidermal cells near the base of the petiole are reactivated and organize shoot meristems. At the same time cells deeper in the tissue of the petiole give rise to root apices. Thus, whole plants are produced by the association of independently initiated shoot and root meristems. A similar type of plantlet initiation can occur from the lamina if a portion of this part of the leaf is explanted separately. Thus, in these cases, cells that in the normal course of events do not express their latent potentialities for further development will do so following injury or removal from the whole plant system.

There are also well-documented cases in which regeneration phenomena resulting from the reactivation of differentiated cells result in development of the more specialized meristem of a shoot or root, rather than producing the rudiment of a whole plant. An interesting example of this type of development is seen in the hypocotyl of *Linum usitatissimum* (flax). Link and Eggers (1946) have shown that buds are initiated by divisions in individual epidermal cells or in groups of these cells. In the intact plant, development of these meristematic centers is arrested at an early stage, but occasionally they may develop as shoots. If, however, the plant is decapitated below the cotyledons, they regularly grow out and, in addition, many more bud primordia are formed.

Similarly, roots are often initiated from mature or partially mature tissues of the stem or even the leaf. These may arise on intact plants, particularly at the nodes, but also in many cases without such localization. Their origin often follows injury and they are perhaps best recognized arising at the base of a severed stem in cuttings. In this way they provide the basis for the commonest method of horticultural propagation. In most cases these adventitious roots arise in the inner tissues of the stem close to the vascular system. Mature or partially mature parenchymatous cells undergo divisions to form a meristematic center which becomes organized as a root meristem and subsequently emerges at the surface of the parent stem. In some cuttings, roots originate from an unorganized callus that arises from the tissues of the stem near the cut end.

The regeneration of shoots or roots or both following injury or the severing of a particular organ is of importance not only from the point of view of horticultural propagation but from the standpoint of survival of the plant under natural conditions. Priestley and Swingle (1929), among many other workers, have investigated the development of shoots and roots that have propagative significance. One of the most interesting and least understood of these regeneration phenomena is the initiation of shoot buds on roots. Many plants initiate shoot buds on severed or damaged roots by the development of shoot meristems from differentiated cells of the internal tissues. Root fragments of such species can be used as propagules for intentional multiplication, and, under natural conditions, survival and even spread are enhanced. Some perennial weeds are extremely difficult to eradicate because their roots possess this property. Moreover, the production of root buds need not depend upon injury (Fig. 17.11), and there are many plants in which the formation of buds on undisturbed root systems provides a mechanism of vigorous spreading. For example, *Populus tremuloides* (aspen poplar) reproduces on the northern prairies only by means of root buds, and large clumps covering many acres are believed to represent the development of original individual plants over periods of several thousand years.

It is thus evident that, under a variety of conditions, differentiated or partly differentiated plant cells are capable of reverting to an embry-

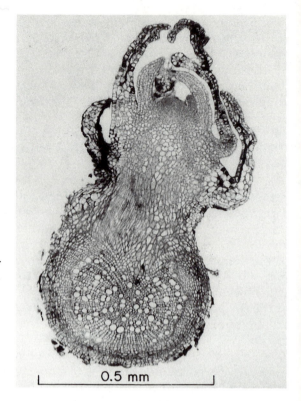

Figure 17.11. *Shoot-bud regeneration from the root of* Viola adunca. *The root is shown in cross section in the lower part of the figure. In the upper part is the shoot with apical meristem, surrounding leaf primordia, scale leaves, and differentiating stem tissues.* ×90. *(M.V.S. Raju et al.,* Can. J. Botany *44:33, 1966.)*

0.5 mm

onic condition and subsequently of expressing a new pattern of differentiation. The conclusion that any cell can express zygotic potentialities if its milieu is adequate to trigger and sustain this development is as valid for cells within a plant as it is for those growing in tissue cultures; the relevancy of the cell culture studies is thus strongly emphasized. However, it is clear that in the normal, intact plant very few cells actually are triggered to express alternative developmental patterns, and the development and maintenance of an organized plant body requires this limitation. The question of regulation, that is, of the integration of the plant body, is thus of prime importance. Although this regulation is far from being elucidated, there are some experiments that hint at the nature of the mechanism.

Developmental control

In horticultural practice it is well known that application of growth regulators of the auxin type often enhances the initiation of roots on cuttings and frequently promotes rooting in cuttings that will not form roots without it. Because of the practical significance of this phenomenon, it is not surprising that much research has been devoted to it. In the present context, however, what is significant is that it implicates a regulatory mechanism in the maintenance of integration in the plant and suggests that when this is disturbed, regeneration may result.

Auxin also influences bud formation in a number of species but with rare exceptions this is an inhibitory effect. Thus, in *Armoracia rusticana* (horseradish) Dore and Williams (1956) found that application of auxin to root slices promotes root initiation and inhibits bud formation, and these two responses increase with increasing concentrations of the hormone. Other substances are known that promote bud formation in various cases. Skoog and Tsui (1948), for example, found a strong bud-promoting effect of adenine on stem segments of *Nicotiana tabacum* (tobacco) in sterile culture. An even more striking effect upon bud formation by the substance kinetin was shown in later experiments by Skoog and Miller (1957) on cultured tobacco stem segments and callus tissues. What emerged clearly from these experiments, however, was the importance of the relative concentrations of interacting substances rather than the absolute concentration of any one substance in regulating organ differentiation. By adjusting the relative concentrations of auxin and kinetin in the culture medium it was possible to induce bud formation, root formation, or the growth of undifferentiated callus, and, within limits, the effect of a change in concentration of one substance could be countered by a corresponding change in the concentration of the other. However, in this and other experimental systems other substances have been found to play a role in controlling shoot and root formation in explanted tissues. For example, in the tobacco system just described, the addition of casein hydrolysate to the culture medium greatly enhanced the bud-promoting effect of kinetin, although alone it had very little effect.

A striking illustration of the interaction among diverse factors in regulating the morphogenetic expression of differentiated cells is provided by recent culture experiments using thin tissue slices from several species of plants (Tran Thanh Van, 1981). These explants, as described in Chapter 10, consist of the stem epidermis and several layers of underlying cortical tissue. Although they show some residual positional effects, the pattern of regeneration in these small (1 mm × 10 mm) pieces can be controlled to a remarkable extent by the hormonal, nutritional, and physical characteristics of the culture system. For example, explants from the inflorescence of tobacco can directly regenerate flower buds, vegetative buds, or roots or can give rise to unorganized callus, depending upon the relative concentrations of auxin and cytokinins within limits imposed by other factors. Flower bud initiation requires a high sugar concentration and is light dependent. Roots arise only when sucrose is supplied instead of glucose. Other factors that influence the hormonal effect are the liquid or semisolid condition of the culture medium and its pH. Most unexpectedly, it has been found that soluble oligosaccharides extracted from plant cell walls can alter the morphogenetic response in specific ways (Albersheim and Darvill, 1985). For example, one type of oligosaccharide preparation causes a shift in response from callus formation to vegetative bud development, whereas a different preparation promotes root formation in a medium on which only callus would be expected. These experiments point to the importance of the chemical and physical environment of the cell in controlling its developmental expression.

In considering the control of these regeneration phenomena, it is important to recognize that it is not an organized root, shoot, or flower that is being evoked. Rather, in all cases, it is a meristem that, by its continued development, gives rise to the particular organ and, as earlier chapters have shown, is capable of doing so more or less autonomously. Moreover, there is evidence in some cases that the meristem may not be determined as to type until some time after initiation. An illustration of this is provided in a study by Bonnett and Torrey (1966) of the initiation of lateral roots and shoot buds on cultured root segments of *Convolvulus arvensis*. At the time of initiation the primordia of both types of organs were morphologically indistinguishable and only subsequently did they become recognizable as root or shoot apices as a result of differences in rate of development and the orientation of cell divisions. Both the number and kind of lateral organs produced on the root segments could be influenced by the constituents of the culture medium, but the implication of the morphological study was that the influences governing initiation of development at a regeneration site and those governing determination as shoot or root meristem were not exerted at the same time.

In considering the nature of the process that allows differentiated cells or those in an unorganized callus to initiate organized development, various investigators have debated whether such a process occurs in a group of cells that collectively form a meristem or whether it must occur in one cell from which the whole meristem is ultimately derived. Histological study has provided evidence for both alternatives in different cases but in fact cannot really answer the question. The actual process might occur in a group of cells that had been derived from one original cell. Fortunately, it has been possible to obtain a definitive answer in several cases through clonal analysis. In several species of *Nicotiana*, mixed callus cultures consisting of genetically different cell lines have given rise to chimeral buds, and periclinal chimeric leaves have been induced to regenerate chimeric shoots (Marcotrigiano, 1986). In these cases, at least, the buds must have arisen from more than one cell or they could not be chimeral.

GENERAL COMMENT

These experiments, and many others of a similar type, have given substance to the statement that any cell can express its latent potentialities if its milieu is such as to trigger and sustain its development. In the organized and integrated plant the suitable cell environment is apparently not often encountered, and therefore differentiated cells tend to remain in a stable mature condition and are mitotically inactive. The development of an organized plant body requires that this be so. However, under a variety of conditions, some of which occur spontaneously and some of which follow injury, the right combination of factors is achieved to trigger and sustain renewed development and previously mature cells may then express their latent potentialities. In doing so, the full developmental potential of the cell may be expressed and an entire plant is produced. In other cases, however, a less complete expression is evoked and only the individual meristems of shoot or root are initiated. Why triggering in some cases should result in the whole developmental course being repeated and in other cases only a portion of this course is not clear and remains an exciting area for future investigation. Possibly the cellular environment during the initial stages of regeneration exerts a determining effect on the fate of the developing structure. If so, it is possible to visualize the meristems of the plant in normal development as representing the partially differentiated derivatives of the zygote, which in the embryo become determined as shoot- or root-forming structures, and are then maintained at this particular level of differentiation by the persistence of a specific set of factors.

The evidence at present available strongly suggests that the factors involved in regeneration are probably numerous and may vary in different cases. Attempts to interpret the phenomena of regeneration in terms of single specific substances or of simple interactions of a few substances have not been notably successful when applied to the whole organism and they do not seem likely to provide meaningful answers to the problems of plant morphogenesis. It seems much more probable that regulation, and consequently regeneration, are mediated through the total environment of the cell, and this conclusion is equally applicable to the zygote and to the differentiated cells in the mature plant body. Thus, although it is not surprising to find embryos normally developing in embryo sacs and archegonia, it is no more surprising to find them developing elsewhere.

REFERENCES

Albersheim, P., and A. G. Darvill. 1985. Oligosaccharins. *Sci. Am.* 253(3):58–64.

Ammirato, P. V. 1983. Embryogenesis. In *Handbook of Plant Cell Culture*, vol. 1, ed. D. A. Evans, W. R. Sharp, P. V. Ammirato, and Y. Yamada, 82–123. New York: Macmillan.

Backs-Hüsemann, D., and J. Reinert. 1970. Embryobildung durch isolierte Einzelzellen aus Gewebekulturen von *Daucus carota. Protoplasma* 70:49–60.

Bonnett, H. T., Jr., and J. G. Torrey. 1966. Comparative anatomy of endogenous bud and lateral root formation in *Convolvulus arvensis* roots cultured *in vitro. Am. J. Botany* 53:496–507.

Crouch, M. L. 1982. Non-zygotic embryos of *Brassica napus* L. contain embryo-specific proteins. *Planta* 156:520–4.

Dore, J., and W. T. Williams. 1956. Studies in the regeneration of horseradish. II. Correlation phenomena. *Ann. Botany* 20:231–49.

Giles, K. L., ed. 1983. *Plant Protoplasts*, Suppl. 16., Int. Rev. Cytol. New York: Academic Press.

Halperin, W. 1966. Alternative morphogenetic events in cell suspensions. *Am. J. Botany* 53:443–53.

Halperin, W., and D. F. Wetherell. 1964. Adventive embryony in tissue cultures of the wild carrot, *Daucus carota. Am. J. Botany* 51:274–83.

Jones, L. E., A. C. Hildebrandt, A. J. Riker, and J. H. Wu. 1960. Growth of somatic tobacco cells in microculture. *Am. J. Botany* 47:468–75.

Kohlenbach, H. W. 1978. Comparative somatic embryogenesis. In *Frontiers of Plant Tissue Culture*, ed. T. A. Thorpe, 59–66. Calgary, Alberta: University of Calgary Press.

——. 1985. Cytodifferentiation. In *Plant Protoplasts*, ed. L. C. Fowke and F. Constabel, 91–104. Boca Raton, Fla.: C.R.C. Press.

Konar, R. N., and K. Nataraja. 1965. Experimental studies in *Ranunculus sceleratus* L. Development of embryos from the stem epidermis. *Phytomorphology* 15:132–7.

Krikorian, A. D., and D. L. Berquam. 1969. Plant cell and tissue cultures: The role of Haberlandt. *Bot. Rev.* 35:58–88.

Link, G.K.K., and V. Eggers. 1946. Mode, site, and time of initiation of hypocotyledonary bud primordia in *Linum usitatissimum* L. *Bot. Gaz.* 107:441–54.

Marcotrigiano, M. 1986. Origin of adventitious shoots regenerated from cultured tobacco leaf tissue. *Am. J. Botany* 73:1541–7.

Muir, W. H., A. C. Hildebrandt, and A. J. Riker. 1958. The preparation, isolation and growth in culture of single cells from higher plants. *Am. J. Botany* 45:589–97.

Naylor, E. 1932. The morphology of regeneration in *Bryophyllum calycinum. Am. J. Botany* 19:32–40.

Priestley, J. H., and C. F. Swingle. 1929. Vegetative propagation from the standpoint of plant anatomy. *U.S. Dept. Agric. Tech. Bull.* 151:1–98.

Skoog, F., and C. O. Miller. 1957. Chemical regulation of growth and organ formation in plant tissues cultured *in vitro. Symp. Soc. Exp. Biol.* 11:118–31.

Skoog, F., and C. Tsui. 1948. Chemical control of growth and bud formation in tobacco stem segments and callus cultured *in vitro. Am. J. Botany* 35:782–7.

Steward, F. C., M. O. Mapes, A. E. Kent, and R. D. Holsten. 1964. Growth and development of cultured plant cells. *Science* 143:20–7.

Steward, F. C., M. O. Mapes, and K. Mears. 1958. Growth and organized development of cultured cells. II. Organization in cultures grown from freely suspended cells. *Am. J. Botany* 45:705–8.

Taylor, R. L. 1967. The foliar embryos of *Malaxis paludosa. Can. J. Botany* 45:1553–6.

Tran Thanh Van, K. M. 1981. Control of morphogenesis in *in vitro* cultures. *Ann. Rev. Plant Physiol.* 32:291–311.

Vasil, V., and A. C. Hildebrandt. 1965a. Growth and tissue formation from single, isolated tobacco cells in microculture. *Science* 147:1454–5.

———. 1965b. Differentiation of tobacco plants from single isolated cells in microcultures. *Science* 150:889–92.

Credits

Figure 2.1: Miller, H. A., and R. H. Wetmore, 1945. *American Journal of Botany*. 32:588–99, Figures 4, 5, 7, 10, 14, 25, 26 and 27.

Figure 2.2: Miller, H. A., and R. H. Wetmore, 1945. *American Journal of Botany*. 32:588–99, Figures 21–24, 28–31 and 36.

Figure 2.4: Randolph, L. F., 1936. *Journal of Agricultural Research*. Figures 4f, 6a, 6c, 6e, 7d and 7k.

Figure 2.5c–d: Lyon, H. G., 1904. *Minnesota Botanical Studies*. 3:275 Plate XXIX, Figure 4 and Plate XXXI, Figure 6.

Figure 2.6e, f: Buchholz, J. T., 1931. *Transactions of the Illinois State Academy of Science*. 23:117, Figures 9a and 9b.

Figure 2.6g: Spurr, A. R., 1948. *American Journal of Botany*. 35:629–641, Figure 21.

Figure 2.7: Ward, M., 1954. *Phytomorphology*. 4:18–26, Figures 1–5, 7 and 8.

Figure 2.8: Schulz, Sister R., and W. A. Jensen, 1968. *American Journal of Botany*. 55:807–19, Figure 11.

Figure 2.9: Pollock, E. G., and W. A. Jensen, 1964. *American Journal of Botany*. 51: 915–21, Figure 43.

Figure 2.10: Schulz, Sister R., and W. A. Jensen, 1964. *Journal of Ultrastructural Research*. 22:376–392, Figure 14.

Figure 3.4: Ward, M., and R. H. Wetmore, 1954. *American Journal of Botany*. 41:428–34, Figures 1–12.

Figure 3.5: Zhou, C., and H. Y. Yang, 1985. *Planta*. 165:225–31, Figures 13 and 14. Heidelberg: Springer-Verlag.

Figure 3.7: Lloyd, F. E., 1902. *Memoirs Torrey Botanical Club*. 8:1, Plate VIII, Figure 4.

Figure 3.8: Finkelstein, R., et. al., 1985. *Plant Physiology*. 78:630–6, Figure 3. Reproduced by permission of ASPP.

Figure 3.9: Marsden, M., and D. W. Meinke, 1985. *American Journal of Botany*. 72:1801, Figure 6.

Figures 4.2 and 4.3: Poething, R. S., and I. M. Sussex, 1985. *Planta*. 165:158, Figures 2 and 3. Heidelberg: Springer-Verlag.

Figure 4.7: Foster, A. S., 1943. *American Journal of Botany*. 30:56–73, Figure 2.

Figure 4.9: Davis, E. L., P. Rennie, and T. A. Steeves, 1979. *Canadian Journal of Botany*. 57:971–80, Figure 1.

Figure 4.11b: Bierhorst, D. W., 1977. *American Journal of Botany*. 64:125–52, Figure 1.

Figure 4.12: Golub, S. J., and R. H. Wetmore, 1948. *American Journal of Botany*. 35:755–767, Figure 6.

Figure 5.1a: Steeves, T. A., M. A. Hicks, J. M. Naylor and P. Rennie, 1969. *Canadian Journal of Botany*. 47:1367–75, Plate I, Figure 3.

Figure 5.1b: Davis, E. L., P. Rennie, and T. A. Steeves: *Canadian Journal of Botany.* 57:971–80, Figure 5.

Figure 5.2: Buvat, R., 155. *Annee Biologique.* 31:595–656, Figure 27.

Figure 5.3: Ball, E., 1960. *Phytomorphology.* 10:377–96, Figures 18, 20 and 23.

Figure 5.4: Steeves, T. A., M. A. Hicks, J. M. Naylor, and P. Rennie, 1969. *Canadian Journal of Botany,* 1969. 47:1367–1375.

Figure 5.5: Sawhney, V. K., P. Rennie, and T. A. Steeves, 1981. *Canadian Journal of Botany.* 59:2009, Figure 5.

Figure 5.8: Satina, S., A. F. Blakeslee, and A. G. Avery, 1940. *American Journal of Botany.* 27:895–905, p. 902, Table I.

Figure 6.1: Wardlaw, C. W., 1947. *Phil. Trans.,* 232:343, Figures 1, 3 and 77. Royal Society, London, B.

Figure 6.3: Smith, R. H., and T. Murashige, 1970. *American Journal of Botany.* 57:562–8, Figures 1b, 1c, 2a and 2c.

Figure 6.4: Pilkington, M., 1929. *New Phytologist.* 28:37–53, Figures 6 and 8.

Figures 7.1 and 7.2: Sussex, I. M., 1955. *Phytomorphology.* 5:253–73, Figures 8b, 9, 10.

Figure 7.4: Steeves, T. A., and W. R. Briggs, 1958. *Phytomorphology.* 8:60–72, Figures 2, 3 and 4.

Figure 7.6: Foard, D. E., 1971. *Canadian Journal of Botany.* 49:1601–3, Plate II, Figures 5c and 6.

Figures 7.10: Snow, M., and R. Snow, 1931. *Phil. Trans.* 221:1–43, Figures 3a, 3c, 5a and 5c. Royal Society, London, B.

Figure 7.11: Wardlaw, C. W., 1949. *Growth.* (Supplement):93–131, Figure 1.

Figure 7.12: Wardlaw, C. W., 1949. *Growth.* (Supplement):93–131, Figure 1.

Figure 7.13: Plantefol, L., 1948. *La Theorie des Helices Foliares Multiples.* p. 194, Figure 7.2.

Figure 8.1: Steeves, T. A., and I. M. Sussex, 1957. *American Journal of Botany.* 44:665–73, Figures 2 and 7.

Figure 8.2: Steeves, T. A. et al, 23 August 1957. "Growth in Sterile Culture of Excised Leaves of Flowering Plants." *Science* 126:350–1, Figure 1. Copyright 1957 by the AAAS.

Figure 8.3: Steeves, T. A., 1961. *Phytomorphology.* 11:346–59, Figure 2.

Figure 8.5, 8.6: Sachs, R., 1969. *Israel Journal of Botany.* 18:21–30, Figures 1a, 1b, 1c, 3a and 3e.

Figures 8.7: Sussex, I. M., 1955. *Phytomorphology.* 5:286–300, Figures 1b, 4, 8, 12, 25 and 26.

Figure 8.8: Hanawa, J., 1961. *Botanical Magazine.* 74:303–9, Figures 1b and 4c. Tokyo: Botanical Society of Japan.

Figure 8.12: Cutter, E.. G., 1964. *American Journal of Botany.* 51:318, p. 319, Figure 1.

Figure 9.1: Steeves, T. A., *J. Linn. Soc. London (Bot.)* 58:401–15. Figures 1 and 2, Plate I, Figure 4.

Figure 9.2: Avery, G. S., 1933. *American Journal of Botany.* 20:565, Figures 1–4, 6, and 21.

Figures 9.3, 9.5, 9.6: Poething, R. S., and I. M. Sussex, 1985. *Planta.* 165:170–84, Figures 2a, 2b, 5b, 9b and 9d. Heidelberg: Springer-Verlag.

Figure 9.4: Avery, G. S., 1933. *American Journal of Botany.* 20:565, Figures 24b, 36–7.

Figure 9.6: White, R. A., and W. C. Dickinson (ed.), 1984. *Contemporary Problems in Plant Anatomy.* in article by S. Poething, p. 235, Figures 3a and 3b.

Figure 9.7: Sharman, B. C., 1942. *Annals of Botany.* 6:245, Figure 22.

Figure 9.9: Kaplan, D. R., N. G. Dengler, and R. E. Dengler, 1982. *Canadian Journal of Botany*. 60:2939, Figure 17.

Figure 9.10: Caponetti, J. D., and T. A. Steeves, 1963. *Canadian Journal of Botany*. 41:545–56, Figure 2.

Figure 9.11: Allsopp, A., 1963. *Journal of the Linnean Society* (Botany). 58:417–27, Figures 1 and 2.

Figure 9.12: Sussex, I. M., and M. E. Clutter, 1960. *Phytomorphology*. 10:87–99, Figures 1, 2, 4 and 6.

Figure 9.13: Foster, A. S., 1932. *American Journal of Botany*. 19:75–99, Plate III, Figures 19–20; Plate IV, Figures 21–3.

Figure 9.14: Mueller, P. A., and N. G. Dengler, 1984. *Canadian Journal of Botany*. 62:1158–70, Figure 14.

Figure 9.15: Schmidt, B. L., and W. F. Millington, 1968. *Bulletin of Torrey Botanical Club*. 95:264–286, Figures 3a and 3d.

Figure 10.1: Bieniek, M. E., and W. F. Millington, 1967. *American Journal of Botany*. 54:61–70, Figures 7, 9, and 10.

Figure 10.4: Cheng, P.C., R. I. Greyson, and D. B. Walden, 1983. *American Journal of Botany*. 70:450, Figures 11 and 22.

Figure 10.5: Steeves, T. A., M. A. Hicks, J. M. Naylor, and P. Rennie, 1969. *Canadian Journal of Botany*. 47:1367–75, Plate I, Figure 5.

Figure 10.6: Krishnamoorthy, H. N. and K. K. Nanda, 1968. *Planta*. 80:43–51, Figure 5a. Heidelberg: Springer-Verlag.

Figure 10.7: Tran Thanh Van, K. 1973. *Planta*. 115:87–92, Figure 2. Heidelberg: Springer-Verlag.

Figure 10.8: Hicks, G. S., and I. M. Sussex, 1971. *Bot. Gaz.* 132:350–363, Figures 15 and 17. By permission of University of Chicago Press.

Figure 10.10: Sawhney, V. K., and R. I. Greyson, 1973. *American Journal of Botany*. 60:514, Figure 23.

Figure 11.1 a,b: Garrison, R., 1949. *American Journal of Botany*. 36:205, Figures 1–2.

Figure 11.5: Sachs, R. M., C. F. Bretz, and A. Lang, 1959. *American Journal of Botany*. 46:376–84, Figure 1B.

Figure 11.6: Titman, P. W., and R. H. Wetmore, 1955. *American Journal of Botany*. 42:364–72, Figure 6.

Figure 11.7: Hallé, F., R. Oldeman, and P. Tomlinson, 1978. *Tropical Trees and Forests: An Architectural Analysis*. p. 33, Figures 8a–8f. Heidelberg: Springer-Verlag.

Figure 11.9: Halle, F., R. Oldeman, and P. Tomlinson, 1978. *Tropical Trees and Forests: An Architectural Analysis*, p. 87, Figure of Chamberlain's model, p. 95, Figure of Rauh's model, p. 91, Figures of Aubreville's model and Nozeran's model, p. 91. Heidelberg: Springer-Verlag.

Figure 11.10: Halle, F., R. Oldeman, and P. Tomlinson, 1978. *Tropical Trees and Forests: An Architectural Analysis*, p. 271, Figures 73(a–f). Heidelberg: Springer-Verlag.

Figure 12.1: Raju, M. V. S., T. A. Steeves, and R. T. Coupland, 1963. *Canadian Journal of Botany*. 41:579, Figure 6.

Figure 12.2: McArthur, I. C. S. and T. A. Steeves, 1969. *Canadian Journal of Botany*. 47:1377, Plate III, Figures 14, 16, and 18.

Figure 12.3: Goodwin, R. H., and C. J. Avers, 1956. *American Journal of Botany*. 43:479–87, Figure 6.

Figure 12.7: Clowes, F. A. L., 1959. *Biological Reviews*. 34:501, Figure 2, Cambridge Philosophical Society.

Figure 14.13: Falconer, M. M., and R. W. Seagull, 1985. *Protoplasma.* 125:190–8, Figure 1.

Figure 15.2: Strasburger, E., 1921. *A Textbook of Botany.* p. 152, Figure 176.

Figure 15.3: Bailey, I. W., 1920. *American Journal of Botany.* 7:417, Plate XXIX, Figure 53.

Figure 15.4: Bailey, I. W., 1920. *American Journal of Botany.* 7:417, Plate XXIX, Figure 50.

Figure 15.5a: Strasburger, E., 1921. *A Textbook of Botany.* London: Macmillan, 1921, p. 154, Figure 179.

Figure 15.5b: Bailey, I. W., 1920. *American Journal of Botany.* 7:417, Plate XXVII, Figure 21.

Figure 15.6: Eames, A. J., and L. H. MacDaniels, 1947. *An Introduciton to Plant Anatomy.* pp. 188 and 189; Figures 87c, and 88c. New York: McGraw-Hill.

Figure 15.8: Bannan, M. W., 1955. *Canadian Journal of Botany.* 33:113–36, Figures 1, 18 and 26.

Figure 15.9a: Barghoorn, E. S., 1940. *Bulletin of Torrey Botanical Club.* 67:303–28, Figure 8.

Figure 15.9b: Barghoorn, E. S., 1940. *American Journal of Botany.* 27:918–28, Figure 17.

Figure 15.9c: Barghoorn, E. S., 1940. *American Journal of Botany.* 27:918–28, Figure 13.

Figure 15.10: Hejnowicz, Z., and J. A. Romberger, 1973. *American Journal of Botany.* 60:209–22, Figure 1.

Figure 16.1: Siebers, A. M., 1971. *Acta Botanica Neerlandica.* 20:211–220, Figure 1.

Figure 16.2: Brown, C. L., and K. Sax, 1962. *American Journal of Botany.* 49:683–91, Figures 1, 14–15.

Figure 16.3: Zimmerman, M. (ed.), 1964. *The Formation of Wood in Forest Trees.* New York: Academic Press in article by C. L. Brown, p. 396, Figure 14.

Figure 16.5: Thair, B. W., and T. A. Steeves, 1976. *Canadian Journal of Botany.* 54:361–73, Figure 13.

Figure 17.2: Steward, F. C. et al, 3 January 1964. "Growth and Development of Cultured Plant Cells." *Science* 143:20–7, Figure 2. Copyright 1964 by the AAAS.

Figure 17.3: Vasil, V., and A. C. Hildebrandt, 12 November 1965. "Differentiation of Tobacco Plants from Single, Isolated Cells in Microcultures." *Science* 150:889–892, Figure 2. Copyright 1965 by the AAAS.

Figure 17.4: Halperin, W., 1966. *American Journal of Botany.* 53:443–53, Figures 7–9.

Figure 17.5: Backs-Hüseman, D. and J. Reinert, 1970. *Protoplasma.* 70:49–60, Figure 3.

Figure 17.6: Halperin, W., 1966. *American Journal of Botany.* 53:443–53, Figures 17–23.

Figure 17.7: Kao, K. N., and M. R. Michyluk, 1980. *Ziet. Pflanzenphysiologische* 96:135, Figures 3, 6 and 7.

Figure 17.8: Konar, R. N., and K. Nataraja, 1965. *Phytomorphology.* 15: 132–137, Figures 1b, 1c, 1i and 1j.

Figure 17.9: Taylor, R. L., 1967. *Canadian Journal of Botany.* 45:1553–6, Figures 1, 2 and 3.

Figure 17.11: Raju, M. V. S. et al., 1966. *Canadian Journal of Botany.* 44:33, Plate I, Figure 2.

Author index

Subject index